Biostatistics in
Clinical Medicine

Joseph A. Ingelfinger, M.D.
Assistant Professor of Medicine, Boston University
School of Medicine: Associate Director, Section of
General Internal Medicine, Boston City Hospital

Frederick Mosteller, Ph.D.
Roger I. Lee Professor of Mathematical Statistics
and Chairman, Department of Health Policy and
Management, School of Public Health,
Harvard University

Lawrence A. Thibodeau, Ph.D.
Senior Manager, Price Waterhouse, New York,
New York

James H. Ware, Ph.D.
Professor of Biostatistics
School of Public Health, Harvard University

Biostatistics in Clinical Medicine

MACMILLAN PUBLISHING CO., INC.
NEW YORK

COLLIER MACMILLAN CANADA, INC.
TORONTO

COLLIER MACMILLAN PUBLISHERS
LONDON

Macmillan Publishing Company
866 Third Avenue, New York, New York 10022

Collier Macmillan Canada, Inc.

Collier Macmillan Publishers · London

Library of Congress Cataloging-in-Publication Data

Biostatistics in clinical medicine.

 Includes bibliographies and index.
 1. Medical statistics. 2. Medicine, Clinical—
Statistical methods. 3. Biometry. I. Ingelfinger,
Joseph A. [DNLM: 1. Biometry. 2. Medicine. 3. Prob-
ability. WA 950 B6165]
RA409.B48 1987   616' .0028   86-18116
ISBN 0–02–359721–6

Printing: 2 3 4 5 6 7 8   Year: 7 8 9 0 1 2 3 4

To
Howard H. Hiatt
who started us, and to
Franz J. Ingelfinger
who guided us along the way

Preface

This book prepares physicians to use and understand applications of probability and statistics for the care of the individual patient. Although probability and statistics are usually offered as research tools for the study of masses of patients, we emphasize their use in problems of diagnosis, treatment, and follow-up. Each principal chapter contains at least one detailed clinical problem, and quantitative methods lead to interpretations and solutions.

The book consists of 12 principal chapters. These chapters have three themes: diagnostic testing and clinical decision analysis (Chapters 1–3), day-to-day variability and the interpretation of clinical data (Chapters 4–10), and the interpretation and application of clinical trials (Chapters 11–12).

In Chapters 1 and 2 we describe a probabilistic approach to diagnosis and indicate how the results of diagnostic testing influence estimates of probability of disease. Chapter 3 introduces the techniques of clinical decision analysis and demonstrates how a probabilistic approach can be used in the decision-making process for patient care. Chapters 4 through 10 introduce the basic ideas of statistical inference and apply them to problems of clinical practice. In Chapter 4 we discuss the day-to-day variability of signs and symptoms and introduce probability distributions as a tool for interpreting observations subject to variability. Chapters 5 and 6 describe methods for summarizing clinical observations as means, medians, and rates, and for interpreting these summaries through confidence intervals and hypothesis tests. In Chapter 7 we approach the problem of inference from a different perspective, explaining how P values are computed, and describing their use in clinical medicine. Chapters 8 and 9 show how chi-square tests and regression analysis can be used to study relationships between two variables. Chapter 10 discusses life-table analysis and its application in summarizing the survival experience of patients suffering from life-threatening chronic disease.

Chapters 11 and 12 describe the evaluation and application of clinical trials. In Chapter 11 we give criteria for evaluating the design, conduct, and analysis of clinical trials. Chapter 12 discusses the application of a clinical trial to the single patient, especially the issues of generalizability and sub-group analysis.

The book also includes three supplementary chapters labeled with a chapter number and a letter (Chapters 5A, 9A, 9B) that more formally discuss the statistical background for the material in the preceding chapter. The principal chapters, however, can be read without referring to these supplements.

The chapters on multiple regression analysis (Chapter 9B) and life-table analysis (Chapter 10) are available for the first time in this second edition. We prepared this new material in response to the increasing reliance on these techniques in the medical literature.

We have also included frequently used statistical tables in Appendix III so that the book can be used as a reference. The tables have special lists on each page intended to speed the reader to the table being sought. Each page lists all the tables at the top, and shows the table on that page in boldface type. The reader can thus tell whether it is necessary to leaf backward or forward to find the desired table.

Problems for practice appear at the end of each chapter. In Appendix II short answers are given for the odd-numbered problems. A Solutions Manual containing solutions to all problems in this book is available to course instructors on written request to the publisher. Some additional problems are given in Appendix I.

Documentary references appear at the end of each chapter and we have occasionally included additional readings to supplement these references.

The organization of the material is shown in the accompanying figure. Material assumed in each of the principal chapters is determined by reading down from Chapter 1. Chapters 1, 4 through 7, and 11 and 12 contain the ideas that are central to the book, as well as much of the material clinicians will find most useful. Chapter 3 addresses a popular and important topic but is not required reading for later chapters. Chapters 2, 8, 9, and 10 can be included or omitted, depending on interest and time limitations. A reader with some previous exposure to elementary probability may wish to skim or omit Chapters 1 and 4.

Above all, we have based each chapter on clinical examples motivating the quantitative solutions. Thus the reader is exposed to both clinical and quantitative material. For example, the chapters discussing diagnostic tests provide extensive information concerning the serum digoxin concentration and its value in diagnosing digitalis toxicity. The chapter on blood pressure measurement contains important information about blood pressure variability of value to physicians who diagnose and treat patients with high blood pressure. With this approach, we believe that this book not only teaches

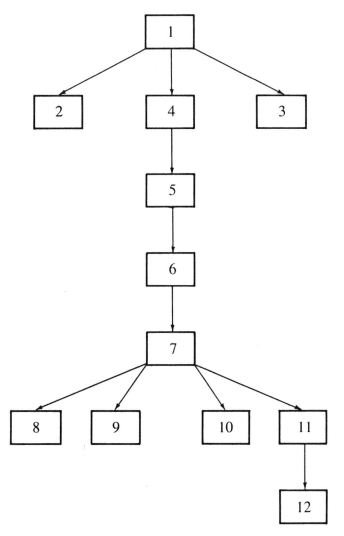

Diagram of the background assumed in each of the 12 main chapters. Read down from Chapter 1 to identify the chapters whose content is assumed. Thus, Chapter 8 assumes familiarity with the probability and statistical content of Chapters 1 and 4 through 7.

probability and statistics for the care of the individual patient, but also provides some of the information required to use these disciplines effectively in clinical medicine.

In a book for professionals, some pacts between readers and authors can maintain better relations. First, although most readers of this work will be physicians, they will be at various stages of their careers, from beginning study through experienced practitioners. Consequently, we sometimes give readers more explanation than they need. The more experienced reader will understand that the less experienced may be grateful for this extra help, rather than that the authors intend to insult the learning of the experienced.

The second understanding flows from the role of this book as an introduction to quantitative methods in clinical medicine. At the beginning, we cannot afford to clutter the reader with many complications. Yet the experienced physician may often be able to add further issues that should be considered in connection with the diagnoses, diseases, and treatments discussed here. These issues will enrich a classroom discussion, but they might drag down the initial explanation of a method by their weight and numbers. Thus both readers and authors should and do appreciate that further possibilities can be treated profitably. Initial oversimplification often gets things started.

Third, happily medicine and knowledge change, so some of the treatments and discussions are sure to need updating as soon as this work appears. We hope that the reader will be tolerant of such obsolescence.

Fourth and finally, some readers have more mathematical preparation than others. We have not leaned heavily on such preparation, but we have tried to write so that readers with more preparation will not be misled in their generalizations. The reader with more mathematical preparation needs to be tolerant of a presentation planned for the less well prepared.

J.A.I.
F.M.
L.A.T.
J.H.W.

Boston, Massachusetts

Acknowledgments

We are grateful to John Bailar, M.D., Ph.D., who drafted early versions of some of this material. He also gave us many suggestions both about the overall plan and the details of the writing.

George A. Lamb, M.D., joined us in teaching classes using the earliest versions of this material and generously provided advice, both oral and written.

Peter Goldman, M.D., encouraged us in preparing the volume. The ideas of and discussions with Lincoln E. Moses contributed greatly to the shaping of Chapter 12.

Nina Leech had primary responsibility for preparing the manuscript, and at various times Patricia Houghton and Ellen Collins aided. Their patient and good-natured support through numerous revisions was an essential part of our efforts to strengthen the clarity and organization of the material. Nina Leech, in particular, devoted many months to the task of creating a finished manuscript from early drafts of individual chapters. We are deeply appreciative of her contribution to the book.

Contents

Biostatistics in
Clinical Medicine

CHAPTER **1**

Diagnostic Testing: Introduction to Probability

OBJECTIVES

Sensitivity and specificity of tests
Predictive values of tests
Probability
The 2 × 2 table
Conditional probability
Product and addition rules for probability
Independence
Bayes theorem
Multiple tests

Much of the information gathered by clinical history, physical examination, laboratory analysis, and radiographic evaluation helps us diagnose a patient's condition. For each patient, the physician considers a number of diagnostic possibilities. The physician adjusts the relative likelihood of these possibilities in the light of newly acquired diagnostic information. But exactly how much should test results alter the probability of disease? Does a positive acid-fast smear guarantee that the patient has active tuberculosis? Does a toxic digoxin concentration inevitably signify digitalis intoxication? Obviously not, as neither test is absolutely accurate, yet such positive test results do tend to increase the probability that the disease is present. Our somewhat idealized view holds that the disease is either present or absent, and that our information is imperfect. The probability of the disease numerically measures our degree of belief in its presence. The disease is in the patient; its probability is in our minds.

Sensitivity and specificity are two ways to describe test accuracy.

Sensitive test. If all persons with the disease have "positive" tests, we say that the test is sensitive to the presence of the disease.
Specific test. If all persons without the disease test "negative," we say that the test is specific to the absence of the disease.

When a test is sensitive and specific, no ambiguities in interpretation arise.

1

These ideal conditions rarely hold. Nevertheless, diagnostic tests frequently yield outcomes that are consistent with the patient's true state.

This chapter explains how to use new diagnostic information to modify the probability of disease when the test is not perfectly accurate.

1-1.   Digitalis Toxicity and Serum Digoxin Concentration

Clinical Problem 1-1.   *Is the Patient Digitalis Toxic?*

Mr. R., a 50-year-old man with congestive heart failure taking digoxin, has been admitted to the hospital. On admission, he has a serum digoxin concentration of 2.5 ng/ml. Is Mr. R. digitalis toxic?

Background

The digitalis glycosides, digoxin and digitoxin, can be used to great benefit in patients with congestive heart failure because they produce slower heart rate and an increased force of myocardial contraction. Digitalis therapy can, however, result in digitalis intoxication, a serious complication. Some degree of intoxication occurs in 5 to 30% of hospitalized patients taking digitalis.

Diagnosing digitalis intoxication may be difficult because many of the symptoms of intoxication, for example, premature ventricular contractions, nausea, and fatigue, are also symptoms of congestive heart failure. The serum concentration of digoxin and digitoxin, measurable since 1968, provides a useful diagnostic test for toxicity.

Beller *et al.* (1971) examined the relation between serum digitalis concentrations and digitalis toxicity. From among 931 consecutive admissions to the Boston City Hospital, they investigated all 135 patients who were taking either digoxin or digitoxin. On admission, each patient's serum digitalis concentration was recorded. Serial electrocardiograms provided follow-up information on each patient. These 135 patients compose all patients receiving a digitalis preparation admitted to the hospital during the study period. We could refer to those patients as being the **population** of all patients taking digoxin or digitoxin admitted to the hospital during the study period.

Although examining this specific population may be instructive, we usually wish to project our experience with the studied patients to other, similar patients. In such projections, we postulate a population of patients and we refer to the patients that we have studied as a **sample** from this population. Then we view our sample as providing information about the population sampled. We will consider the 135 patients studied by Beller *et al.* as a sample from the population of all hospitalized patients taking a digitalis preparation.

Table 1-1.   Rhythm disturbance acceptable for consideration of digitalis intoxication

Ectopic ventricular rhythms
  Multifocal ventricular ectopic beats
  Unifocal ventricular ectopic beats, but > 5 per minute
  Ventricular bigeminy or trigeminy
  Ventricular tachycardia
Nonparoxysmal atrioventricular junctional tachycardia (> 80 per minute)
  Atrioventricular junctional escape rhythms and atrioventricular junctional exit block
Atrioventricular dissociation with ventricular rate exceeding atrial rate
Atrial fibrillation with ventricular response < 50 per minute if accompanied by ectopic ventricular beats
Mobitz type 1 (Wenckebach) second-degree atrioventricular block
Paroxysmal atrial tachycardia with atrioventricular block
Sinoatrial exit block or sinus arrest

Source: Reprinted by permission from G. A. Beller, T. W. Smith, W. H. Abelmann, E. Haber, and W. B. Hood, *N. Engl. J. Med.*, **284**:989–97, 1971.

Implications for Clinical Problem 1-1

    Let us relate the toxic state of the patient to the serum digoxin or digitoxin concentration. Using the clinical criteria shown in Tables 1-1 and 1-2, the investigators classified 34 patients as definitely toxic, 9 as possibly toxic

Table 1-2.   Criteria for digitalis intoxication

No digitalis toxicity
  Absence of rhythm disturbance
  Arrhythmia remits when digitalis continued or dosage increased
  Arrhythmia present **on** or **off** digitalis
Definite digitalis toxicity
  Complete resolution of arrhythmia with discontinuance of digitalis or reduction in dosage of drug
  Any of the following in a patient who dies before resolution of rhythm disturbance, and in whom no other etiology for the arrhythmia was apparent:
    Ventricular bigeminy, multifocal premature ventricular beats, runs of ventricular tachycardia
    Nonparoxysmal atrioventricular junctional tachycardia, atrioventricular junctional exit block, or atrioventricular junctional escape rhythm
    Paroxysmal atrial tachycardia with atrioventricular block
    Mobitz type 1 second-degree atrioventricular block or atrioventricular dissociation
Possible digitalis toxicity
  Partial resolution of rhythm disturbance when digitalis reduced in dosage or discontinued
  Digitalis discontinued but patient discharged or signs out of hospital before serial tracings obtained
  Concomitant electrocardiographic evidence of acute myocardial infarction or changes of acute ischemia, although rhythm disturbance disappears when digitalis discontinued

Source: Reprinted by permission from G. A. Beller, T. W. Smith, W. H. Abelmann, E. Haber, and W. B. Hood, *N. Engl. J. Med.* **284**:989–97, 1971.

Table 1-3. 2 × 2 table from the data of Beller *et al.* on digoxin toxicity[a]

		Toxicity		Row total
		D+	D−	
Serum digoxin level	T+	25[b]	14	39
	T−	18[c]	78	96
Column total		43[d]	92	135 Grand total

[a] D+, toxic or possibly toxic;
D−, not D+ (nontoxic);
T+, serum digoxin level greater than 1.7 ng/ml or digitoxin level greater than 25 ng/ml;
T−, not T+.
[b] 24 definitely and 1 possibly toxic.
[c] 10 definitely and 8 possibly toxic.
[d] 34 definitely and 9 possibly toxic.
Source: From the data of Beller *et al.* (1971).

(adding to 43), and 92 as nontoxic. A digoxin level of 1.7 ng/ml and a digitoxin level of 25 ng/ml are essentially equivalent, and we regard higher concentrations than these as positive tests, labeled T+. The investigators reported positive tests in 24 of the 34 definitely toxic, 1 of the 9 possibly toxic, and 14 of the 92 nontoxic patients. These findings produce the results shown in Table 1-3.

In view of the findings cited above, Mr. R.'s elevated digoxin level of 2.5 ng/ml suggests toxicity because 25 of 39 patients with elevated digoxin levels were definitely or possibly toxic. In the following sections we discuss how the results of the investigation of Beller *et al.* can be used to evaluate the accuracy of serum digoxin level as a test for digoxin toxicity. This information, used with rules for combining probabilities, provides a quantitative estimate of the probability of toxicity in Mr. R.

1-2.   Determining the Accuracy of a Diagnostic Test

The 2 × 2 Table

One way to summarize test results in patients with and without a disease lays out the data in a 2 × 2 table (read: two-by-two table). Table 1-3 is the 2 × 2 table summarizing the results reported by Beller *et al.* The basic table consists of four cells arranged in two columns and two rows. The column, row, and grand totals which appear at the end of each column or row are called marginal totals because they appear on the margins of the table.

The column factor, disease, separates patients by presence, D+, or absence, D−, of disease. (In this study the disease is digitalis toxicity diagnosed according to the criteria in Tables 1-1 and 1-2.) The row separates patients by positive, T+, or negative, T−, test results. Beller *et al.* define positive and negative tests by selecting, somewhat arbitrarily, a **critical value** of the serum digoxin (and serum digitoxin) concentration. Values greater than the critical value are labeled positive (toxic), smaller values negative (nontoxic). As mentioned above, Beller *et al.* selected a digoxin concentration of 1.7 ng/ml as the critical value separating positive from negative test results.

Estimating Probabilities from the 2 × 2 Table

The 2 × 2 table provides an easy format for computing the proportion of patients with any particular diagnosis and/or test result. By looking at the right-hand marginal totals of Table 1-3, one can see that 39 of the 135, or 29%, of the patients had a positive test. Looking at the first row, one can see that 25 of 39, or 64%, of the patients with a positive test had toxicity. Each cell in the table corresponds to an event. The upper left-hand corner, for example, corresponds to the event

"the patient has a positive test and is diseased"

In this group of 135 patients, 25 satisfy this event. The T + row corresponds to the event

"the patient has a positive test"

and in this group of 135 patients, 39 satisfy this event. Similar meanings correspond to the total cells, the other row, and to the columns.

To speak of the probability of an event occurring, we write $P(A)$ meaning the probability of event A, often read as "P of A." For example, $P(T+)$ is shorthand for the probability that a randomly drawn patient gives a positive test. [Note that $P(T+)$ means that we draw a patient and the patient gives a positive test. This is different from the probability that a particular patient gives a positive test. The latter probability would arise if we repeatedly tested a single patient, getting sometimes positive and sometimes negative results.]

The probability of the event A is written $P(A)$.

In the study of Beller *et al.* and the present analysis, we use the exact proportion diseased in the small sample of patients actually studied to estimate the probability of disease in future patients. More generally, let the sample size be designated by n, as is commonly used. Further, let the number of patients or items in the sample satisfying the event A be denoted by a. Then we say that "the estimated probability of event A is a/n." For shorthand, we write

$$\hat{P}(A) = \frac{a}{n}$$

where the circumflex ("hat") on P emphasizes that this is an estimate rather than the true probability $P(A)$.

> When we estimate the probability $P(A)$ from a sample, we use the symbol $\hat{P}(A)$ to represent this estimate.

The estimate a/n plays two roles:

1. *Population interpretation.* If the population consists of the n patients or items under discussion, of whom a satisfy the event A, then the exact probability of event A when an item is randomly drawn from this population is a/n. Equivalently, it is the proportion of items satisfying event A.
2. *Sample interpretation.* From a population, a sample of n items is randomly drawn, of which a satisfy the event A. Then a/n is an estimate of the proportion of items in the population that satisfy event A.

From Table 1-3 one can make the following estimates of the probabilities of disease and toxicity in patients taking digoxin who are admitted to the hospital.

Positive test: $\hat{P}(T+) = \dfrac{39}{135} = .29$

Negative test: $\hat{P}(T-) = \dfrac{96}{135} = .71$

Diseased: $\hat{P}(D+) = \dfrac{43}{135} = .32$

Not diseased: $\hat{P}(D-) = \dfrac{92}{135} = .68$

Note that

$$\hat{P}(T+) + \hat{P}(T-) = 1.0$$

as all patients have either a positive test or a negative test but not both. Similarly,

$$\hat{P}(D+) + \hat{P}(D-) = 1.0$$

Conditional Probability

Consider the probability that a toxic patient will have a positive test. From Table 1-3 we see that 25 of 43, or 58%, of the toxic (D+) patients tested T+. Thus we say: "An estimate of the probability of T+ given D+ is .58." A shorthand notation for this is

$$\hat{P}(T+|D+) = .58$$

In this notation, events that are given as occurring are written to the right of the vertical bar, here D+; to the left of the bar, we write the event whose probability is desired. Note that "conditioning" events, to the right of the bar, may be under our control (such as treatment) or they may not (such as disease).

> The probabilities of events occurring given that certain conditions hold are called **conditional probabilities.** More generally, we write the conditional probability that event A occurs given that event B occurs as $P(A|B)$.

Actually, all probabilities are conditional on their population. When the population has been defined separately, we do not write it in. Thus $P(D+)$ means the probability that a patient in the given population is diseased. If we had a name for the population such as B, where B indicates all patients entering a hospital with specific symptoms, we could write $P(D+|B)$ for the probability that one of these patients was diseased. Thus $P(T+|D+)$ means that we focus on the subpopulation of diseased patients and give the proportion of those who test positively. Secretly, we had an original unnamed population here, part of whom are the D+'s, in our example the population described by Beller *et al.*'s protocol.

Sensitivity and Specificity

One measure of the test's accuracy for detecting a disease is the probability that a diseased patient tests positive. This conditional probability, $P(T+|D+)$, is referred to as the **sensitivity** of the test. A second measure of test accuracy is the probability that a nondiseased patient tests negative. This conditional probability, $P(T-|D-)$, is referred to as the **specificity** of the test.

> The sensitivity of a test is the probability $P(T+|D+)$. The specificity is $P(T-|D-)$.

EXAMPLE. Using the information in Table 1-3, we can compute estimates of the sensitivity and specificity of the serum digoxin concentration as a test for digitalis toxicity. (Recall that 1.7 ng/ml has been used as the critical value to separate positive and negative tests.)

By definition,

Sensitivity $= P(T+|D+)$

Using Table 1-3, we get the estimate

$$P(T+|D+) = \frac{25}{43} = .58$$

Similarly, the estimate of specificity is

$$\hat{P}(T-|D-) = \frac{78}{92} = .85$$

Often we can change the sensitivity and specificity of a test by changing the critical value. If 1 ng/ml had been chosen to divide positive and negative test results, the test may have been more sensitive because more toxic patients might have had positive results. At the same time, the specificity may have decreased because additional nontoxic patients may also give positive tests.

This feature is shown in Figure 1-1, which is a histogram depicting digoxin concentrations from a simulated sample of 21 toxic and 62 nontoxic patients. (Beller *et al.* did not publish the individual levels of the toxic and nontoxic patients, so we simulated data with characteristics similar to those reported.) The number of individuals is indicated by the vertical bars of the histogram. Among these, 18 toxic patients have digoxin concentrations greater than 1.0 ng/ml, and 24 nontoxic patients have serum digoxin concen-

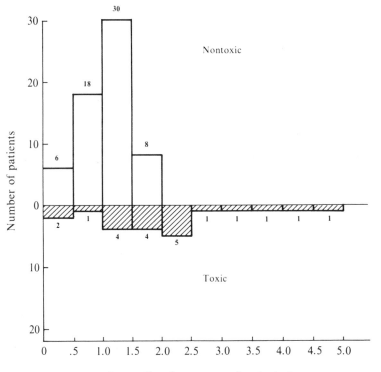

Figure 1-1. Histograms of serum digoxin concentrations for 62 clinically nontoxic patients (above the axis) and 21 clinically toxic patients (shaded and below the axis). No nontoxic patients had concentrations higher than 2.0. (Frequencies are shown at the tip of each bar.)

trations less than 1.0 ng/ml. Choosing 1 ng/ml as the critical value yields a test with a higher estimated sensitivity, 18 out of 21, or .86, but this gain is paid for by a loss in estimated specificity, 24 out of 62, or .39. Whatever critical value is chosen to separate toxic and nontoxic patients, test sensitivity and test specificity can be estimated. Recall that using a critical value of 1.7 ng/ml resulted in an estimated sensitivity of .58 and an estimated specificity of .85. So although we have gained in sensitivity, .58 to .86, we have lost specificity, .85 to .39. Whether such a trade-off is desirable hinges on benefits and losses associated with correct and erroneous diagnoses as well as the probability of disease in our patients.

Test sensitivity and specificity may change from one sample of patients to the next, especially if important characteristics of the sampled patients differ. For example, toxic and nontoxic infants receiving digoxin have concentrations higher than adults. The sensitivity and specificity of the serum digoxin concentration as a test for toxicity, using 1.7 ng/ml as the critical value, are different in infants, as shown by the following estimates based on two investigations (Hayes *et al.*, 1973; Krasula *et al.*, 1974):

Sensitivity: $\hat{P}(\text{T}+|\text{D}+) = \dfrac{12}{12} = 1.0$

Specificity: $\hat{P}(\text{T}-|\text{D}-) = \dfrac{21}{60} = .35$

Although the probability of disease may vary in different clinics, the sensitivity and specificity of a test do not depend on the underlying probability of disease. This means that test sensitivity and specificity determined by other investigators can be applied in a clinical setting where disease is more common or less common.

> The sensitivity and specificity of a test do *not* depend on the underlying disease probability.

Predictive Value

The positive and negative predictive values of a test also help us evaluate tests. Although sensitivity and specificity do not depend on the proportion of diseased patients, the predictive values depend both on test accuracy and on the frequency of disease in the sample of patients tested.

> We define the **predictive value of a positive test** by $P(\text{D}+|\text{T}+)$, and the **predictive value of a negative test** by $P(\text{D}-|\text{T}-)$.

Based on the series of Beller *et al.* (1971) in Table 1-3, we computed:

Predictive value of

Positive test: $\hat{P}(D+|T+) = \dfrac{25}{39} = .64$

Negative test: $\hat{P}(D-|T-) = \dfrac{78}{96} = .81$

On admission to the hospital, one would estimate that Mr. R. had a .32 chance of toxicity, because $\hat{P}(D+) = 43/135 = .32$. When we find that his digoxin concentration exceeds 1.7 ng/ml, we change our estimate to .64. Thus a positive test has doubled the probability of toxicity, although we are still far from certainty.

1-3.   Revising the Estimated Probability of Disease

Clinical Problem 1-2.   *Using the Digoxin Concentration to Revise the Probability of Digitalis Toxicity*

Mrs. Q. is a 60-year-old woman with congestive heart failure who is taking digoxin. On admission to the hospital, she complains of fatigue. Her electrocardiogram shows ventricular premature contractions which had not been noted previously. Digoxin toxicity is a possibility, as about 60% of such patients are toxic. How will knowledge of Mrs. Q.'s digoxin concentration influence the probability of digoxin toxicity?

Combining Probabilities

Based on clinical presentation, one estimates for Mrs. Q. that $P(D+$ for Mrs. Q.$) = .6$ and so $P(D-$ for Mrs. Q.$) = .4$. If the sensitivity and specificity of the serum digoxin test are estimated, then $P(D+|T+)$ and $P(D-|T-)$, the revised probabilities of disease given a positive or negative test, can be computed. As we now explain, the computation depends on the product rule for joint probabilities and the addition rule for marginal probabilities.

Consider the 2 × 2 table in Figure 1-2, which symbolizes the results of a study investigating the relationship between a disease and a test. Here a, b, c, and d are the numbers of observations falling in each cell. The row, column, and grand totals are shown in the margins. A **marginal probability** is the probability of a single (marginal) event, for example,

$$P(T+) = \frac{a + b}{a + b + c + d}$$

A **joint probability** is a probability of two (or more) events occurring simultaneously (jointly). For example, the probability that a patient would

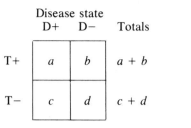

                              Disease state
                               D+   D−   Totals

Test outcome

             T+   a   b   a + b

             T−   c   d   c + d

             Totals   a + c b + d   a + b + c + d = n

Figure 1-2. A 2 × 2 table of counts.

be both diseased and have a positive test is denoted by $P(T+$ and $D+)$, which equals $a/(a + b + c + d)$.

A fundamental rule for computing a joint probability is the product rule:

> Product Rule: For any two events A and B, the joint probability of A and B occurring is equal to the product of the conditional probability of A given B times the probability of B.
>
> $$P(A \text{ and } B) = P(A|B)P(B)$$

EXAMPLE. Using the product rule and the 2 × 2 table of Figure 1-2, we will show that

$$P(D+ \text{ and } T+) = P(D+|T+)P(T+)$$

First note that

$$P(D+|T+) = \frac{a}{a + b}$$

which is the number of $(T+, D+)$ patients divided by the total number of $T+$ patients. Similarly,

$$P(T+) = \frac{a + b}{a + b + c + d}$$

It follows that

$$P(D+|T+)P(T+) = \frac{a}{a + b} \times \frac{a + b}{a + b + c + d} = \frac{a}{a + b + c + d}$$

We have already noted that

$$P(T+ \text{ and } D+) = \frac{a}{a + b + c + d}$$

Therefore,

$$P(T+ \text{ and } D+) = P(D+|T+)P(T+)$$

This completes the proof. Furthermore, because we can substitute A for D+ and B for T+ throughout, we have also proved the general product rule given in the box.

EXAMPLE. Use the product rule and the data in Table 1-3 to find the probability that a patient will be both toxic and have an elevated digoxin concentration. (We will treat the numbers as if we have a whole population, so we can compute P's instead of $\hat{P}$'s. Actually, we ordinarily have estimates rather than population proportions, but we do not always make the distinction.)

SOLUTION. Writing out the product rule gives

$$P(D+ \text{ and } T+) = P(D+|T+)P(T+)$$

$$= \frac{25}{39} \times \frac{39}{135} = \frac{25}{135}$$

CHECK. By direct calculation,

$$P(D+ \text{ and } T+) = \frac{a}{a + b + c + d}$$

$$= \frac{25}{135}$$

Independence of events is an important concept related to the product rule.

Independence: Two events, A and B, are **independent** if

$$P(A|B) = P(A)$$

This condition states that the probability of A is not altered by the occurrence of B. When A and B are independent, the product rule becomes

---

Product rule with independence:

$$P(A \text{ and } B) = P(A)P(B)$$

Another fundamental rule for combining probabilities is the addition rule. This time joint probabilities combine to give a marginal probability. Essentially, the addition rule rests on the idea that either the event "B" occurs or the event "not B" occurs. So all possibilities are covered by the two events B or not B.

> Addition rule: For any two events A and B, the marginal probability of A equals the joint probability of A and B plus the joint probability of A and not B.
>
> $$P(A) = P(A \text{ and } B) + P(A \text{ and not } B)$$

PROBLEM: Addition rule: Using the 2×2 table of Figure 1-2, show that

$$P(T+) = P(T+ \text{ and } D+) + P(T+ \text{ and } D-)$$

Now using the product and addition rules for combining probabilities, values for sensitivity and specificity, and the probability of disease, we can construct a 2×2 **probability table.** A 2×2 probability table has the same structure as the 2×2 tables of counts that we have been using; the difference is that the entries in the probability table are probabilities, not counts. For example, in Figure 1-2 in the upper left cell where the count a is the **number** of $(T+, D+)$ patients, a probability table would have the joint probability, $P(T+ \text{ and } D+)$. In Figure 1-2, $P(T+ \text{ and } D+)$ equals $a/(a + b + c + d)$. Similarly, the marginal total in Figure 1-2 of $a + b$, the number of $T+$ patients, would be replaced by the marginal probability $P(T+)$, which is $(a + b)/(a + b + c + d)$.

To see how to get such a probability table, let us construct one for Mrs. Q. We have been told that for Mrs. Q.,

$$P(D+ \text{ for Mrs. Q.}) = .60$$

(This did not come from our table but from outside information.) Further, using the sensitivity and specificity of the serum digoxin test reported by Beller *et al.,* we have, from Table 1-3,

$$P(T+|D+) = \frac{25}{43} = .58$$

and

$$P(T-|D-) = \frac{78}{92} = .85$$

Recall that these characteristics of a test do not depend on the probability of disease. Using the product rule and these three probabilities, we can compute two joint probabilities. They are, for Mrs. Q.,

$$P(T+ \text{ and } D+) = P(T+|D+)P(D+) = .58(.6) = .348 \approx .35$$

and

$$P(T- \text{ and } D-) = P(T-|D-)P(D-) = .85(.4) = .34$$

Figure 1-3A shows a 2×2 probability table with these two joint probabilities and the marginal probabilities for $D+$ and $D-$ entered.

A

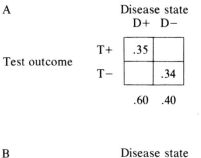

Disease state
D+ D−

Test outcome

	D+	D−
T+	.35	
T−		.34

.60 .40

B

Disease state
D+ D−

Test outcome

	D+	D−	
T+	.35	.06	.41
T−	.25	.34	.59

.60 .40

Figure 1-3. Filling in the 2 × 2 probability table for Mrs. Q.: (A) partial; (B) complete.

To complete the 2 × 2 probability table, we need only to recall the addition rule for probabilities. Thus, since

$$P(D+) = P(T+ \text{ and } D+) + P(T- \text{ and } D+)$$

and since

$$P(D+) = .60$$

then

$$P(T+ \text{ and } D+) = .35$$

$$P(T- \text{ and } D+) = .60 - .35 = .25$$

Using the addition rule to compute the other missing joint probability and marginal probabilities, we get Figure 1-3B.

To compute the positive and negative predictive value for a test, we need a method for computing conditional probabilities from marginal and joint probabilities. We can obtain the needed representation by rewriting the product rule.

Provided that $P(B) \neq 0$, we can write the conditional probability $P(A|B)$ as

$$P(A|B) = \frac{P(A \text{ and } B)}{P(B)}$$

Solution to Clinical Problem 1-2

With the formula just developed, we can use the table shown in Figure 1-3*B* to compute the predictive value of a positive or negative test when $P(D+) = .6$. They are, for Mrs. Q.,

$$P(D+|T+) = \frac{P(T+ \text{ and } D+)}{P(T+)} = \frac{.35}{.41} = .85$$

$$P(D-|T-) = \frac{P(T- \text{ and } D-)}{P(T-)} = \frac{.34}{.59} = .58$$

Thus a positive test increases the estimated probability that Mrs. Q. is toxic from .60 to .85, and a negative test increases the probability that Mrs. Q. is not toxic from .40 to .58.

Some clinicians would question the value of the serum digoxin level in this case. They would withhold digoxin whether the probability of toxicity was .85 following a positive test or .42 ($= 1 - .58$) following a negative test.

Bayes Theorem

Constructing a 2×2 probability table offers a direct way to compute the predictive value of a positive or negative test. Another way gives identical results and avoids the construction of the complete 2×2 probability table. The method is based on Bayes theorem.

Bayes theorem: If $P(B) \neq 0$, then

$$P(A|B) = \frac{P(B|A)P(A)}{P(B|A)P(A) + P(B|\text{not } A)P(\text{not } A)}$$

Noting that the denominator on the right is, by the addition rule, $P(B)$, we can also state Bayes theorem as

$$P(A|B) = \frac{P(B|A)P(A)}{P(B)}$$

EXAMPLE. Mrs. Q. has a probability of toxicity of .6, $P(D+$ for Mrs. Q.$) = .6$. How would this estimate of the probability of toxicity change if she had an elevated digoxin concentration, or if she had a therapeutic (not elevated) concentration?

SOLUTION. By Bayes theorem,

$$P(D+|T+) = \frac{P(T+|D+)P(D+)}{P(T+|D+)P(D+) + P(T+|D-)P(D-)} \qquad (1\text{-}1)$$

All the quantities on the right are known once we note that $P(T+|D-)$ is equal to $1 - P(T-|D-)$. Substituting into equation (1-1) numerical values from Table 1-3 for sensitivity and specificity, and the postulated $P(D+$ for Mrs. Q.) $= .60$, we get

$$P(D+|T+) = \frac{\left(\frac{25}{43}\right)(.60)}{\left(\frac{25}{43}\right)(.60) + \left(1 - \frac{78}{92}\right)(.40)} = .85$$

Similarly, we can compute the predictive value of a negative test as

$$P(D-|T-) = \frac{P(T-|D-)P(D-)}{P(T-|D-)P(D-) + P(T-|D+)P(D+)}$$

Using the conditional probabilities from Table 1-3 and the postulated $P(D-$ for Mrs. Q.) $= .40$, we get

$$P(D-|T-) = \frac{\left(\frac{78}{92}\right)(.4)}{\left(\frac{78}{92}\right)(.4) + \left(1 - \frac{25}{43}\right)(.6)} = .57$$

which differs slightly from the earlier result, .58, because of rounding errors in the first calculation.

Thus we can use Bayes theorem to compute the conditional probabilities that we computed from the 2×2 probability table in Figure 1-3.

Bayes theorem is more than simply a short cut rule. For disease and test situations, Bayes theorem tells us how sensitivity and specificity, two properties of the test, combine with the probability of disease to give the predictive value of a positive or negative test.

Sampling Patient Populations

Beller *et al.* wanted to know the probabilities of disease and positive tests, and the sensitivity and specificity, for their population of patients. To estimate these they took a sample consisting of a consecutive series of patients so that they would not introduce selection effects by their own picking and choosing of patients for consideration. Although it is true that the patient mix can change from one hospital to another or even with seasons or with years, the hope is that the estimates of the probabilities obtained from this sample will apply to the population of admissions found in other hospitals in the near future. (Discussing how one might keep up with the changing times and populations would take us beyond the scope of this chapter.) Sometimes tests are evaluated in a sample of patients who are chosen because they either clearly have the disease or they clearly do not. Of course, a most important group of patients to include in evaluating a test are patients who (at least initially) present a confusing diagnostic picture.

1-4. Testing for Pancreatic Cancer

This section discusses the planning and interpretation of diagnostic procedures when several are available to diagnose a particular condition. The background article, "A prospective comparison of current diagnostic tests for pancreatic cancer" (DiMagno *et al.*, 1977), provides information on several tests used to diagnose carcinoma of the pancreas for a selected group of patients. The results help us to evaluate various testing strategies.

> **Clinical Problem 1-3.** *Choosing a Strategy for the Diagnostic Workup*
>
> A 60-year-old widow of one year has a six-month history of upper abdominal pain radiating to her back and a weight loss of 15 lb. Although her symptoms may relate to a depression following the death of her husband, further diagnostic tests investigate the possibility of pancreatic cancer. How should the diagnostic workup proceed?

Background

DiMagno *et al.* describe their experiences with several diagnostic tests used for the diagnosis of pancreatic cancer*:

> Patients were referred for this study if they had abdominal pain, weight loss, or jaundice, but otherwise normal findings on physical examination and routine laboratory tests. . . . From this group . . . subjects (were selected) if they were more than 35 years old and had been experiencing two or more of the following problems, each for 18 months or less: pain in the upper abdomen, pain extending to the back, or night awakening with pain; loss of more than 10 percent of ideal body weight; obstructive jaundice; or unexplained pancreatitis (in patients beyond 50 years of age).

Patients so selected were evaluated by the following tests:

1. *ANG.* Pancreatic angiography
2. *ERP.* Endoscopic retrograde pancreatography (a procedure in which radiopaque dye is injected into the pancreatic duct after a duodenoscope has been used to visualize and catheterize the pancreatic duct)
3. *US.* Ultrasonography of the pancreas
4. *PFT.* Pancreatic function testing (a test in which pancreatic output of enzymes is measured following a dose of cholecystokinin to stimulate pancreatic output)

It was expected that US and PFT could detect pancreatic disease but not distinguish between pancreatic carcinoma and chronic pancreatitis,

---

* All extracts in this chapter from DiMagno *et al.* (1977) are reprinted by permission from E. P. DiMagno, J. R. Malagelada, W. F. Taylor, and V. L. Go, *N. Engl. J. Med.*, **297**:737–42, 1977.

whereas ANG and ERP could diagnose pancreatic carcinoma in distinction
to chronic pancreatic inflammation. Of the 70 patients initially entered into
the study, only 42 had all four diagnostic tests successfully performed. ERP,
US, and PFT were attempted but unsuccessful in 8, 5, and 3 patients,
respectively. DiMagno *et al.* report:

> At the completion of testing, the final diagnosis was established by laparotomy
> in 68 cases and by liver biopsy in two; it is these cases that form the data base
> for this report. Of the 70 patients, 30 had pancreatic cancer, seven had pancrea-
> titis, nine had non-pancreatic but intra-abdominal neoplasms (three lymphoma,
> four carcinoma of small intestine, one ovarian cancer, and one uterine cancer
> with retroperitoneal metastasis), and 24 had miscellaneous non-pancreatic,
> non-neoplastic disease.

Before examining the specific test results, it is worth noting the excellent choice of study patients. All subjects had signs or symptoms raising the
clinical suspicion of pancreatic disease, but the diagnosis was not always
obvious from signs and symptoms alone.

Table 1-4. Four test results and surgical diagnoses in 42 patients who had
all four tests performed[a]

Combination	Test results					Surgical diagnosis			
						PANCREATIC DISEASE		OTHER	
	ANG	ERP	US	PFT	Totals	CANCER	INFLAM-MATORY	CANCER	NON-CANCER
A	−	−	−	−	12				12
B	−	−	−	+	4			2	2
C	−	−	+	−	2			1	1
D	−	+	−	−	0				
E	+	−	−	−	3				3
F	−	−	+	+	0				
G	−	+	−	+	1	1			
H	+	−	−	+	0				
I	−	+	+	−	0				
J	+	−	+	−	1			1	
K	+	+	−	−	2	2			
L	−	+	+	+	2		2		
M	+	−	+	+	2		2		
N	+	+	−	+	1	1			
O	+	+	+	−	2		2		
P	+	+	+	+	10	10			
Totals					42	14	6	4	18

[a] Possible combinations of test results for the four tests are listed on the left and the distribution
of patients to the various outcomes is given in the four columns headed "Surgical diagnosis."
ANG, ERP, US, PFT, +, and − represent angiography, endoscopic retrograde pancreatography,
ultrasonography, pancreatic function test, and positive and negative test results, respectively.

Source: Reprinted by permission from E. P. DiMagno, J. R. Malagelada, W. F. Taylor, and
V. L. Go, *N. Engl. J. Med.,* **297**:737–42, 1977.

Table 1-5.   Sensitivity, specificity, and predictive values of
four tests for detecting pancreatic cancer[a]

	Test			
	ANG	ERP	US	PFT
Sensitivity: $\hat{P}(T+|D+)$	.93	1.00	.71	.86
(14 patients)				
Specificity: $\hat{P}(T-|D-)$	.71	.86	.68	.78
(28 patients)				
Predictive values				
Positive: $\hat{P}(D+|T+)$	.62	.78	.53	.60
(21, 18, 19, and 20 patients)				
Negative: $\hat{P}(D-|T-)$	.95	1.00	.83	.91
(21, 24, 23, and 22 patients)				

[a] The number of patients used to estimate each probability are given in
parentheses. For predictive values, the order of the patient numbers and
test columns correspond.

Table 1-4 shows the test results and the diagnosis in the 42 patients who
had all four tests performed. Although it is difficult to appreciate the performance of each test in this table, such a presentation of data allows the
reader to evaluate different strategies involving one or more tests.

Table 1-5 presents the sensitivity, specificity, and predictive value of
positive and negative tests for the four tests. ERP is overall the most accurate test for the diagnosis of pancreatic carcinoma as judged by all four performance measures. Assuming our patient to be similar to the patients reported by DiMagno *et al.*, a positive ERP would imply a 78% chance of pancreatic carcinoma and a negative ERP would provide strong evidence that
our patient did not have carcinoma of the pancreas. (Although the estimated
predictive value of a negative test is 1.00, based on a sample of 24, sampling
fluctuations could reduce the estimate on another occasion.)

Solution to Clinical Problem 1-3

A skilled endoscopist is required to perform an ERP and the test is often
accompanied by some patient morbidity. For these reasons, DiMagno *et al.*
recommend that

> ultrasonography, a noninvasive test, be performed first. If the ultrasound gives
> negative results, a pancreatic-function test, a procedure of negligible morbidity,
> should be performed next. Positive results from either or both of these proce-
> dures should be followed up with endoscopic retrograde pancreatography.

Figure 1-4 gives the algorithm corresponding to the test strategy. Using
the data in Table 1-4, we can compute its sensitivity and specificity to pancreatic cancer. To compute the sensitivity, we add the number of patients

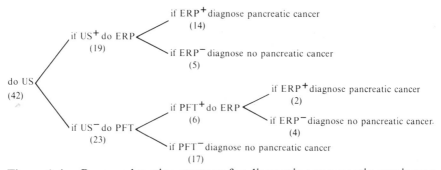

Figure 1-4.   Proposed testing strategy for diagnosing pancreatic carcinoma. The plus (+) or minus (−) appended to the abbreviated test name denotes a positive or negative test result. The number of patients in each category is noted in parentheses.

with pancreatic cancer who had positive test outcomes—namely those with ERP+ and US+, or those with ERP+, US−, and PFT+ (these give combinations I, L, O, P, G, and N of Table 1-4)—with cases $0 + 0 + 0 + 10 + 1 + 1 = 12$, and divide by 14, the number of pancreatic cancer patients. The resulting probability is .86. Similarly, specificity, the probability of diagnosing no pancreatic cancer given no pancreatic cancer, is computed by dividing 24, the number of noncancer patients with negative test outcomes, by 28, the total number of patients without pancreatic cancer. The resulting probability is .86. Thus the strategy involving three tests is not as accurate as performing ERP alone, which had a sensitivity of 1.00 in the 14 patients with pancreatic cancer and the same specificity, .86. The sequential test strategy avoids performing ERP in 14 patients.

A final judgment of which diagnostic workup is better, ERP alone or the sequential test strategy proposed by the authors, hinges not only on the number of correct diagnoses but also on the "cost" of missing a diagnosis, the "benefit" of correct diagnosis, and the cost and morbidity of the test procedures used. These considerations are addressed further in Chapter 3.

In the DiMagno *et al.* example, no test strategy involving multiple tests leads to a more accurate separation of patients into those with and without pancreatic carcinoma than ERP alone. This is not surprising in view of the accuracy of ERP reported.

Multiple testing can provide additional information by allowing the identification of several groups of patients, each with a different probability of pancreatic carcinoma. Table 1-6 shows how performing all 4 tests on the group of patients studied by DiMagno *et al.* separates the subjects into 5 groups with different probabilities of pancreatic carcinoma based on the number of positive test results obtained on each patient. Among the 12 people with 0 positive tests and the 9 with 1 positive test, none had pancreatic cancer. The 10 patients with all 4 tests positive all had pancreatic cancer. Those with 2 and 3 positive tests had a .75 and .14 probability of pancreatic cancer, respectively, the order being different from what we might anticipate.

Table 1-6. Number and proportion of patients with pancreatic cancer, by the number of positive tests

	Number of positive tests				
	0	1	2	3	4
Number of patients (42 total)	12	9	4	7	10
Number with pancreatic cancer	0	0	3	1	10
Proportion with pancreatic cancer	.00	.00	.75	.14	1.00

Source: Reprinted by permission from E. P. DiMagno, J. R. Malagelada, W. F. Taylor, and V. L. Go, *N. Engl. J. Med.* **297:**737–42, 1977.

An alternative method of grouping is shown in Table 1-7. Here we have divided the patients into those with 1 or more, 2 or more, 3 or more, and 4 positive. (You are asked to fill in the probabilities for 2 or more tests in Problem 1-5.) As our test requirements become more stringent, sensitivity decreases in favor of specificity; simultaneously, negative predictive value decreases in favor of positive predictive value.

Simply counting the number of positive results, as in Table 1-6, gives the tests equal weight. We already know that ERP is the best test and that they are not all equally strong. Suppose that our patient has a positive ERP but negative US, PFT, and ANG. Using Table 1-6, one might say that pancreatic carcinoma was unlikely in such patients because the observed proportion is .00 in 9 patients with 1 positive test. However, by reviewing Table 1-4, we see that no patient had the specific pattern of test results (combination D). Our doubts about using this method for evaluating these contradictory test results are heightened because ERP is the most sensitive of the four tests. Ideally, we would find other cases with the same combination of results and determine the diagnosis in these instances. Such information is often difficult to find, as we have just observed in Table 1-4.

Table 1-7. Sensitivity, specificity, and predictive values of the number of positive tests out of 4 for diagnosing pancreatic cancer[a]

	Number of positive tests			
	1 OR MORE	2 OR MORE[b]	3 OR MORE	4
Sensitivity: $\hat{P}$(positive|D+) (14 patients)	1.00	*	.79	.71
Specificity: $\hat{P}$(negative|D−) (28 patients)	.43	*	.79	1.00
Predictive values				
Positive: $\hat{P}$(D+|positive) (30, *, 17, and 10 patients)	.47	*	.65	1.00
Negative: $\hat{P}$(D−|negative) (12, *, 25, and 32 patients)	1.00	*	.88	.88

[a] The number of patients used to estimate each probability is given in parentheses. For predictive values, the order of patient numbers and positive tests numbers correspond.

[b] For the values indicated by asterisks, see Problem 1-5.

A review of pancreatic imaging techniques by Van Dyke *et al.* (1985) pooled information from several studies of patients with various pancreatic diseases, including neoplasm. Table 1-8 shows the number of individuals, the percent of individuals on whom a satisfactory test could not be performed (the failure rate), and the sensitivity and specificity of US, ERP, ANG, and computed tomography.

Table 1-8. Failure rate, sensitivity and specificity of four pancreatic imaging methods

	Patients studied	Failure rate (%)	Sensitivity	Specificity
Ultrasound	908	14	.78	.89
Endoscopic retrograde pancreatography	376	6	.94	.97
Angiography	100	11	.98	.95
Computed tomography	300	1	.92	.90

ERP has a high sensitivity and specificity, as does ANG. Both these tests are invasive and accompanied by substantial morbidity. Computed tomography is a good screening test as it is not invasive, has a low failure rate, and a relatively good sensitivity and specificity.

The review provided no information about testing strategies involving two or more imaging techniques.

Since few, if any, tests are perfect discriminators of disease, conflicting results are always possible when more than one test is performed. The more tests ordered, the more likely that at least one contradictory result will be reported. In the DiMagno study, about half the patients had contradictory test results. As shown in Table 1-6, patients with multiple test results which are all in agreement usually can be regarded as having a fairly accurate diagnosis, whereas the diagnosis in patients with contradictory results should be regarded as less certain.

Medicine could use many more studies that compare the simultaneous performances of several tests, such as that of DiMagno *et al.*, especially in the light of the wide use of multichannel tests. Without such studies we cannot calculate the additional benefits received when several tests are used for diagnosis.

1-5. Summary

Sensitivity, $P(T+|D+)$, and specificity, $P(T-|D-)$, are measures of test accuracy that do not depend on the overall probability of disease. The predictive values of a test, $P(D+|T+)$ and $P(D-|T-)$, depend both on test accuracy and the probability of disease. For any given probability of disease, Bayes theorem provides a rule by which the predictive values can be computed from the sensitivity and specificity when they are known. Testing accuracy is not the only test characteristic important in planning a patient workup; cost and the risk of morbidity must also be considered.

Problems

1-1. Let A and B indicate two events, with

$$P(A|B) = .2$$

$$P(\text{not } A|\text{not } B) = .4$$

and

$$P(B) = .3$$

  a. Use Bayes theorem to compute $P(B|A)$.
  b. Construct the 2×2 probability table with columns (A, not A) and rows (B, not B). Compute $P(B|A)$ directly from the table.

1-2. Two tests T_1 and T_2 are available for the detection of condition D with the characteristics.

$$P(T_1+|D+) = .90 \qquad P(T_1-|D-) = .80$$

and

$$P(T_2+|D+) = .80 \qquad P(T_2-|D-) = .90$$

What properties of condition D and its treatment might influence which test you prefer? For example, does it matter if condition D is early cancer of the cervix and treatment is biopsy, or if D is cancer of the pancreas and treatment is surgery?

1-3. Suppose that given D+ or D−, tests 1 and 2 (Problem 1-2) are independent. Compute the sensitivity and specificity of the procedure "diagnose disease if and only if both tests are positive."

1-4. Narain *et al.* (1971) investigated the AFB smear as a diagnostic test for pulmonary tuberculosis. AFB smears and AFB cultures were obtained from 6453 residents of the Bangalore district, southern India. There were 142 positive smears and 180 positive cultures. In 98 instances, the smear was negative and the culture positive.
  a. Construct a 2×2 table showing the relation between smear results and culture results.
  b. What is the sensitivity and specificity of the AFB smear as a test for tuberculosis? (Assume that the culture results are the "gold standard" for the diagnosis of tuberculosis.)
  c. An elderly alcoholic comes into the hospital with a cough. Chest x-ray shows a right upper lobe infiltrate. Use your clinical skills to "guess" the probability of tuberculosis in this patient. You obtain a sputum sample and smear it for AFB. It is positive. What is the probability of active tuberculosis in light of this test result?

1-5. Consider the following strategy for the diagnosis of pancreatic cancer using four tests (ERP, US, PFT, ANG). Refer to the data of DiMagno *et al.*
  (i) If two or more of the tests are positive, diagnose pancreatic cancer.
  (ii) If more than two of the tests are negative, diagnose no pancreatic cancer.
  a. Set up a 2×2 table for this strategy, cross-classifying diagnosis with pancreatic cancer for the data shown in Table 1-4.
  b. Compute the sensitivity, specificity, and predictive values of this strategy and fill in Table 1-7.

    c. How does this strategy compare to ERP alone?

    d. Are the results consistent with what we expected in Table 1-7?

1-6. Read "Ultrasonic and radiographic cholecystography" (Bartrum *et al.*, 1977) and use it to respond to the following questions:

    a. What are the values of the sensitivity and specificity of the tests? (Exclude the indeterminate results.)

    b. Do false negative rate and false positive rate relate to sensitivity and specificity?

    c. What is the "gold standard" for diagnosis?

    d. Does an indeterminate ultrasonogram or cholecystogram carry any diagnostic information?

References

Bartrum, R. J.; Crow, H. C.; and Foote, S. R. (1977). Ultrasonic and radiographic cholecystography. *N. Engl. J. Med.*, **296**:538–41.

Beller, G. A.; Smith, T. W.; Abelmann, W. H.; Haber, E.; and Hood, W. B. (1971). Digitalis intoxication: a prospective clinical study with serum level correlations. *N. Engl. J. Med.*, **284**:989–97.

DiMagno, E. P.; Malagelada, J. R.; Taylor, W. F.; and Go, V. L. (1977). A prospective comparison of current diagnostic tests for pancreatic cancer. *N. Engl. J. Med.*, **297**:737–42.

Hayes, C. J.; Butler, V. P.; and Gersony, W. M. (1973). Serum digoxin studies in infants and children. *Pediatrics*, **52**:561–68.

Krasula, R.; Yanagi, R.; Hastreiter, A. R.; Levitsky, S.; and Soyka, L. F. (1974). Digoxin intoxication in infants and children: correlation with serum levels. *J. Pediatr.*, **84**:265–69.

Narain, R.; Subba Rao, M. S.; Chandrasekhar, P.; and Pyarelal (1971). Microscopy positive and microscopy negative cases of pulmonary tuberculosis. *Am. Rev. Respir. Dis.*, **103**:761–73.

Van Dyke, J. A.; Stanley, R. J.; and Berland, L. L. (1985). Pancreatic imaging. *Ann. Intern. Med.*, **102**:212–17.

Additional Reading

Emerson, J. and Colditz, G. Uses of statistical analysis in the *New England Journal of Medicine*. In Bailar, J. and Mosteller, F., eds. (1986) *Medical Uses of Statistics*, chapter 2. Waltham, MA. *New England Journal of Medicine*.

CHAPTER 2

Diagnostic Testing: Likelihood and Odds

OBJECTIVES

Prior probability

Odds

The likelihood ratio as a measure of test accuracy

Posterior probabilities: adjusting prior probabilities and prior odds with new diagnostic information

The binomial distribution

Using the binomial distribution to estimate the likelihood ratio

Abnormal results from multichannel tests

This chapter reemphasizes the two steps involved in interpreting a diagnostic test. In step 1 the physician estimates the **prior probability** of disease. The prior probability is the probability of disease given the history, physical findings, and previous diagnostic test results for the individual patient, as well as knowledge about disease frequency in various populations of patients. In step 2 the physician modifies this probability based on the new test result and information about test accuracy to obtain the **posterior probability** of disease. The posterior probability is the probability that the patient has the disease, given both the initial information and the test result.

In Chapter 1 we used Bayes theorem to compute the posterior probability. Here, we introduce the notion of odds, a concept related to probability. We also introduce the likelihood ratio as a measure of test information and describe a simple method for computing the posterior odds in favor of disease from the prior odds and the likelihood ratio.

Much of the clinical material in this chapter deals with the diagnosis of the hemophilia carrier state. We introduce the binomial distribution to compute the probability distribution for the number of diseased males in affected families. We also apply the binomial distribution to the interpretation of multichannel diagnostic tests.

2-1. Using Measures of Factor VIII Activity and Factor VIII Antigen to Detect Carriers

Background

Genetic hemophilia is a bleeding disorder caused by a genetic defect in the activity of coagulation factor VIII, also called antihemophilic globulin or AHG. It is the most common genetic disorder affecting coagulation factors. The disease is sex linked, so that overt disease appears in males whose X chromosome has the abnormal gene and, at least theoretically, in females whose two X chromosomes are affected. (In fact, females with hemophilia are rare, and in them the disease may have a more complex genetic mechanism.) Females with one affected X chromosome are called "carriers"; they do not have overt hemophilia but can transmit the affected gene to their offspring. Since the disease is determined by a single gene in males, males cannot be carriers without having overt clinical hemophilia. The disease can also occur as a result of a new mutation.

In about 30% of hemophilia cases, no family history of hemophilia can be elicited. Such cases occur because of random variation in passage of the gene (it can be passed, unrecognized, through a succession of females before appearing in a male descendant), or because of incomplete family histories, or because of new mutations. It is estimated that 15% of cases with no family history, or 4.5% of total cases, arise from a new mutation.

For many years, daughters of known carriers knew only that they had a 50% chance of being a carrier. When methods for measuring the level of factor VIII were developed, they showed that carriers frequently had lower factor VIII activity than normals. However, the distributions of factor VIII activity in carriers and normals were not sufficiently different, and the assay for factor VIII activity was a poor test for the carrier state. Around 1970, investigators discovered that hemophiliacs produce normal amounts of a molecule that is immunologically identical to factor VIII but with no factor VIII activity. The hemophilic gene of carriers produces about half their antigen, and this half has no factor VIII activity. Hence the ratio of factor VIII activity to factor VIII antigen tends to be lower in carriers than in normals, thus providing a basis for diagnostic testing.

Rizza *et al.* (1975) compared the ratio of factor VIII activity to factor VIII antigen for 34 normals and 34 obligatory carriers. Defining a positive test as a factor VIII ratio of .8 or lower, we obtain the test results shown in Table 2-1 for the 68 study participants. The estimated sensitivity of this test is

$$\hat{P}(T+|D+) = \frac{32}{34} = .94$$

and the estimated specificity is

$$\hat{P}(T-|D-) = \frac{28}{34} = .82$$

Table 2-1. Factor VIII
ratio test results for 34
normals and 34 obligatory
hemophilia carriers[a]

		Carrier state D+	D−	
Test result	T+	32	6	38
	T−	2	28	30
		34	34	68

[a] D+, hemophilia carrier;
D−, normal woman (noncarrier);
T+, positive test (factor VIII ratio
≤ .8);
T−, negative test (factor VIII ra-
tio > .8).

Source: Rizza *et al.* (1975).

In the following pages, we utilize these data to modify the probability of disease for individual patients based on their factor VIII ratios.

Clinical Problem 2-1. *Is the Patient a Carrier of Hemophilia?*

Ms. R.P. is 24 years old. One of her brothers and a maternal uncle
have hemophilia; her father and another brother are free of the disease. (She does not have hemophilia.) Her factor VIII antigen is
110% normal with an activity of 57% normal, giving a factor VIII
ratio of .52 and a positive test result. What is the probability that
she is a carrier?

Solution Using Bayes Rule

Ms. R.P.'s family history clearly indicates that her mother was a carrier. If we had no laboratory results, we would say that she has a 50-50
chance (probability .5) of being a carrier. This is the prior probability. A
factor VIII ratio of .52 is a positive test result, and Table 2-1 shows that, for
carriers,

$$\hat{P}(T+|D+) = \frac{32}{34} = .94$$

while for noncarriers,

$$\hat{P}(T+|D-) = \frac{6}{34} = .18$$

Applying Bayes theorem as in Chapter 1, we compute

$$\hat{P}(D+|T+) = \frac{\hat{P}(T+|D+)P(D+)}{\hat{P}(T+|D+)P(D+) + \hat{P}(T+|D-)P(D-)}$$

$$= \frac{(.94)(.50)}{(.94)(.50) + (.18)(.50)}$$

$$= .84$$

Given the outcome of the diagnostic test, Ms. R.P. has probability .84 of being a hemophilia carrier. This is the posterior probability.

Although Bayes theorem provides a reliable method for modifying disease probabilities based on diagnostic test results, the method seems unintuitive to many and is difficult to remember. When we remember the sensitivity and specificity of a test, there is no quick way to apply these values to a prior probability of disease. Perhaps this is why many people remember the predictive value of a test, a value that applies only at one specific prior probability of disease (see Chapter 1). When we perform the same calculation in terms of the **odds** in favor of disease, the corresponding calculations are quite simple and easy to remember. We define the odds in favor of disease and demonstrate the procedure in the next section.

2-2.   Using Odds Instead of Probability

> We define the odds in favor of A (against not A) as:
>
> $$\text{Odds in favor of } A = \frac{P(A)}{P(\text{not } A)} = \frac{P(A)}{1 - P(A)} \qquad (2\text{-}1)$$

For example, if $P(A) = 2/3$, then the odds in favor of A have the ratio 2 to 1, or 2. The probability, $P(A)$, is expressed in terms of the odds as

> $$P(A) = \frac{\text{odds}}{1 + \text{odds}} \qquad (2\text{-}2)$$

Some simple examples:

> The odds in favor of heads when a coin is tossed have a ratio of 1 to 1, or 1.
> The odds in favor of rolling a 6 on any throw of a fair die have a ratio of 1 to 5 or 1/5, equal to 0.2.
> The odds against rolling a 5 are 5 to 1, or 5.
> The odds in favor of drawing an ace from an ordinary deck of playing

cards are 1 to 12, or 1/12. The odds against drawing an ace are 12 to 1, or 12.

If B denotes a test result or other new information, the **posterior odds** in favor of A are defined as:

$$\text{Posterior odds in favor of } A = \frac{P(A|B)}{P(\text{not } A|B)}$$

$$= \frac{P(A|B)}{1 - P(A|B)}$$

Recall that

$$P(A|B) = \frac{P(A \text{ and } B)}{P(B)} = \frac{P(A)P(B|A)}{P(B)} \tag{2-3}$$

and

$$P(\text{not } A|B) = \frac{P(\text{not } A \text{ and } B)}{P(B)} = \frac{P(\text{not } A)P(B|\text{not } A)}{P(B)} \tag{2-4}$$

Then the posterior odds in favor of A can be computed from the ratio of expressions (2-3) and (2-4). We obtain:

$$\text{Posterior odds in favor of } A = \frac{P(A|B)}{P(\text{not } A|B)} = \frac{P(A)P(B|A)}{P(\text{not } A)P(B|\text{not } A)}$$

$$= \frac{P(A)}{P(\text{not } A)} \times \frac{P(B|A)}{P(B|\text{not } A)} \tag{2-5}$$

Thus we have the fundamental relationship:

$$\begin{array}{c|c} \begin{array}{c}\text{Posterior odds} \\ \text{in favor of } A\end{array} = \begin{pmatrix}\text{prior odds in} \\ \text{favor of } A\end{pmatrix} \times \begin{pmatrix}\text{likelihood} \\ \text{ratio}\end{pmatrix} & (2\text{-}6) \end{array}$$

Equations (2-5) and (2-6) show that the posterior odds can be written as the prior odds multiplied by what is called the likelihood ratio. The likelihood ratio is the ratio of the conditional probabilities of B given the conditions A and not A, namely $P(B|A)/P(B|\text{not } A)$.

The prior odds are a ratio of probabilities, the prior probabilities for and against disease. The likelihood ratio is also a ratio of probabilities. When interpreting a positive test, it is the ratio obtained when the probability of a positive test given disease is divided by the probability of a positive test given no disease. Equivalently, it is the sensitivity divided by $1 -$ specificity. This is the likelihood ratio for a positive test, abbreviated LR$+$. The likelihood ratio for a negative test, abbreviated LR$-$, is the probability of a negative test given disease divided by the probability of a negative test given no disease. Thus the LR$-$ is $1 -$ sensitivity divided by the specificity. To sum up:

$$LR+ = \frac{\text{sensitivity}}{1 - \text{specificity}}$$

$$LR- = \frac{1 - \text{sensitivity}}{\text{specificity}}$$

For the test based on the factor VIII ratio, we estimate

$$LR+ = \frac{\frac{32}{34}}{\frac{6}{34}} = \frac{32}{6} = 5.3$$

$$LR- = \frac{\frac{2}{34}}{\frac{28}{34}} = \frac{2}{28} = .07$$

These two numbers summarize all the diagnostic information in Table 2-1.

Solution to Clinical Problem 2-1 Using Odds

We can apply this approach to obtain the odds that Ms. R.P. is a carrier. Let A represent the carrier state and B represent a positive test. The prior odds in favor of A are .5/.5, or 1, and the $LR+$ is 5.3, so

Posterior odds in favor of $A = 1 \times 5.3 = 5.3$

The odds are 5.3 to 1 in favor of Ms. R.P. being a carrier.
    To convert odds back to probabilities, recall that

$$P(A) = \frac{\text{odds}}{1 + \text{odds}} \tag{2-7}$$

Thus, to convert the odds in favor of A to the posterior probability of A, we divide the odds by 1 plus the odds. In the example above, the posterior odds of being a carrier given a positive test result are 5.3. Then

$$\hat{P}(\text{Ms. R.P. is a carrier}|T+) = \frac{5.3}{1 + 5.3} = \frac{5.3}{6.3} = .84$$

which agrees with the result we got through Bayes theorem. As in Chapter 1, the circumflex above P indicates that this probability is estimated, in this instance because $LR+$ was estimated based on a sample of 68 patients.
    Suppose that Ms. R.P.'s test had been negative. How would this change her posterior probability of disease? The prior odds in favor of disease are still 1, but $LR-$ equals .07. So

Posterior odds in favor of $A = 1 \times .07 = .07$

and

$$\hat{P}(\text{Ms. R.P. is a carrier}|T-) = \frac{.07}{1 + .07}$$
$$= .065$$

These calculations demonstrate an important feature of diagnostic testing when we have two disease states and two test outcomes:

> The effect of a diagnostic test on the posterior probability of disease depends on only two quantities, LR + and LR −.

Clinical Problem 2-2.   *Using Family History to Compute the Carrier Probability*

Ms. Cook has a maternal uncle who is a hemophiliac, indicating that her grandmother is a carrier. However, her mother had no sons. Ms. Cook's factor VIII ratio is .97, a negative test. Estimate the probability that she is a carrier.

Solution

Since Ms. Cook's grandmother was a carrier, her mother was a carrier with probability .5, and Ms. Cook's prior probability of being a carrier is .25. However, the test information, expressed as LR − for a negative test, remains unchanged. Since LR − equals .07, we can apply formula (2-6) and get the odds in favor of being a carrier as

Posterior odds = prior odds × LR −

$$= \frac{.25}{.75} \times .07 = .023$$

To compute the posterior probability that Ms. Cook is a carrier, we have

$$\hat{P}(\text{Ms. Cook is a carrier}|T-) = \frac{\text{odds}}{1 + \text{odds}} = \frac{.023}{1.023} = .022$$

or about .02.

Clinical Problem 2-3.   *Computing the Carrier Probability When There Is No Family History of Hemophilia*

Ms. K.I. is a 22-year-old woman whose neighbor's newborn son has been discovered to have hemophilia. Ms. K.I.'s family has no history of hemophilia. Her factor VIII ratio is .65, a positive test. What is the chance that she is a carrier?

Solution: Using Prior Probabilities Based on Population Prevalence

The prevalence of carriers in the general population is believed to be about 3 per 10,000 women. Thus the prior probability for Ms. K.I. is

$$\hat{P}(\text{Ms. K.I. is a carrier}) = .0003$$

and the prior odds for disease are

$$\frac{.0003}{.9997}$$

The occurrence of hemophilia in a neighbor does not affect this estimate.

The estimated likelihood ratio corresponding to a positive test is 5.3 once again. Consequently, the posterior odds are:

$$\text{Posterior odds} = \frac{.0003}{.9997} \times 5.3$$

$$= .0016$$

and the estimated posterior probability that Ms. K.I. is a carrier is

$$\hat{P}(\text{Ms. K.I. is a carrier} | T+) = \frac{.0016}{1.0016}$$

$$= .0016$$

The posterior probability that Ms. K.I. is a carrier is still small, as it would be even with a likelihood ratio of 100. While a factor VIII ratio of .65 is rare in a normal woman, the occurrence of the carrier state in the absence of any family history is even more unusual. This woman is almost certainly normal.

Why use odds instead of the more familiar probabilities? First and most important, we can quickly modify prior odds using diagnostic information. The calculations require only that we recall $LR+$ and $LR-$, and then they can be performed without pencil and paper. Second, the likelihood ratio becomes a meaningful and intuitive measure of test accuracy. For instance, an $LR+$ of 4 indicates that persons with the disease are 4 times as likely to test positive as those without the disease. Finally, we can sometimes multiply the likelihood ratios of several diagnostic tests to measure their diagnostic value when performed jointly. Since this calculation depends on the independence of the several tests, we discuss it later in the chapter.

2-3. Hemophilia and Family History: Is the Patient a Hemophilia Carrier?

Clinical Problem 2-4. *Computing the Carrier Probability When There Are Normal Sons*

Ms. G.R. is a 42-year-old woman who has 2 sons, neither of whom has hemophilia. However, her brother is a hemophiliac, indicating that her mother is a carrier. Ms. G.R. has a factor VIII ratio of .70. She is concerned that she is a carrier.

Table 2-2. Distribution of children's genotypes for various parental genotypes assuming no mutations

Mother's genotype	Father's genotype	
	NORMAL	DISEASED
NORMAL	All normal	All males normal All females carriers
CARRIER	Half of males normal Half of males diseased Half of females normal Half of females carriers	Half of males normal Half of males diseased Half of females carriers Half of females diseased
DISEASED	All males diseased All females carriers	All diseased

Solution by Computing the Odds of Disease

A factor VIII ratio of .70 is a positive test result, corresponding once again to a likelihood ratio (LR +) of 5.3. On the other hand, computing the prior odds that Ms. G.R. is a carrier is somewhat more difficult. Since her mother is a carrier, we might say that Ms. G.R.'s probability of being a carrier is .5 and that her odds of being a carrier are .5/.5, or 1. But this calculation ignores the fact that she has 2 normal sons. Certainly, this information must affect the probability that she is a carrier. To understand how Ms. G.R.'s 2 normal sons influence her odds, we need to compute the probability that families of carriers with 2 sons will have no hemophiliacs.

Table 2-2 gives the expected proportion of hemophiliacs and carriers among children born to each possible union of parental genotypes. For instance, we expect half of the male offspring of a normal father and a carrier mother to be normal and half to be hemophilic. Of course, the proportion of normals will vary in individual families. The halves just mentioned are long-run averages.

A family with 1 male child could have either 0 diseased or 1 diseased child with equal probability. If such a family were to add a male child, it would be diseased or not diseased (normal), leading to 4 possible family patterns, as shown in Table 2-3. Since the 4 family types occur with equal frequency, a family of 2 male children will have 0, 1, or 2 diseased offspring

Table 2-3. Possible outcomes for families with 2 male offspring of a normal father and carrier mother

Family	First male	Second male	No. diseased males in family
a	Normal	Normal	0
b	Diseased	Normal	1
c	Normal	Diseased	1
d	Diseased	Diseased	2

with the following probabilities:

Number diseased	0	1	2	
Probability	1/4	2/4	1/4	(2-8)

This is called the **probability distribution** for the number of diseased children in families with a normal father, a carrier mother, and 2 male children.

> The probability distribution for any variable gives the probability for each possible outcome, in this instance, 0, 1, or 2 diseased children.

How does this help us with Ms. G.R.? If she were a carrier, then the probability that neither of her 2 sons would be hemophiliacs is .25. Also, if she were not a carrier, the probability of 2 normal sons is 1.0, if we neglect the small probability of mutation. Moreover, the probability that Ms. G.R. receives a defective gene from her mother, a carrier, is .5. Now we can use our rule for combining prior odds with a likelihood ratio to compute the posterior odds that Ms. G.R. is a carrier, given her sons' hemophilia status. In this instance, the prior odds, based on her mother's carrier status, are .5 to .5, or 1. The "diagnostic test" is the outcome for her children, in this instance, 2 normal sons. Thus we compute the likelihood ratio for the data "2 normal sons." The posterior odds that Ms. G.R. is a carrier, given that neither of her sons had hemophilia, is computed from equation (2-6) as

$$\text{Posterior odds} = \text{prior odds} \times \text{likelihood ratio}$$

$$= \frac{P(\text{Ms. G.R. is a carrier})}{P(\text{Ms. G.R. is normal})}$$

$$\times \frac{P(\text{two normal sons}|\text{Ms. G.R. is a carrier})}{P(\text{two normal sons}|\text{Ms. G.R. is normal})}$$

$$= \frac{.50}{.50} \times \frac{.25}{1.00} = .25$$

Odds of .25 convert to a probability of $.25/1.25 = .20$. Therefore, knowing that Ms. G.R. had two healthy sons reduced the probability that she is a carrier from .50 to .20. This probability, .20, that Ms. G.R. is a carrier now becomes the prior probability of disease to be used with the results of diagnostic testing. Combining this prior probability with the likelihood ratio based on a positive test, we have the posterior odds that Ms. G.R. is a carrier:

$$\text{Posterior odds} = \text{prior odds} \times \text{likelihood ratio}$$
$$= .25 \times 5.3 = 1.33$$

Converting to posterior probability, we have $1.33/2.33 = .57$, the posterior probability that Ms. G.R. is a carrier.

Comment on Multiple Tests for the Same Disease

The foregoing example has suggested that we can successively update the odds of disease by multiplying by the likelihood ratio computed from new information. This is correct if the new information is statistically independent from that previously incorporated or, equivalently, if the likelihood ratio is computed conditionally on the previous information. However, in many cases, the independence of two or more tests has not been investigated.

Refinements

By reporting the factor VIII ratio as well as a positive or negative test result, we have suggested that the value of the factor VIII ratio is more informative than a positive or negative test result. This is sometimes true. Methods have been developed for computing likelihood ratios from the numeric value of a laboratory measurement, but their description is beyond the scope of this book.

2-4.   Binomial Distribution When Outcomes Are Equally Likely

Preliminaries

From Table 2-3 we deduced the probabilities of each family type (0, 1, 2 diseased) and ultimately used one of these probabilities to modify the probability that Ms. G.R. is a carrier. Suppose, however, that Ms. G.R. had 3, 4, or more sons and we were interested in the number of diseased sons. How would we compute the probability distribution? In general, how would we compute the probability that d sons would have hemophilia in a family with n sons? The binomial distribution provides a method of computing these probabilities.

To provide some insight into the binomial distribution, we can construct the distribution of the number of diseased males in families with a normal father and a carrier mother, and 3 or 4 sons. We have already given the distribution of the number of diseased males in families with 2 males in tabulation (2-8). To find the probabilities for the number of hemophiliacs among 3 male children, start with Table 2-3. Half the time, the third son will not be diseased. The contribution of this outcome to the probability distribution for families with 3 sons is

Number diseased	0	1	2	3
Probability contribution	$\frac{1}{8}$	$\frac{2}{8}$	$\frac{1}{8}$	0

(2-9)

We obtain the probability contribution by multiplying the probabilities in distribution (2-8) by ½.

The other half of the time, the additional child is diseased. These outcomes add 1 to the number diseased in tabulation (2-8) and multiply the probabilities by ½ to give the following probability contribution:

Number diseased	0	1	2	3
Probability contribution		$\frac{1}{8}$	$\frac{2}{8}$	$\frac{1}{8}$

(2-10)

Adding contribution (2-9) and contribution (2-10) gives

Number diseased	0	1	2	3
Probability	$\frac{1}{8}$	$\frac{3}{8}$	$\frac{3}{8}$	$\frac{1}{8}$

(2-11)

Tabulation (2-11) gives the probability distribution for the number of diseased males out of 3 male offspring when the father is normal and the mother is a carrier. Table 2-4 gives the distribution for families having 1, 2, 3, or 4 male children when the father is normal and the mother is a carrier.

Binomial Formula for Equally Likely Outcomes

When we deal with 50-50 situations, as we have for male offspring of a normal father and carrier mother, a special formula applies. Since each male child is either diseased or not diseased, in n males there are $2 \times 2 \times \cdots \times 2 = 2^n$ possible families. This accounts for the powers of 2 in the denominators of the probabilities in Table 2-4. The numerators are called binomial coefficients. They give the number of birth orders in which exactly d diseased males can occur among the n male offspring. The symbol for such a coefficient is $\binom{n}{d}$, often read "n choose d." Note that no fraction bar appears between the n and the d. To evaluate this quantity, we need the idea of a factorial. A **factorial** is the product of descending integers. For instance, "5 factorial" is the number $5 \times 4 \times 3 \times 2 \times 1$, or 120. We use the shorthand symbol

$$5! = 5 \times 4 \times 3 \times 2 \times 1$$

By convention, $0! = 1$.

Table 2-4. Probabilities of each number of diseased males among sons of a normal father and a carrier mother

Number	Number diseased				
of sons	0	1	2	3	4
1	$^1/_2$	$^1/_2$			
2	$^1/_4$	$^2/_4$	$^1/_4$		
3	$^1/_8$	$^3/_8$	$^3/_8$	$^1/_8$	
4	$^1/_{16}$	$^4/_{16}$	$^6/_{16}$	$^4/_{16}$	$^1/_{16}$

> The formula for the binomial coefficient $\binom{n}{d}$ is
>
> $$\binom{n}{d} = \frac{n!}{d!(n-d)!}$$

If $n = 4$ and $d = 2$, we get

$$\binom{4}{2} = \frac{4!}{2!2!} = \frac{4 \times 3 \times 2 \times 1}{2 \times 1 \times 2 \times 1} = 6$$

This is the numerator of the probability of 2 diseased males in a family of 4 males. The denominator is $2^4 = 16$. The ratio gives the probability 6/16, which agrees with the value 3/8 in Table 2-4.

In general, the probability of d diseased males among n males in this special inheritance problem is

$$P(d \text{ diseased}|n \text{ males}) = \frac{\binom{n}{d}}{2^n}$$

EXAMPLE. Find the probability that in a family of 6 males with a normal father and a carrier mother, exactly 2 are diseased.

SOLUTION. We can apply the formula for $n = 6$ and $d = 2$. Then

$$P(d = 2|n = 6) = \frac{\binom{6}{2}}{2^6} = \frac{\frac{6!}{2!4!}}{64}$$

$$= \frac{15}{64}$$

2-5.   General Binomial Formula

Clinical Problem 2-5.   *What Are the Chances of Tay-Sachs Disease?*

Mr. and Mrs. R. are both known to be carriers of Tay-Sachs disease. From such parents, each child has (independently) the probability .25 of being born with Tay-Sachs disease. The disease is fatal in early childhood. It is possible to diagnose Tay-Sachs in early pregnancy and then abort the pregnancy if necessary. However, Mr. and Mrs. R. are opposed to abortion of any kind. What is the probability distribution of the number of children with Tay-Sachs disease if they have 2 children?

Solution: Computing Binomial Probabilities

Using the probability of a diseased child (.25), we can compute the probabilities of each of the possible events, 0, 1, or 2 diseased. The first child is diseased with probability .25; similarly, the second child is diseased with probability .25. Thus the probability that both are diseased is

$$P(2 \text{ diseased} | 2 \text{ offspring}) = (.25)(.25)$$
$$= .0625$$

For the event "1 child diseased," we need to compute the probabilities for the two events "first diseased, second not" and "first not, second diseased." These probabilities are $.25(1 - .25) = .1875$ and $(1 - .25).25 = .1875$, respectively. The probabilities are equal, because of the constant probability of disease and the independence of the disease state between the offspring. Adding these two probabilities, we have

$$P(1 \text{ diseased} | 2 \text{ offspring}) = .1875 + .1875$$
$$= 2(.1875) = .375$$

Finally, we want to compute the probability of 0 diseased children out of 2 offspring. We might simply note that the probabilities of the mutually exclusive, exhaustive events, 0, 1, and 2 diseased children, must add to 1, then solve for $P(0 \text{ diseased} | 2 \text{ offspring})$ by subtraction. Alternatively, following our previous arguments, we note that the probability of a healthy child is $1 - .25 = .75$. It follows that the probability of two healthy children is

$$P(0 \text{ diseased} | 2 \text{ offspring}) = (.75)(.75)$$
$$= .5625$$

Thus for Mr. and Mrs. R., the distribution of diseased children is given by:

Number diseased	0	1	2	
Probability	.5625	.3750	.0625	(2-12)

We computed the probabilities above by establishing the probability of disease for each child and the number of offspring and then doing a complete examination of the possibilities. Although this is a reasonable method for a small number of possibilities, it quickly becomes unmanageable as the number increases. We use a general formula for the binomial distribution in such cases. The binomial probabilities depend on three quantities: d, the number of positives (in our example, children with Tay-Sachs); n, the number of trials (in our example, 2 offspring); and π, the probability of positive for a single trial (in our example, the probability of Tay-Sachs disease, .25).

> Then the probability of d successes is
> $$P(d|n, \pi) = \binom{n}{d} \pi^d (1 - \pi)^{n-d} \qquad (2\text{-}13)$$
> for $d = 0, 1, \ldots, n$.

Formula (2-13) is comprised of three factors:

$\binom{n}{d}$, the binomial coefficient

π^d, the probability of d positives in d trials

$(1 - \pi)^{n-d}$, the probability of $n - d$ negatives in $n - d$ trials

Comment on Notation

The use of the symbol π to denote the probability of a positive response is one of several instances in which we use a Greek letter to denote an unknown quantity. Another frequently used symbol for the probability of a positive response is p. Since we subsequently use $\bar{p}$ to denote the **observed** proportion of positive responses, we use π for the **expected** proportion to avoid confusion.

EXAMPLE.   Find the probabilities corresponding to $d = 0$, 1, and 2 when $n = 2$ and $\pi = .25$ by using the general binomial formula.

SOLUTION.   Applying formula (2-13) with $n = 2$ and $\pi = .25$ gives

$$P(d|n = 2, \pi = .25) = \binom{2}{d} (.25)^d (.75)^{2-d}$$

The components are, for $d = 0$:

$$\binom{2}{0} = \frac{2!}{0!(2-0)!} = 1$$

$$(.25)^0 = 1$$

$$(.75)^{2-0} = (.75)^2 = .5625$$

So putting them together gives

$$P(d = 0|n = 2, \pi = .25) = 1(1)(.5625)$$
$$= .5625$$

Similarly, for $d = 1$:

$$\binom{2}{1} = \frac{2!}{1!(2-1)!} = \frac{2 \times 1}{1 \times 1} = 2$$

$$(.25)^1 = .25$$

$$(.75)^1 = .75$$

So

$$P(d = 1|n = 2, \pi = .25) = 2(.25)(.75)$$
$$= .375$$

Finally, for $d = 2$:

$$\binom{2}{2} = \frac{2!}{2!(2-2)!} = 1$$

$$(.25)^2 = .0625$$

$$(.75)^0 = 1$$

So

$$P(d = 2|n = 2, \pi = .25) = 1(.0625)(1)$$
$$= .0625$$

The general binomial formula gives the same probability distribution for the number of children with Tay-Sachs disease out of 2 offspring for Mr. and Mrs. R. that we obtained with a direct argument.

Clinical Problem 2-6. *Interpreting Results from Multichannel Tests*

At the Beth Israel Hospital, an automated clinical chemistry analyzer is used to give 18 routinely ordered chemical determinations on one order [glucose, blood urea nitrogen, creatinine, sodium, potassium, chloride, bicarbonate, calcium, phosphate, uric acid, total protein, albumin, bilirubin, glutamic oxoloacetic transaminase (SGOT), lactic dihydrogenase (LDH), creatine phosphokinase (CPK), alkaline phosphatase, and iron]. The normal values for these 18 tests were established by the concentrations of these chemicals in the sera of a large sample of healthy volunteers. The normal range was defined so that an average of 3% of the values found in these healthy subjects fell outside.

Using the binomial formula, compute the probability that a healthy subject will have normal values on all 18 tests. Also compute the probability of 2 or more abnormal values.

Solution: Applying the General Binomial Formula

For this binomial problem, we have $n = 18$ and $\pi = .03$. The binomial formula becomes

$$P(d|n = 18, \pi = .03) = \binom{18}{d} (.03)^d(.97)^{18-d}$$

For $d = 0$, no abnormal values, we have

$$P(d = 0|n = 18, \pi = .03) = \binom{18}{0} (.03)^0(.97)^{18}$$

The values for the three terms in this expression are

$$\binom{18}{0} = \frac{18!}{0!(18 - 0)!} = 1$$

$$(.03)^0 = 1$$

$$(.97)^{18} = .58$$

So

$$P(d = 0|n = 18, \pi = .03) = .58$$

That is, we expect to observe normal values on all 18 tests for only 58% of healthy subjects.

To determine the probability of 2 or more abnormal tests, we could compute the individual probabilities of 2, 3, . . . , up to 18. However, it is easier to observe that the probability of 2 or more abnormal results is simply 1 minus the probability of 0 or 1 abnormal results. We have already computed the probability of 0 as .58. The probability of 1 abnormal test result is

$$P(d = 1|n = 18, \pi = .03) = \binom{18}{1} (.03)^1(.97)^{17}$$
$$= 18(.03)(.97)^{17}$$
$$= .32$$

Thus the probability of 2 or more abnormal test results is

$$P(d \geq 2|n = 18, \pi = .03) = 1 - (.58 + .32)$$
$$= 1 - (.90)$$
$$= .10$$

We would expect 2 or more abnormal test results about 10% of the time in healthy subjects.

2-6.   When Does the Binomial Distribution Apply?

Requirements

To get a binomial distribution, we have five requirements:

1. *Two categories.* The outcome of each trial results in one of two categories arbitrarily labeled success and failure. (Examples are: alive and dead, accident and no accident, cure and no cure, head and tail—in general two possibilities, exactly one of which must occur.)

2. *Independence.* The trials are independent. Outcomes on any set of trials do not change the probabilities on the others. (When a coin is tossed, a succession of heads does not change the probability of a head on the next toss.)
3. *Fixed π.* We have the same probability of success on every trial. (The chance that the next baby born in a hospital is female is just about constant.)
4. *Fixed n.* The number of trials is fixed and chosen in advance. (We want to know what will happen in a family of, say, 6 males. The binomial does not deal with, say, number of trials to first success where sample size depends on outcome.)
5. *Size of d.* We are interested in the value of d, not some simpler question such as whether d is equal to 0.

When these conditions are satisfied, the binomial distributions give the probability associated with each possible number of successes, d.

To sum up, for n independent trials, each with probability of success π, the binomial distribution is the set of paired values

$$d, P(d|n, \pi)$$

Thus it associates each value of d with its probability.

We have illustrated such sets of pairs repeatedly for the hemophilia and Tay-Sachs problems and also for multiple diagnostic tests.

EXAMPLE.   Check the first four requirements for the binomial distribution of the number of abnormal results when 18 laboratory results are obtained for normal subjects.

SOLUTION.   The four requirements are:

1. *Two categories.* The categories are normal and abnormal.
2. *Independence.* The test results are probably not independent, as we explain below.
3. *Fixed π.* The normal range was intended to exclude 3% of normal values for each test. Since the normal range was estimated from a large sample of normal subjects, the actual probability should vary only a little from test to test.
4. *Fixed n.* Complete results consist of 18 tests for each subject.

Independence simply means that the value of one test is not related to other values. This is not true in this situation. For instance, subjects with low albumin are likely to have low total protein, since total protein is the sum of albumin and globulin. However, the adequacy of the binomial distribution for describing the distribution of the number of abnormal test results can be evaluated by comparing the **theoretical** distribution, obtained from calculations using the binomial distribution, to the **observed** distribution of

abnormal results for a sample of normal subjects. Among 82 normal employ-
ees at Beth Israel Hospital, 52 of 82 (64%) had all normal tests, 19 of 82
(23%) had 1 abnormal test, and 11 of 82 (13%) had 2 or more abnormal results. The corresponding probabilities from the binomial distribution were
58%, 32%, and 10%. The observed and theoretical distributions are not too
different for these data. Thus we feel comfortable in using the binomial distribution as an approximation in this problem.

2-7.   Summary

To compute the probability of disease given diagnostic test results, we first
estimate the prior odds that the patient has the disease based on previously
available information. Next we compute the likelihood ratio for the observed
test result, either LR+ or LR−. The prior odds in favor of disease are then
revised to compute the posterior odds in favor of disease using the formula

Posterior odds = (prior odds) × (likelihood ratio)

Using odds makes calculations easier and the likelihood ratio provides an
intuitively meaningful measure of test accuracy.

The binomial formula can sometimes be used to compute the prior probability of disease. More generally, the binomial formula is used to compute
the probability of observing d positives in n trials, where the probability of a
positive result in any one trial is π. The binomial formula can be used to adjust the probability that a patient is a carrier of a genetic disease based on
family history. It also can provide the probability of d abnormal test results
among n independent tests. In each application, we can review the five requirements for the binomial distribution to determine whether the distribution applies.

Problems

2-1. Using arguments which parallel those given in Section 2-3, give the distribution
of the number of hemophilic males out of 5 male offspring when the father is
normal and the mother is a carrier.

2-2. If 10% of a population has a certain disease, what are the odds that a person selected at random from this population would have the disease?

2-3. Explain the relation between test sensitivity and specificity and the likelihood
ratio for a negative test result.

2-4. Diagnosis of diabetes mellitus by measurement of the urinary glucose. Background: About 75% of patients with diabetes will have glucose in their urine one
to two hours after having eaten a meal; 5% of normals will have glucosuria under
similar circumstances.
   a. Patient S is a 21-year-old without a family history of diabetes who has glucosuria on a routine urine sample obtained one hour after he drank a Coke.

The prevalence of diabetes in asymptomatic 21-year-old subjects without family history of diabetes is about .1%. What is the chance that he is diabetic?

b. Patient E is a 20-year-old who comes to your office complaining of fatigue, polydipsia, polyurea, and weight loss. He also has glucose in his urine. He has eaten within the past two hours. What are the chances that he is diabetic? (*Note:* In this problem, as in clinical practice, you will need to estimate the prior probability of diabetes in the patient using your clinical knowledge and your assessment of the signs and symptoms.)

2-5. Diagnosis of streptococcal pharyngitis. Background: About 10% of young adults (aged 11 to 30) with sore throats have streptococcal pharyngitis, as indicated by a positive throat culture. Crawford *et al.* (1979) investigated several clinical features of pharyngitis and a new diagnostic test, a Gram stain of the pharyngeal exudate. The accompanying table shows the sensitivity and specificity of the Grain stain, several signs, and history of exposure to a family member with streptococcal pharyngitis.

	Fever ≥ 38.3°C	Cervical adenopathy	Pharyn- geal exudate	History of exposure	Positive Gram stain
Sensitivity	.33	.73	.45	.18	.73
Specificity	.89	.55	.78	.92	.96

a. Compute the LR+ and LR− for each test.

b. A 20-year-old with a sore throat has a fever of 39°C. What is the probability of streptococcal pharyngitis?

c. A 20-year-old man with a sore throat has a sister with streptococcal pharyngitis. He has a fever of 39°C. What are the odds and probability of streptococcal pharyngitis?

d. Which of the tests do you think are independent?

e. Which is the best test according to these data?

2-6. Suppose that in the general population of men aged 60 to 64 years, 60% have a resting systolic blood pressure that exceeds 120 mm Hg. Discuss the probability that 0, 1, 2, or 3 of a group of 3 men in this age range will have systolic pressure over 120 if the 3 men:

a. Are randomly selected from the entire general population.

b. Are randomly selected from patients in the hospital hypertension clinic.

c. Are triplets, randomly selected from among all sets of triplets in the general population.

2-7. Treatment Y causes a toxic reaction in 25% of persons to whom it is given. What is the probability that 0, 1, 2, 3, or 4 of four patients chosen at random will have a toxic reaction?

2-8. Mrs. A. has mild diabetes controlled by diet. Her morning urine sugar test is negative 80% of the time and positive (+) 20% of the time. (It is never graded higher than +).

a. At her regular visit she tells her physician that the test has been + on each of the last 5 days. What is the probability that this would occur if her condition has remained unchanged? Does this observation give reason to think that her condition **has** changed?

b. Is the situation different if she observes, between visits, that the test is + on 5 successive days and telephones to express her concern?

Problems for Classroom Discussion

Estimate the diagnostic probabilities for the following "cases." Explicitly supply your own estimates for the prior probability or odds of disease and the sensitivity and specificity or likelihood ratio of the tests involved in each situation. For example, explain whether you are estimating

$P(\text{CPK abnormal} | \text{health})$

or

$P(\text{exactly 1 of 18 tests abnormal} | \text{health})$

or

$P(\text{at least 1 of 18 tests abnormal} | \text{health})$

All examples deal with lab results from a multichannel Auto Analyzer.

CASE A.   A 30-year-old man has a normal history and physical examination. His routine laboratory test reveals 17 normal values and a modestly elevated CPK. What are the chances that he has had a recent myocardial infarction?

CASE B.   A 50-year-old man has had 6 hours of chest pain that is somewhat atypical of myocardial infarction. Seventeen of his laboratory values are normal but he has a modestly elevated CPK. What are the chances that he has had a recent myocardial infarction?

CASE C.   You are reviewing laboratory results for your colleague. One patient has a modestly elevated CPK and 17 normal values. How do you interpret these results?

CASE D.   A 20-year-old man has a history of hepatitis exposure two weeks ago. He says that he may be slightly fatigued. He has 17 normal values and a modestly elevated SGOT. What are the chances that he has hepatitis?

References

Crawford, G.; Brancato, F.; and Holmes, K. K. (1979). Streptococcal pharyngitis: diagnosis by Gram stain. *Ann. Intern. Med.*, **90**:293–97.

Rizza, C. R.; Rhymes, I. L.; Austen, D. E. G.; Kernoff, P. B. A.; and Aroni, S. A. (1975). Detection of carriers of haemophilia: a "blind" study. *Bri. J. Haematol.*, **30**:447–56.

CHAPTER 3

Management of Patients with Urinary Tract Symptoms: Decision Trees

OBJECTIVES

The idea of decision trees

How to use a decision tree for making decisions

Decision trees show systematically what information is required

Missing probabilities have to be estimated

Weights express the relative importance of outcomes

Sensitivity testing can show when decisions would be changed by changes in probabilities and changes in the weights

Any systematic way of making decisions is equivalent to assigning probabilities and weights; and so a decision process has implicit probabilities and weights, or ranges of them, even when they are not explicity displayed

A 15-year-old boy has right lower quadrant pain that has increased over two days. You think acute appendicitis is possible. Should you recommend immediate appendectomy, or should the patient be observed? When should penicillin be prescribed immediately for patients with sore throat, and when should antibiotic therapy await results of a throat culture? Physicians face such situations daily, and treatments are prescribed even though diagnoses are not certain and therapy is not universally effective. Indeed, many clinical decisions are made in the face of uncertainty, that is, on the basis of diagnoses and anticipated results which are probabilities, not certainties.

This chapter introduces the decision tree, a device to aid decision making when the diagnosis is unclear and the therapy risky.

> A decision tree is a schematic method for examining decision problems which assists in identifying, organizing, and analyzing the information that bears on a decision.

For the clinician, a decision tree can help (1) to point out the relevant background information, (2) to explore the need for more diagnostic information, and (3) to expose the consequences of treatment when the diagnosis is uncertain or the treatment has risks for the patient.

To apply the method of decision trees to a clinical problem, the physician has to enumerate the possible consequences of the decision, estimate the probability of each possible outcome, and assess the desirability or "utility" of each outcome. Although decision trees do not usually take every possible special situation into account, they may offer much useful guidance.

This chapter presents the decision tree in a highly structured form, but the essence of the logical process can be applied without formal analysis to many simple, common, but important clinical problems.

3-1. Choice of Treatment When Urinary Tract Infection Is Probable and Gonorrhea Is Possible

Background: Treating Urinary Tract Infections

Physicians tentatively diagnose female patients with acute onset of dysuria and increased frequency of urination as having a urinary tract infection (UTI) if their urine contains more than five white blood cells per microscopic field. A urine specimen is sent for bacterial culture to confirm the diagnosis and discover the antibiotic sensitivity of any organism identified. Pending results of these tests (which take up to four days for completion) patients are started on an antibiotic. The usual choices are a sulfa antibiotic, ampicillin, or tetracycline. Each of the three antibiotics has an 85% chance of curing a UTI. Sulfa has been the traditional choice at many clinics. The frequency of side effects with sulfa is low, and it is less expensive than the other two antibiotics.

Clinical Problem 3-1. *A Woman with Urinary Tract Symptoms*

A physician made the following comments:

One evening at the clinic, I was asked to see a woman with signs and symptoms of UTI. I did not immediately prescribe sulfa because I thought there was a slight chance that she might have gonorrhea, which can produce signs and symptoms mimicking a urinary tract infection. Gonorrhea can be successfully treated with ampicillin or tetracycline but not with sulfa. Perhaps then ampicillin or tetracycline would be a more sensible choice of initial antibiotic.

One recommended treatment for gonorrheal infection (GC) is 3.5 g of ampicillin plus 1 g of probenecid given all at once. Such therapy cures 90 to 95% of gonorrheal infections. However, it is not known how effective ampicillin would be for GC if it were given by the same schedule as for UTI (250 mg four times a day for ten days). What is the wise course of action?

Constructing the Decision Tree

1. *Listing outcome.* To help resolve this problem, we lay out a decision tree showing each possible course of action and each possible outcome. Then by making the best choices when we have a choice, we can increase the probability of getting desirable outcomes. We can formulate the problem as a decision tree as shown in Figure 3-1. Let us focus on the choice between sulfa and ampicillin administered in doses appropriate for UTI. (The 3.5-g ampicillin dose is considered later.) The initial choice of antibiotic is represented as a small square at the far left, with the decision between sulfa and ampicillin distinguishing the two branches leading away from the square.

In general, squares in decision trees represent decisions made by the physician. Each branch may branch again into two or more outcomes. Any branch point is called a **node** in the diagram.

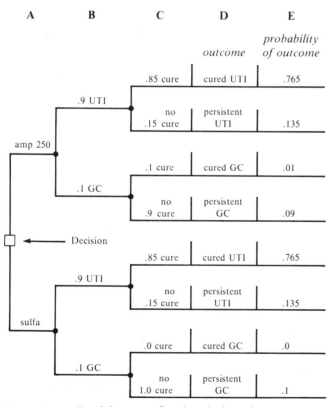

Figure 3-1. Decision tree for the choice of treatment—sulfa against ampicillin; amp 250 = ampicillin, 250 mg QID × 10 days; sulfa = sulfa, 1 g QID × 10 days.

In this example, all the other nodes in the tree represent branch points over which the physician has no control. These are indicated by circles. More generally, circles represent points where chance or nature "makes" the choices.

We have indicated the stages in the tree by letters at the top of the diagram for easy reference. The branching in column B arises because the patient is considered to have either a urinary tract infection (UTI) or gonorrhea (GC). The final branches in column C catalog the outcome of the treatment given the disease and the specified therapy. Again the outcomes are subject to chance. Column D gives the final state of the patient, and column E the probability of the state given the treatment.

2. *Assigning probabilities.* After listing the possible outcomes of treatment in branches stemming from each of the possible therapeutic strategies, one has to estimate how often each outcome will occur. The physician is considering two diseases: UTI and GC. Suppose that on the average, 9 out of 10 patients with signs and symptoms similar to those of this patient would have UTI and then 1 of 10 would have GC. We have translated the physician's "a slight chance that she might have gonorrhea" into a probability of .1. A shorthand way of writing "the probability of UTI is .9" is

$$P(\text{UTI}) = .9$$

Suppose also that the probability that UTI will be cured with either antibiotic is 85%:

$$P(\text{antibiotic cures UTI}) = .85$$

Similarly, assume that sulfa never cures GC, so that

$$P(\text{sulfa cures GC}) = 0$$

We will also assume that ampicillin given in doses appropriate for a UTI will cure some GC, so

$$P(\text{ampicillin cures GC}) = .1$$

Note that for each chance branching point, the probabilities shown on all branches emanating from the point must add to 1 because one of the events must happen. For instance, in column B,

$$P(\text{UTI}) + P(\text{GC}) = 1$$

and in column C,

$$P(\text{cure}) + P(\text{no cure}) = 1$$

These sum to 1 because the possibilities at each branch point are the only ones that can occur and do not overlap. Obviously, each patient has either cure or no cure, but not both; not so obviously, we assume that UTI and GC do not occur in the same patient at the same time.

As we have seen in Chapter 1, probabilities which vary depending on the conditions that hold are called conditional probabilities. Thus we say: "The probability of a UTI given urinary symptoms is .9." The shorthand notation for this is:

$P(\text{UTI}|\text{urinary symptoms}) = .9$

We often have two or more events or conditions on the same side of the vertical bar. To represent the probability that a patient with UTI who is treated with ampicillin will be cured is .85, we can write

$P(\text{cure}|\text{UTI, ampicillin}) = .85$

where the comma means that **both** UTI and ampicillin treatment are assumed true or present in the patient. They are the givens in this example. Similarly,

$P(\text{cure}|\text{GC, ampicillin}) = .1$

Note that the "conditioning" events listed on the right of the vertical bar may be under the control of the physician (such as treatment) or they may be random (such as disease). We may not know that the conditioning event occurs for the individual patient. In spite of such ignorance, we may still be able to make useful calculations employing conditional probabilities.

3. *Determining the expected outcomes.* Figure 3-1 shows in column E the probability of each outcome (cure or no cure) for each treatment–disease pair. We get these column E probabilities by multiplying the unconditional probability of the disease shown in column B by the conditional probability of outcome shown in column C.

For example, to get the proportion of all patients who would be treated with ampicillin and are cured of UTI, we multiply the probability of having UTI by the probability of a cure using ampicillin. We need a little care in justifying this step. We know that if A and B are events, we can write the probability that both A and B occur as

$P(A \text{ and } B) = P(A, B) = P(A)P(B|A)$

[Note that $P(A, B, C)$ would be the probability that A, B, and C occur, and so on.] In our problem, A corresponds to having UTI, and B corresponds to cure of UTI. Where does the treatment with ampicillin come in? We can handle this in either of two equivalent ways. We can regard ourselves as dealing only with the ampicillin population so that all the probabilities are implicitly conditional on the ampicillin treatment. In this approach we write for ampicillin:

$P(\text{UTI, cure}) = P(\text{UTI})P(\text{cure}|\text{UTI})$

where the formulas suppress (omit) the condition that we are talking about—ampicillin treatment. Or we can make the discussion more explicit by writing a more general formula for combining conditional probabilities:

$P(A, B|C) = P(A|C)P(B|A, C)$

(Note that the A, C following the final vertical bar means that both A and C occur.) In this notation we make more explicit the universe of people and events. Thus, in our example, C is the people who come to the hospital with these symptoms and are treated with ampicillin. We would probably still simplify the notation and focus on the ampicillin, letting the hospital source of the patients be understood. Then we could write

$$P(\text{UTI, cure}|\text{ampicillin}) = P(\text{UTI}|\text{ampicillin})P(\text{cure}|\text{UTI, ampicillin})$$

This rule seems obvious when the probabilities are interpreted as rates. If 90% of all patients have UTI, and if 85% are cured by ampicillin, then $100 \times .9 \times .85$ of patients, or 76.5% have UTI and can be cured using ampicillin. We note that

$$P(\text{UTI}|\text{ampicillin}) = P(\text{UTI}|\text{sulfa}) = P(\text{UTI})$$

These formulas say that the probability of UTI does not depend on what medication is administered.

Under the foregoing hypotheses, the tree shows that among 1000 patients treated with ampicillin, 765 would be cured of UTI and 10 would be cured of GC. Among 1000 patients, sulfa therapy would cure 765 UTI patients and 0 GC patients. Going beyond the details, the tree demonstrates that the outcome with ampicillin is at least as good as that with sulfa whether the patient has UTI or GC. No matter what the exact probability of GC and no matter what the precise efficacy of ampicillin as a cure for GC at this dosage, ampicillin therapy does no worse than sulfa. Thus, in this case, ampicillin must be considered a better treatment than sulfa.

The **sure-thing principle** is at work in this example. Ampicillin cures as many UTIs as sulfa, but in addition cures some GC, and so, barring side effects, ampicillin is better. This simple reasoning can often be used, and this is one ground for saying that such analyses can sometimes be made easily by the physician.

> The sure-thing principle says that even if two policies of action give the same results for some situations, if one gives better results for all the rest of the situations, it is the better policy.

3-2. Payoff Functions: Choosing the Better Strategy

Choosing a Payoff Function

We analyze decisions with a decision tree to help choose the action with the best payoff. Therefore, in addition to a list of the possible strategies and outcomes with their associated probabilities, we need the idea of a payoff function. In our UTI-GC example, the payoff was percent cured. It was especially simple because we did not have to face a trade-off between percent cured of UTI and percent cured of GC. Some problems have multiple payoff functions and their resolution requires further assumptions on the part of the decision maker.

Some examples of payoff functions are percent cured, average time to cure, and percent alive at five years. We want to maximize percent cured, minimize time to cure, and maximize percent alive at five years, if other complications do not intervene. Other situations may call for minimizing the number of toxic events, minimizing costs of treatment, or even minimizing the probability of malpractice suits. When we have a single payoff function, the decision analysis is oriented toward choosing a strategy that optimizes the value of the payoff function.

When we deal with several payoff functions simultaneously, we cannot expect to optimize more than one at the same time. If ampicillin cured 700 UTI patients and 100 GC patients per 1000 treated and sulfa cured 900 UTI patients and 0 GC patients per 1000 treated, we would need to form a payoff function to help decide which treatment is preferable.

Assigning Values

In some problems we may wish to assign values to the outcomes to compute the average performance of a given strategy. The payoff function might be the average obtained. To illustrate, in our UTI-GC example, it may be that curing one case of GC is worth more in health terms than curing one UTI. We could assign values to the four outcomes: cured UTI, persistent UTI, cured GC, and persistent GC. For example, these outcomes might be assigned the values 1, 0, 2, and -1, respectively, in units on some arbitrary scale.

Then we obtain the average value of treatment by weighting each value of cure or no cure by its corresponding probability. For ampicillin,

$$1(.765) + 0(.135) + 2(.01) + (-1)(.09) = .695 \text{ units}$$

Similarly for sulfa, the average value of treatment is

$$1(.765) + 0(.135) + 2(0) + (-1)(.1) = .665 \text{ units}$$

Again ampicillin is the preferred treatment because .695 exceeds .665.

Maximizing the Expected Value

We have just illustrated an application of the concept of **average** or **expected value.** Let us define it more generally:

> Average or expected value: If scores $x_1, x_2, \ldots, x_k$ have probabilities $\pi_1, \pi_2, \ldots, \pi_k$, then the average or expected value of the score is the sum of their products
>
> $$x_1 \pi_1 + x_2 \pi_2 + \cdots + x_k \pi_k$$

Using average outcome to compare and select actions necessitates the assignment of numerical values, called utilities, to each outcome of each action. If we had added the percentage of cures from the two diseases to get our payoff value, 77.5% cured for ampicillin and 76.5% cured for sulfa, this would be equivalent to assigning a value of 1 for cure of either disease, and 0 for failure to cure.

When we applied the sure-thing principle, we were not forced to make comparative assignments because the percentage cured of UTI was the same for both diseases. We merely needed to know that cures were beneficial and lack of cure was not.

Although enumerating actions and outcomes and assigning probabilities may be difficult, for many problems the assignment of values (utilities) to outcomes may at first seem impossible. Consider, for example, treatments that may produce several types of toxicity. What are the relative values of cure versus blindness? versus death? Questions of this type are not intended to discourage such analyses but rather to encourage exploration and caution. For example, if a small change in the values assigned to the outcomes changes the decision on the preferred action, we should proceed with caution. In our UTI-GC example, the choice of ampicillin is insensitive to the assignment of values, provided that a cure is considered worth more than no cure. Since it is, we can apply the sure-thing principle.

It is important to remember that observations and testing also have associated costs in dollars, time, pain, or increased risk of toxicity. In a complete decision analysis each such cost will be reflected in the value assigned to the outcome.

Values Are Implicit in Any Decision

Scoring may not be easy, but can it be avoided? For example, suppose that a patient with a malignant neoplasm of the femur can be treated with surgical amputation or by radiation, and that we estimate the probabilities of cure to be 45% and 40%, respectively. The decision tree approach would require that life with one leg, life with two legs, and death be given scores reflective of their relative value. The physician may consider such scoring to be a hopeless task and abandon the tree. Yet the physician in consultation with the patient still makes the decision. The decision that is reached implicitly rates the uncertainties and outcomes of radiation against the uncertainties and outcomes of surgery.

If in the present example of neoplasm of the femur, the physician elects surgery, he or she is acting as though a 45% chance of survival without a leg is considered "better" than a 40% chance with a leg. One way to see how any decision implies a scoring of life with amputation is to assign the value of death (with or without the leg) the value 0. If we assign survival with both legs the value 1, the expected score from not operating is:

Expected score without operation
$$= 1 \times P(\text{survival with leg}) + 0 \times P(\text{death with leg})$$
$$= 1 \times .4 + 0 = .4$$

Now let us assign some as yet undetermined score, V, to stand for the value of survival without the leg. The expected score for operating is:

Expected score with operation
$= V \times P(\text{survival without leg}) + 0 \times P(\text{death without leg})$
$= V(.45) + 0 = .45V$

Although someone may object to scoring amputation against life and death, few would object to scoring two outcomes, one against the other, on a scale that is otherwise undefined. Assigning death the value 0 and life (with two legs) the value 1 chooses the origin and unit for the outcome scale, not the value of life and death. A difficulty comes in putting a third outcome on this scale. In the present case, we are discussing the feasibility of determining a numerical score for life with amputation on a scale where death is 0 and life with two legs is 1. It seems plausible that there is some value, say V, presumably between 0 and 1, which is the appropriate score for life with only one leg. The difficulty arises in that it seems impossible to determine the value of this score in the individual case. Hence one might argue that all scoring should be abandoned.

Suppose that the physician abandons scoring and decides in favor of operation: The following value judgment has been made:

$.45V \geq .40$

so

$$V \geq \frac{8}{9}$$

The decision implies that life with one leg has a score of 8/9 or greater on a scale where death is scored 0 and life with two legs is scored 1. Similarly, a decision not to operate implies that V is less than 8/9 on this scale. Despite the objection that we cannot analyze such decisions quantitatively, making a decision actually has implications about the value, so one should know what the decisions imply in this numerical language. If we are surprised by the implications, we may need to rethink the values.

Although we are often reluctant to assign values in clinical decision making, decisions actually have implications about values, so we should consider these implications quantitatively.

3-3. Using More Complex Decision Trees

Because decision trees are most useful in pointing out the consequences of decisions on outcomes which are relatively frequent, omitting a frequent event or often-used therapeutic strategy may constitute a major flaw in the tree. When we detect such omissions, they can often be incorporated into a new, improved, and perhaps more complex tree. In this section we expand our original tree to include the possibility of skin rash following antibiotic therapy.

Background: Side Effects of Sulfa and Ampicillin

About 5% of patients treated with ampicillin develop a drug rash. Sulfa causes rash less often and it occurs in about 1% of treated patients (see Arndt and Jick, 1976). In addition, let us suppose that a patient who developed a skin rash will stop taking the antibiotic before cure has occurred, so that

$P(\text{cure}|\text{rash}) = 0$

Table 3-1 shows all the probability estimates for this problem.

We have assumed that the probability of rash given ampicillin does not depend on whether the patient has UTI or GC. Thus the probability is the same whether UTI is a conditioning event or not:

$P(\text{rash}|\text{ampicillin, UTI}) = .05 = P(\text{rash}|\text{ampicillin, no UTI})$

When this equality holds, recall that we say that the events "rash" and "UTI" are independent. Events that do not affect a probability, such as UTI and no UTI, in this example, are often omitted from the notation and we can replace the probabilities by $P(\text{rash}|\text{ampicillin})$.

Constructing and Examining the Decision Tree

Figure 3-2 shows the decision tree under the hypotheses outlined in Table 3-1. Table 3-2 summarizes outcomes of Figure 3-2, assuming that 10,000 people have been treated with each strategy.

Scrutiny of the outcome probabilities in Figure 3-2 or Table 3-2 shows that the sulfa is more likely to cure UTI and that it produces fewer rashes. Ampicillin seems better in terms of curing GC. But one should formally score the severity of the various clinical outcomes before comparing the hypothetical results of sulfa and ampicillin treatment.

Expected Value When the Undesirable Outcomes Are Weighted Equally

If one assumes that persistent GC, persistent UTI, and sulfa and ampicillin skin rashes are all equally undesirable and assigns to each occurrence a score of − 1, then from the bottom panel of Table 3-2, we find the total score for ampicillin to be − 3138 and for sulfa − 2527. The difference (ampicillin − sulfa) is

$-3138 - (-2527) = -611$

This rating assigns a score of − 2 to those patients with persistent dis-

Table 3-1. Summary of
probabilities from Figure 3-2

$P(\text{UTI}) = .9$

$P(\text{GC}) = .1$

$P(\text{rash}|\text{ampicillin}) = .05$

$P(\text{rash}|\text{sulfa}) = .01$

$P(\text{cure}|\text{rash}) = .0$

$P(\text{cure}|\text{UTI, ampicillin, no rash}) = .85$

$P(\text{cure}|\text{UTI, sulfa, no rash}) = .85$

$P(\text{cure}|\text{GC, ampicillin, no rash}) = .1$

$P(\text{cure}|\text{GC, sulfa}) = .0$

ease and rash, a score of -1 to those with only persistent disease, and 0 to those who are cured. The expected score for ampicillin is worse with this scoring system.

Expected Value When the Undesirable Outcomes Are Weighted Unequally

But is this scoring reasonable? Suppose that we consider persistent GC more serious than the other undesirable outcomes and score it -2. Using the bottom panel of Table 3-2, we compute the ampicillin treatment average value as

$$(-1)(500) + (-1)(1733) + (-2)(905) = -4043$$

Table 3-2. Outcome of treating 10,000 patients with signs and symptoms of UTI assuming that 10% of patients have gonorrhea

Outcome	Amp 250	Sulfa
Cured UTI	7,268	7,574
Cured GC	95	0
Persistent UTI	1,283	1,337
Persistent UTI plus rash	450	90
Persistent GC	855	990
Persistent GC plus rash	50	10
Total	10,000[a]	10,000[a]
Overlapping[b] totals		
Number with rash	500	100
Number with persistent UTI	1,733	1,427
Number with persistent GC	905	1,000
Total	3,138	2,527

[a] Because of "rounding off" fractions, the total adds to 10,001.

[b] The totals are overlapping because one patient may be counted more than once (e.g., a patient with rash and persistent UTI appears twice).

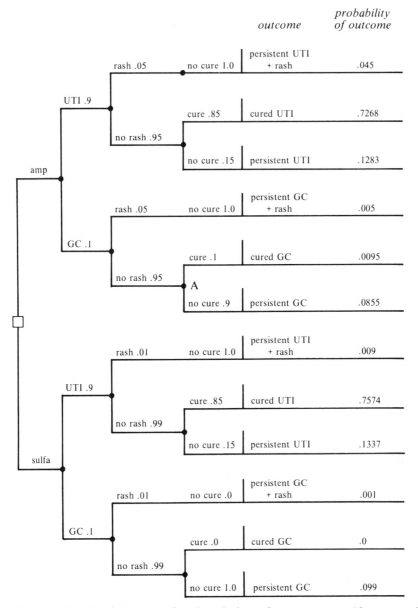

outcome *probability of outcome*

Figure 3-2. Decision tree for the choice of treatment, sulfa or ampicillin (amp), including rash as a toxic event.

The sulfa average value is

$$(-1)(100) + (-1)(1427) + (-2)(1000) = -3527$$

Thus sulfa has the better average score and with these scores would be preferred.

One can show that only if persistent GC is assigned a score more extreme than -7.5 would the outcome favor ampicillin. For this multiplier, the ampicillin average minus the sulfa average equals .5 and ampicillin would be slightly preferred.

In many analyses using the decision tree, as in this one, the decision implied by the tree is not sensitive to modest changes in outcome weights. Because of this insensitivity, the decision indicated by the outcomes presents little problem in this example. If the outcomes are sensitive to the weights, the decision will be known to depend on the value system of the physician or patient. The decision tree will help organize the thinking and make clear what events create the sensitivity (a new meaning for the term).

Sensitivity Testing

The branches of a decision tree are of two kinds: those where a decision is made and those where a chance outcome occurs. For optimum use of the tree, probabilities must be known or estimated for each of the chance branches.

When the exact probabilities are not known, we use a technique known as sensitivity testing to determine whether the decision implied by the tree is influenced by slight changes in probability estimates or whether a broad range of estimates would lead to the same decision. This, of course, is the same idea as varying the scores to see if the decision changes. In the end, both issues may need to be explored.

> In sensitivity testing, we vary the probabilities, weights, or other quantities in a problem to see whether changes in them would change the decisions.

1. Is our decision sensitive to $P(\text{cure}|GC, \text{ampicillin})$ when $P(GC) = .1$ and undesirable outcomes are weighted equally? No one has investigated the efficacy of low-dose ampicillin (250 mg four times a day for ten days) in the treatment of GC, although the lowest cure rate reported with any dose schedule of ampicillin is about 90%. This percentage might suggest that low-dose ampicillin therapy was effective in 50% of GC infections rather than the 10% used above. Would this alter our decision, assuming that the other probabilities shown in Table 3-1 still hold?

Using the tree of Figure 3-2 as a model, one can compute the probabilities associated with the ampicillin outcomes by inserting $P(\text{cure}|\text{ampicillin}, GC) = .5$ at chance node A (see Figure 3-3). (The sulfa branch is not changed.) Now the outcome probability associated with cured GC equals .0475 (instead of .0095) and associated with persistent GC equals .0475 (instead of .0855). When we compute the scores based on 10,000 patients, we get

$$\text{Ampicillin score} = -500 \quad -1733 \quad -475 \quad = -2708$$
$$\text{Sulfa score} = -100 \quad -1427 \quad -1000 = -2527$$

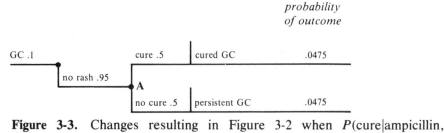

Figure 3-3. Changes resulting in Figure 3-2 when $P(\text{cure}|\text{ampicillin}, \text{GC}) = .5$.

In spite of the fact that ampicillin would now cure 4275 patients with GC, the choice of sulfa as the preferred drug would not change. Under the other assumptions used in this decision tree, the therapeutic strategy is relatively unrelated to the exact probability that ampicillin cures GC.

2. Is the decision sensitive to $P(\text{GC}|\text{urinary symptoms})$ when $P(\text{cure}|\text{ampicillin}, \text{GC}) = .1$ and undesirable outcomes are weighted equally? In a Memphis, Tennessee, emergency room about 25% of the patients with signs and symptoms of UTI actually have GC (Curran *et al.*, 1975). Table 3-3 shows the outcome of treatment by sulfa or low-dose ampicillin.

Table 3-3 assigns a score of -1 to each bad outcome, based on a decision tree with the same probabilities as those of Table 3-1, except that the probability of GC is .25 and that of UTI is .75. In this situation, ampicillin treatment would cure 238 patients with gonorrhea and sulfa would cure 0. Ampicillin still is associated with 400 extra rashes and fails to cure 255 extra persistent UTIs. The ampicillin score from the bottom panel of Table 3-3 is -4206 and the sulfa score -3789. Thus sulfa is preferred.

3. Is the decision sensitive to outcome weights when $P(\text{GC}) = .25$ and $P(\text{cure}|\text{ampicillin}, \text{GC}) = .1$? Although ampicillin still is associated with 417

Table 3-3. Outcome of treating 10,000 patients with signs and symptoms of UTI assuming that 25% of patients have gonorrhea

Outcome	Amp 250	Sulfa
Cured UTI	6,056	6,311
Cured GC	238	0
Persistent UTI	1,069	1,114
Persistent UTI plus rash	375	75
Persistent GC	2,137	2,475
Persistent GC plus rash	125	25
Total	10,000	10,000
Overlapping totals		
Number with rash	500	100
Number with persistent UTI	1,444	1,189
Number with persistent GC	2,262	2,500
Total	4,206	3,789

extra undesirable events, the decision implied by the tree becomes much more sensitive to the weighting of the clinical outcome. In the Memphis study, two of six patients with inadequately treated gonorrhea were hospitalized within the ensuing weeks for acute complications of GC. In contrast, ampicillin rash and persistent UTI rarely lead patients to be hospitalized. Suppose that a physician chose to represent the greater risk by assigning a score of -3 to persistent GC compared to drug rash or persistent UTI which were given a score of -1. Then, using the bottom panel of Table 3-3,

$$\text{Ampicillin score} = (-1)(500) + (-1)(1444) + (-3)(2262) = -8730$$

$$\text{Sulfa score} = (-1)(1000) + (-1)(1189) + (-3)(2500) = -8789$$

This scoring would slightly favor ampicillin at Memphis.

4. Is the decision sensitive to $P(\text{cure}|\text{ampicillin, GC})$ when $P(\text{GC}) = .25$ and undesirable outcomes are weighted equally? If the probability of GC is high, it also becomes more important to know the exact cure rate for the proposed ampicillin treatment. If one assumes that the probability of ampicillin curing GC is .5, ampicillin therapy will cure 1188 patients with GC while sulfa will still cure none. The number of persistent UTI and skin rashes remains the same as in Table 3-3. If each unwanted outcome is weighted equally, ampicillin would be the treatment of choice.

Expanding the Tree Again: Should a GC Culture Be Obtained?

If patients with urinary symptoms have a 25% frequency of gonorrheal disease, neither of the two treatment strategies discussed so far may be appropriate. One might consider obtaining a sample for GC culture from every female patient with signs or symptoms of UTI, as is already done for many males with signs or symptoms of UTI. The decision then might be between the two strategies: strategy A—immediate therapy with sulfa as for UTI, with a change if culture results were positive for gonorrhea, or strategy B—treatment with a large single dose of ampicillin plus probenecid, which is effective against gonorrhea (probability of cure .95), with continued ampicillin at 250 mg four times a day for 10 days for possible UTI.

Strategy A, culture and sulfa treatment, now makes use of a diagnostic test. The outcome of therapy will depend, therefore, on the accuracy of this test and the feasibility of the planned antibiotic switch. Formulating a tree for this decision will show how these factors can be incorporated in the analysis.

Pruning the Tree

Figure 3-4 shows the tree for this decision. The probabilities for each chance node are not included on the diagram itself but are listed in Table 3-4. The tree has been ''pruned'' in that the risk of rash from sulfa has not been considered. Among other things, including rash from sulfa makes the tree

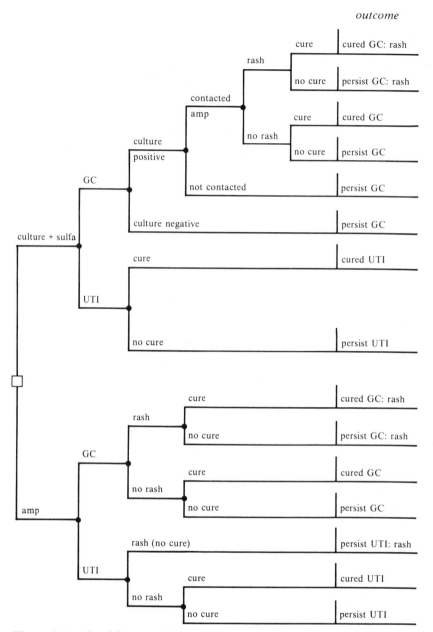

Figure 3-4. Decision tree for the strategies culture plus sulfa against ampi-cillin (amp) as treatments of patients with urinary symptoms.

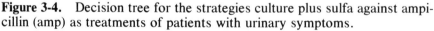

very bushy (complicated). Usually, it is best to consider all possible branches and prune those to which the decision is insensitive.

We have also pruned the tree by not including branches that have conditional probabilities of 0. For example, $P(\text{cure}|\text{GC, sulfa}) = 0$, and this branch has not been included.

Table 3-4. Summary of probabilities for tree comparing sulfa and culture against ampicillin as treatment of patients with urinary symptoms

$P(GC) = .25$
$P(UTI) = .75$
$P(\text{cure}|\text{UTI, sulfa}) = .85$
$P(\text{cure}|\text{GC, sulfa}) = .0$
$P(\text{culture positive}|\text{GC}) = .8$
$P(\text{contacted}|\text{positive culture}) = .9$
$P(\text{rash}|\text{ampicillin}) = .05$
$P(\text{cure}|\text{ampicillin, GC}) = .95$ (independent of rash)
$P(\text{cure}|\text{UTI, sulfa}) = .85$
$P(\text{cure}|\text{ampicillin, UTI, no rash}) = .85$
$P(\text{cure}|\text{ampicillin, UTI, rash}) = .0$

Many of the probabilities listed in Table 3-4 are those that we have used in our previous examples. We have estimated

$$P(\text{culture positive}|\text{GC}) = .8$$

by referring to Isselbacher *et al.* (1980), which states that cultures for gonorrhea will be positive in approximately 80% of women with GC. We have assumed that no women without GC will have a positive culture. We have also assumed, based on clinical experience, that about 90% of patients with a positive culture can be successfully contacted and adequately treated.

Using 3.5 g of ampicillin with probenecid in one initial dose cures GC about 95% of the time. It is even effective in the presence of rash, as the rash develops after the antibiotic has been taken (we will assume that rash will interfere with the treatment of UTI).

Comparing Treatment Strategies

Table 3-5 shows that the ampicillin strategy cures 665 more cases of gonorrhea, but results in 410 more rashes and 319 more persistent urinary tract infections than sulfa when 10,000 patients are treated.

If we assign a score of -1 to each occurrence of rash, persistent UTI, or persistent GC, then

$$\text{Ampicillin score} = -500 \quad -1444 \quad -125 = -2069$$
$$\text{Sulfa score} = \quad -90 \quad -1125 \quad -790 = -2005$$

and sulfa still has a better score. The difference between the scores is small and the decision is obviously sensitive to our estimates of probability and our outcome scores. Indeed, ampicillin would have a higher average if any of the following changes occurred:

1. A decrease in P(contacted|positive culture)
2. A decrease in culture sensitivity
3. An increase in P(GC|urinary symptoms)
4. An increase in the (negative) score assigned to persistent GC

Opposite changes in the estimates above tend to favor sulfa therapy. Although the analysis suggests that it is difficult to choose the best strategy, it is reassuring that both strategies appear about equivalent and neither strategy will be greatly inferior.

So far, we have treated these analyses as suggesting one strategy for all patients with symptoms of UTI. In fact, the analyses can be used to tailor the treatment to the individual patient. The tree has enabled us to identify several probabilities (some of which are listed above) to which the decision is sensitive. Therefore, the physician might try to individualize estimates of these probabilities rather than use the "average" of the entire patient population. One might sensibly vary

P(GC|urinary symptoms) or P(contact|positive culture)

based on history or even change

P(culture positive|GC)

by performing several cultures.

Thus the tree has underscored those estimates of probability that most directly effect therapeutic decisions. This enables physicians to focus on those aspects of the patient's history and physical which will allow good estimates of those factors that play the largest role in determining the expected outcome of therapy for that individual patient.

Table 3-5. Outcome of treating 10,000 patients with signs and symptoms of UTI with large doses of ampicillin or sulfa and a culture for gonorrhea

Outcome	Amp	Sulfa and culture
Cured UTI	6,056	6,375
Cured GC	2,256	1,625
Cured GC plus rash	119	85
Persistent UTI	1,069	1,125
Persistent UTI plus rash	375	0
Persistent GC	119	785
Persistent GC plus rash	6	5
Total	10,000	10,000
Overlapping totals		
Number with rashes	500	90
Number with persistent UTI	1,444	1,125
Number with persistent GC	125	790
Total	2,069	2,005

Other authors have applied decision trees to the management of urinary tract infections. Carlson and Mulley (1985) compared single- and multiple-dose regimens of two different antibiotics, amoxicillin and trimethoprim-sulfamethoxazole, in the treatment of women with dysuria and pyuria. Single-dose regimens were found preferable to multiple-dose regimens, primarily because of the higher incidence of side effects with the latter treatment. Trimethoprim-sulfamethoxazole was preferable to amoxicillin because of its greater effectiveness in bacterial infections and its possible effectiveness in nonbacterial urethritis.

The authors considered the morbidity of symptoms due to side effects equivalent to the morbidity of symptoms due to infection. Single-dose trimethoprim-sulfamethoxazole minimized the expected number of days with symptoms from either side effects or infection and was therefore judged to be optimal treatment.

Decision trees are usually applied to complex situations where actual observations are impractical. There are several clinical trials, however, comparing different treatment of urinary tract infections. Schultz et al. (1984) compared single- and multiple-dose regimens of trimethoprim-sulfamethoxazole in women with dysuria. The efficacy rate of these two regimens and the incidence of side effects were similar to the estimates used by Carlson and Mulley in their decision tree analysis. Schultz et al., however, considered that side effects were "of minimal clinical significance" and opted for multiple-dose treatment.

It is reassuring in this instance to see that the side-effect frequency and efficacy rate of an actual clinical study echoed the rates used in a decision tree analysis. The two studies differed in their choice of optimal therapy because the authors weighted the severity of side effects differently. Perhaps a survey of patients who have experienced infection and side effects could establish a more empirical weighting for these two morbidities.

3-4. Summary

The decision tree is a formal scheme for analyzing the consequences of therapeutic decisions. It is particularly useful in situations where the diagnoses may be uncertain or the responses to therapy variable.

It requires that the physician enumerate the possible consequences of a decision, estimate the probability of each possible outcome, and assess the desirability of each outcome. Sensitivity testing can show how the expected value of a strategy is changed by modest changes in probability estimates or changes in the weights assigned to the outcomes.

A simple tree can often assist clinical decision making in fairly straightforward situations. Not only can decision trees point out the expected outcome of a given strategy, but they can also focus attention on aspects of the individual patient which may be pivotal in formulating a good therapy.

Problems*

The following case of suspected appendicitis provides the background for Problems 3-1 through 3-3.

Case Description

A 15-year-old boy has right-lower-quadrant pain that has persisted for two days with progressive increase in severity. He is anoretic, but has not experienced nausea or vomiting. He has had two loose bowel movements each day. His temperature is 38°C by rectum, and abdominal examination shows diffuse voluntary guarding, most marked in the right lower quadrant, but no palpable masses. Urinalysis is normal, and the white cell count is 15,000, with a slight shift to the left (more polymorphonuclear leukocytes than usual).

Assume that the physician assesses the probability of acute appendicitis to be .3 and the probability of acute gastroenteritis to be .7 on the basis of these findings. Should immediate operation be performed, or should the patient be observed further?

Assumptions Made

Operative fatality of laparotomy is .1% whether or not the appendix is removed. Survival following this surgery is thus 99.9%.

The exact figure for fatality following untreated appendicitis in a comparable cohort of healthy boys of this age is unknown, but for the purposes of this analysis, we assume a 50% probability of perforation if operation is delayed and fatality rate of 2% for perforation. The fatality rate of appropriately treated appendicitis with perforation in the general population is approximately 4%. We assume that our patient's good general health and age implies a lower rate, which explains our choice of 2%.

3-1. Construct a decision tree using the information given above (consider only survival initially).

3-2. On the basis of the calculations, should immediate operation be performed?

3-3. At what probability of acute appendicitis is delay the appropriate treatment?

The following case of suspected pulmonary embolism provides the background for Problems 3–4 through 3–7.

---

* The case descriptions and assumptions are reprinted with modifications by permission from S. G. Pauker and J. P. Kassirer, *N. Engl. J. Med.*, **293**:229–34, 1975.

Case Description

A 30-year-old asthmatic woman who takes oral contraceptive agents is seen because of right-sided pleuritic chest pain and dyspnea. She had an anaphylactic reaction several years earlier in response to intravenous pyelography. Examination shows diffuse wheezing and an accentuated pulmonic component of the second heart sound. There is no evidence of phlebitis. X-ray study of the chest shows only hyperaeration. A lung scan shows a defect in the right-lung field consistent with asthma or pulmonary embolism. You feel there is a 10% chance of embolism. Should the patient be anticoagulated?

Assumptions Made

Given embolism without treatment, there is a 50% chance of reembolization.

Of patients with reembolization, 50% die.

Long-term anticoagulation is associated with a 5% morbidity rate (requiring hospitalization) and a .01% mortality rate.

With a prior pulmonary embolism and with anticoagulation, there is a 15% chance of reembolization and thus a 7.5% chance of death from reembolization.

3-4. Construct a decision tree for this case.

3-5. On the basis of your calculations, would you anticoagulate this woman?

3-6. If we assume that the patient has a 1% chance of dying if a pulmonary angiogram is performed, and that this test will correctly identify 95% of patients with embolism and 95% of patients without an embolism, that is

$$P(\text{positive test} \mid \text{embolism}) = .95$$

$$P(\text{negative test} \mid \text{no embolism}) = .95$$

would you perform an angiogram?

3-7. Is the decision sensitive to your estimate of $P(\text{embolism})$?

3-8. Construct the decision tree corresponding to a clinical problem of your choice.

References

Arndt, K. A., and Jick, H. (1976). Rates of cutaneous reactions to drugs: a report from the Boston Collaborative Drug Surveillance Program. *JAMA*, **235**:918–22.

Carlson, K. J., and Mulley, A. G. (1985). Management of acute dysuria. *Ann. Intern. Med.*, **102**:244–49.

Curran, J. W.; Rendtorff, R. C.; Chandler, R. W.; Wiser, W. L.; and Robinson, H. (1975). Female gonorrhea: its relation to abnormal uterine bleeding, urinary tract symptoms, and cervicitis. *Obstet. Gynecol.,* **45**(2):195–98.

Isselbacher, K. J.; Adams, R. D.; Braunwald, E.; Petersdorf, R. G.; and Wilson, J. D., eds (1980). *Harrison's Principles of Internal Medicine.* New York: McGraw-Hill.

Pauker, S. G., and Kassirer, J. P. (1975). Therapeutic decision making: a cost-benefit analysis. *N. Engl. J. Med.,* **293**(5):229–34.

Schultz, H. J.; McCaffrey, L. E.; Keys, T. F.; and Nobrega, F. T. (1984). Acute cystitis: a prospective study of laboratory tests and duration of therapy. *Mayo Clin. Proc.* **59**:391–97.

Additional Reading

Bunker, J. P.; Barnes, B. A.; and Mosteller, F., eds. (1972). *Costs, Risks, and Benefits of Surgery.* New York: Oxford University Press.

Weinstein, M. C.; Fineberg, H. V.; Elstein, A. S.; Frazier, H. S.; Newhauser, D.; Neutra, R. R.; and McNeil, B. J. (1980). *Clinical Decision Analysis.* Philadelphia: W. B. Saunders.

CHAPTER **4**

Has the Treatment Helped the Patient? Intrasubject and Intersubject Variability

> **OBJECTIVES**
>
> *Signs and symptoms may vary in an individual patient*
>
> *Among the sources of this variation are biologic change, temporal change, and measurement error*
>
> *Inter- and intrapatient variation should be distinguished*
>
> *Records are important in assessing signs, symptoms, or clinical conditions which are highly variable*
>
> *Mean and median are important measures of position or location*
>
> *Some measures of variation are range, interquartile range, and standard deviation*
>
> *Frequency and relative frequency distributions help in comparing distributions, as do their representations as histograms*

If the clinical course of some illness were always the same in the absence of treatment, and if treatment always had the same effect, it would be easy to determine whether some new treatment was an improvement. We would need only to prescribe the treatment to see whether the outcome was changed. A similar approach is still possible when the course of disease is not precisely uniform. One has to observe a sufficient number of cases of the illness and record the frequency of each possible outcome in the absence of treatment, or in the presence of a standard treatment. (We return in Chapter 7 to what is "sufficient.") Then after giving the new treatment to a sufficient number of additional cases, one can see whether and how the probabilities of various outcomes have changed. This chapter discusses the kinds of variability that may affect clinical observations and shows how frequency distributions are useful in the study of clinical observations that may vary from patient to patient or from time to time.

4-1. Evaluating Treatment of the Individual Patient

Background: Angina Pectoris

Angina pectoris (substernal chest pain typically brought on by exercise and relieved by rest) is a common symptom of coronary vascular disease. Angina can cause substantial morbidity by limiting a patient's activity. For

69

many years, trinitroglycerine (TNG), administered sublingually, has been used to treat angina. Usually, TNG will relieve an attack in one to three minutes, and most patients find TNG helpful for most of their attacks. TNG acts for only five to fifteen minutes; it is impossible to prescribe this drug frequently enough to have day-long prevention of angina. One therapeutic approach uses "long-acting" nitrate preparations, although some authorities question their effectiveness. Patients have also been treated with propranolol (a beta blocker) in an effort to decrease the number of anginal attacks.

Clinical Problem 4-1. *Does Long-Acting Nitrate Therapy Help?*

Mr. Lewis is a 55-year-old man with angina. His attacks typically occur after he has climbed half a flight of stairs or walked a quarter of a mile. He has been having about six attacks each week.

His physician recently prescribed a nitrate preparation, isosorbide nitrate (ISDN). ISDN has a much longer duration of action than TNG, which might give it substantial advantages if it is equally effective. Mr. Lewis called his physician later to say that he had his usual angina halfway up his 14 stairs one hour after taking ISDN. In addition, he experienced headache and palpitations (both being known side effects of ISDN). He wondered whether he should stop the ISDN as he has noted no change in his angina and the drug caused him bothersome side effects.

Gathering Evidence from the Patient

To decide whether ISDN has value for Mr. Lewis, we need some idea of how often his angina attacks occur without treatment. (*Note:* The comparison proposed here is ISDN versus nothing, not ISDN versus TNG as discussed just above.) If Mr. Lewis or his physician kept records of his angina before treatment started, some available information might help answer the question. Mr. Lewis did keep a diary concerning his angina attacks. The diary reads as follows:

August 16: angina on tenth step.
August 18: angina on third step.
August 19: angina on sixth step.
August 20: climbed all 14 stairs without angina.

The diary has some 50 entries for the most recent two months. We summarize the information in the frequency distribution shown in Table 4-1. We converted the data on angina experience before ISDN to a histogram in Figure 4-1. The horizontal scale gives the number of steps completed without angina. Thus if Mr. Lewis got angina on the first step, he completed 0 steps without angina. If he climbed the whole flight without angina, he completed 14 steps.

Figure 4-1 has two vertical scales. The left-hand scale gives the observed frequency or count of the number of times Mr. Lewis climbed the

Table 4-1.   Frequency distribution
of steps before angina for
Mr. Lewis before and after ISDN

Steps climbed without angina	Frequency before ISDN	Frequency after ISDN
0	0	1
1	0	0
2	2	0
3	3	1
4	5	0
5	4	2
6	9	2
7	6	3
8	4	3
9	4	2
10	3	2
11	1	1
12	1	0
13	0	1
14	8	2
Total	50	20

given number of steps without angina. The right-hand scale gives the relative frequency, the proportion of trials out of 50. Usually, we are more interested in the proportion because we may be comparing two sets of data based on different total counts.

Relative frequencies can be used to estimate the probability of observing angina at each step. Thus for Mr. Lewis, we estimate the unknown probability that angina will occur just after the sixth step in the absence of treatment as $9/50 = .18$, or 18%.

Interpreting One Observation

Looking at the data in the second column of Table 4-1 or the histogram in Figure 4-1A, we see that Mr. Lewis's new single observation of six steps without angina does not prove or disprove that the drug has some beneficial effect. Already, though, we can make some estimate as to the possible response Mr. Lewis will get from ISDN. His first experience shows that ISDN does not completely prevent attacks. We do not know whether ISDN has changed the probability of angina after the sixth or any other step. To think about this, suppose for a moment that the probability of angina at or before the seventh step were very low, say, 1 in 100 with ISDN. Then we would have observed a rare event (probability 1%) the very first time Mr. Lewis climbed his stairs after starting the new medication. Faced with this single observation, suppose that we must decide whether to believe that ISDN has reduced the probability of attacks at or before the seventh step to 1%. We have arbitrarily formulated two mutually exclusive conclusions:

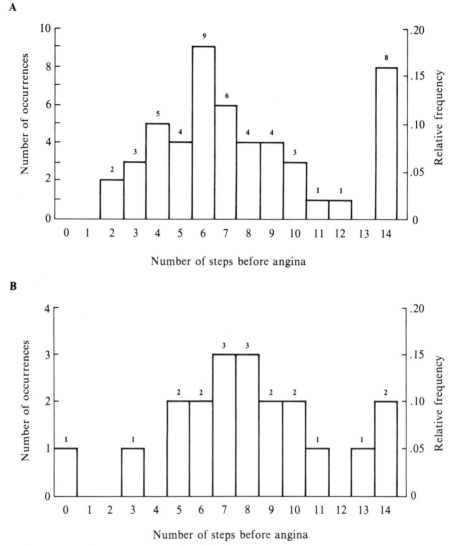

Figure 4-1. Histograms showing the frequency distribution of the number of steps before angina, prior to (*A*) and after beginning (*B*) ISDN therapy, as recorded in Mr. Lewis's diary. The left-hand scale gives the frequency; the right-hand scale gives relative frequency or proportion.

> *Conclusion 1.* With ISDN, angina occurs at or before the seventh step only 1% of the time, and we have observed a rare event; or
> *Conclusion 2.* With ISDN, angina occurs on steps 1 to 7 just as often as without ISDN, and we have observed a commonplace event.

If forced to choose between these extreme conclusions on the basis of this one observation, most people would choose conclusion 2. Thus even one measurement has produced a conclusion about the extent of improvement from ISDN.

Interpreting Several Observations

We can learn more, of course, if we have more than one observation. Only very striking effects of treatment can be demonstrated with a single observation. In Mr. Lewis's case, one observation would be insufficient to show a benefit even if ISDN were completely effective, because about 16% of the time he climbs to the top of his stairs without angina even while taking no medication. In a given situation, the smaller the effect—gain or loss—of some treatment, the more observations will be needed to demonstrate that effect. The technical reason is that the variability of a mean decreases with increasing sample size, and we measure our assurance in terms of the variability. We give more detail on this later.

Since Mr. Lewis had some unpleasant side effects from ISDN, it is a reasonable view that he should not take the drug unless he gets a "fairly large" benefit. What would be "fairly large" is hard to define in any precise way, but his physician believes that if the benefit is large enough to balance the side effects, it should be apparent after 20 or 25 observations. He instructs Mr. Lewis to continue the medication for three weeks and to keep a record of the point at which angina occurs each time he climbs the stairs.

Comparing Outcomes

Mr. Lewis returns with the data recorded in Table 4-1, which shows both the 50 observations before treatment was started and the 20 observations since then. The general shapes of the histograms shown in Figure 4-1*A* and *B* do not appear to differ a great deal. Notice that the scale of measurement for the raw frequencies differs in the two figures, while that for relative frequencies remains the same. The observed fraction of climbs without angina has gone down from 16% to 10%, a loss, and the median number of steps climbed before angina is now 8, whereas before it was 7, a slight gain.

MEDIAN.   The median is the middle value of a set of numbers when they are ordered according to size. If the number of values is odd, it is the middle number. If the number of values is even, it is the average of the two middle numbers.

EXAMPLES.   The median of 4, 5, 5, 7, 8 is 5. The median of 4, 5, 5, 7, 8, 8 is 6.

Even without formal statistical analysis, it seems that Mr. Lewis has had no marked benefit from his ISDN and the continued presence of side effects suggests that it would be prudent to discontinue the medicine. (Chapters 5, 6, and 8 deal with formal ways to test whether two frequency distributions have different properties.)

A second important approach to evaluating whether ISDN is beneficial for Mr. Lewis reviews how other patients respond to the drug. Two points are especially important: the proportion of patients similar to Mr. Lewis who respond, and the degree of improvement for those who do respond. If some patients are almost completely unresponsive, while responders tend to derive large benefits, this two week trial may be enough to conclude that the drug should be stopped. If almost all patients derive some benefit, but the average improvement is small, one might want to reconsider whether just 20 observations is enough to conclude that continued treatment is unwise.

4-2. Evaluation of Antianginal Drugs

Exercise Testing

In one method of measuring antianginal effect, patients perform some physical activity at a controlled rate and we measure the time until the onset of angina. Frequently, patients exercise on a laboratory bicycle ergometer with the work load initially set at 25 watts and increased 25 watts every three minutes.

A Crossover Study

Danahy et al. (1977) used this method to evaluate the effect of TNG and ISDN in 21 patients with angina. Each patient took exercise tests after administration of each drug and placebo. The order of testing was determined by randomization. This approach to comparison of treatments, known as a crossover design, is illustrated further in Section 4-4.

The investigators measured exercise tolerance two minutes after placebo and two minutes after .4 mg of TNG. In a subsequent investigation, they measured exercise tolerance before (control) and 1, 2, 3, and 5 hours after ISDN and placebo. Table 4-2 shows the results. The average, or mean value, of the exercise times begun two minutes after medication was given were 250 seconds for placebo and 356 seconds for TNG (see the next-to-last line of columns 2 and 3 of Table 4-2). On the average, TNG increased exercise tolerance by 106 seconds (356 minus 250). Exercise tolerance after oral ISDN was also increased 144 (399 minus 255), 100, and 55 seconds over the corresponding values in the oral placebo group at 1, 3, and 5 hours post drug administration, respectively. Note that the times for the control period for ISDN averaged 27 seconds higher than those for the control period for the oral placebo.

Placebo Effects

We compare results after treatment with those after the administration of placebo. We could compare 1-, 3-, and 5-hour posttreatment values with the pretreatment values, which might be a better indicator of the pretreatment status of patient than the control values taken at another time. However, there is a possibility that the patient's expectation of benefit might itself affect the outcome of the test, so that we might see posttreatment

Table 4-2. Exercise time to angina pectoris (seconds)

Patient number	SL placebo	SL TNG	Oral placebo				Oral ISDN			
			CONTROL	1 HR	3 HR	5 HR	CONTROL	1 HR	3 HR	5 HR
1	155	431	150	172	118	143	136	445	393	226
2	269	259	205	237	211	207	250	300	206	224
3	408	446	221	244	147	147	250	215	232	258
4	308	349	150	290	205	210	235	248	298	207
5	135	175	87	157	135	105	129	121	110	102
6	409	523	301	357	388	388	425	350	613	514
7	455	488	342	390	441	408	441	504	319	484
8	182	227	215	210	188	189	208	264	210	172
9	141	102	131	125	99	115	154	110	123	105
10	104	231	108	114	136	111	89	145	172	123
11	207	249	228	221	251	206	250	230	264	216
12	198	247	190	199	243	222	147	403	290	208
13	274	397	234	249	267	241	231	510	370	316
14	191	251	218	194	297	223	224	432	291	212
15	155	401	199	329	197	176	132	733	492	303
16	458	766	406	431	448	328	417	743	566	391
17	188	199	191	168	168	159	213	250	150	180
18	258	566	277	264	276	251	490	559	557	439
19	437	552	424	512	560	478	406	651	624	554
20	115	237	234	232	281	237	229	327	280	321
21	200	387	227	199	223	227	265	565	501	517
Mean	250	356	226	255	217	235	253	399	317	290
SD	115	158	85	100	118	101	113	192	163	138

Source: Reprinted by permission of the American Heart Association from D. T. Danahy, D. T. Burwell, W. S. Aronow, and R. Prakash, Sustained hemodynamic and antianginal effect of high dose oral isosorbide dinitrate, *Circulation,* **55:**381–87, 1977.

changes that were a result of the testing procedure, not the drug. Such "placebo effects" are nearly universal in clinical investigations. They usually tend to favor any active procedure over a control regimen. Thus we accept the penalty of having slightly less appropriate control points to avoid the potentially serious bias caused by placebo effects. The mean values in Table 4-2 do suggest a small improvement in the oral placebo data at 1 and 5 hours after the control point.

Variability of Patient Response

The results in Table 4-2 show that the time to angina is highly variable despite the fact that experimental conditions were rigidly controlled. (For instance, patients reported to the exercise laboratory in a fasting state at 8 a.m. Fasting was continued for the duration of each testing, and no smoking was allowed during or for 8 hours prior to each testing.) The variation that each subject exhibits during testing is best seen by examining exercise tests that were done under the same conditions in the same patient. There are two "control" observations for each patient. In no patient are the two values identical, and the difference ranges up to 213 seconds (patient 18). We see a noticeable difference between the means of the two control observations (226 with placebo, 253 with ISDN). This would ordinarily be a cause for concern, as one would expect patients in the untreated state to respond in similar fashion whether they were to be given ISDN or placebo. It is possible that one or more of the patients with high differences are in different clinical states at the time of the two examinations (e.g., with an undetected myocardial infarct after the ISDN test but before the placebo test). We have no information about this possibility.

Table 4-2 also shows that, 1 hour after the administration of ISDN, exercise time was prolonged by 144 seconds as compared with placebo (399 seconds minus 255 seconds). This does not tell us whether some patients respond a lot and some a little, or whether most patients respond to about the same degree. Nor does this difference, by itself, tell us whether the drug causes exercise tolerance to be markedly different from usual or whether it just prolongs exercise a small amount as compared with the usual variation in exercise tolerance.

The data in Figure 4-2 do suggest that some patients responded well to ISDN, whereas others did not respond at all. Figure 4-2 shows a histogram of the difference between the exercise tolerance time 1 hour after oral ISDN and the comparable value after placebo. For patient 1, it went up from 172 to 445 seconds. For 8 patients in order of difference (patients 4, 5, 3, 9, 6, 11, 10, and 8), the 1-hour exercise tolerance changed by less than 60 seconds; in 3 of these it went up, and in 5 it went down. The remaining 13 patients showed improvement under ISDN treatment ranging from 63 to 404 seconds. The lower part of Figure 4-2 shows that when the same patients were treated by TNG, 12 (patients 9, 2, 17, 7, 3, 5, 4, 11, 8, 12, and 14) showed changes of 60 seconds or less, while the remaining 9 all improved by 114 to 308 seconds. Patient 1 improved by 276 seconds from 155 to 431.

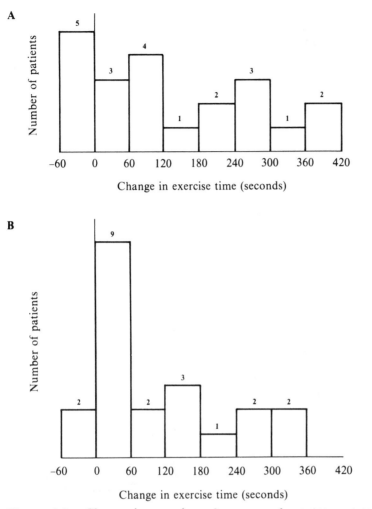

Figure 4-2. Change in exercise tolerance under treatment (treated value minus control value, in seconds): (*A*) 1 hour after oral ISDN; (*B*) after TNG.

Although this analysis suggests that some patients were marked responders and some poor responders, it does not show whether the same pattern of variation in response to ISDN would be observed if the same experiment were repeated with the same subjects. On the one hand, it may be that some patients are in a disease state that is always helped by ISDN and some are never helped. Under this hypothesis, the marked responders in Figure 4-2 would have a similar marked response had they been tested on another day and the variation observed would reflect intersubject variation. On the other hand, it may be that a patient's responsiveness to ISDN varies from day to day just as a patient's exercise tolerance without medicine can vary from day to day. If this occurs, we might find about the same proportion of responders each day, but they would not always be the same patients.

Table 4-3.   Responsiveness of patients to ISDN and TNG

TNG	*ISDN*	
	RESPONDER	NONRESPONDER
RESPONDER	Nos. 1, 13, 15, 16, 18, 19, 20, 21	Nos. 6, 10
NONRESPONDER	Nos. 7, 12, 14, 17	Nos. 2, 3, 4, 5, 8, 9, 11

To put the possibilities in other words, the data are consistent with the following conditions:

1. Some patients are always helped a great deal by ISDN, or
2. All patients are sometimes helped by oral ISDN, or
3. Some combination of these.

This matter cannot be resolved without obtaining repeated sets of measurements under the same treatment for some patients.

There is some evidence that response to nitrates (oral ISDN or sublingual TNG) is reasonably constant from day to day. If a change in exercise tolerance of 60 seconds or less is taken to mean "nonresponder," the patients may be classified as shown in Table 4-3. In all, 15 patients were classified the same way (as responders or nonresponders) after each of the two treatments, while only 6 responded to one but not the other. It is possible, of course, that some patients change from responder to nonresponder and back again over a period of weeks or months, not days. If so, the longer the interval between testing the two drugs, the less the two responses would tend to be the same.

The analysis above suggested that oral ISDN substantially improves the exercise tolerance of some patients, but not of all. Although exercise tolerance measured in the laboratory is not the same as angina following climbing stairs at home, one might consider them sufficiently similar to conclude on the basis of his diary data that Mr. Lewis is probably not one of those patients whose exercise tolerance is markedly benefited by oral ISDN. That is, the estimated prior probability of no benefit (less than 60 seconds) is 9/21, or 43%. Mr. Lewis's further data encourages us to raise his posterior probability of being a nonresponder.

4-3.   Biologic, Temporal, and Measurement Variation

Importance of Variation in Interpreting Outcomes

After the unsuccessful attempt to control Mr. Lewis's angina with ISDN, he went without treatment. His clinical state was apparently unchanged for five months, at which time he told his physician that the angina had recently begun to appear more often, and on a lower step than before. His physician must now consider three broad kinds of reasons for the change. First, Mr. Lewis may have suffered a biologic change (his coronary disease may have worsened). Second, Mr. Lewis's angina may be temporarily worse for no apparent reason, just as exercise tolerance is higher on some days than on others. Finally, Mr. Lewis may have become a more (or less) accurate observer or reporter of his angina.

> Generally, clinical observations are subject to three sources of change, which may be called biologic, temporal, and measurement variation.

In evaluating the status of Mr. Lewis, his physician was at first concerned with whether ISDN caused a biologic variation in his anginal pattern. This evaluation was made difficult because angina has a great deal of temporal (day-to-day) variation and perhaps some measurement variation as well. In discussing the trial of ISDN reported by Danahy *et al.*, we asked whether patients who responded well to ISDN were biologically different from the poor responders in some permanent way (interpatient variation), whether the observed variation in response to ISDN might reflect only the day-to-day variation of patient's responsiveness (intrapatient variation), or whether some combination of these was at work.

When a series of observations is made on different individuals, the variation in responses is due to both intersubject variation (secondary to biologic, temporal, or measurement differences between the subjects) as well as intrasubject variation (also due to biologic, temporal, and measurement variation within a subject). To distinguish the contribution of each source to the overall variation, a series of separate observations on separate persons will not do. One has to study the same individuals more than one time to see whether the individual frequency distributions are similar to each other, and hence to the frequency distribution for the population. For instance, obviously patients' heights vary widely, and the source is intersubject variation in height. Body temperature in patients at the outpatient clinic also varies. However, it is likely that most of that variation is due to intrasubject variation in body temperature, since unusual temperatures may be a symptom associated with going to the clinic.

Implications for Patient Care

A number of principles follow from recognizing that a clinical observation is subject to biologic, temporal, and measurement variation, and that each of these sources may be reflected in intra- as well as intersubject variability.

1. In conditions that have large temporal and/or measurement variation, therapeutic efficacy or other biologic changes may be difficult to detect even with large numbers of well-controlled observations.
2. The "normal range," as determined by observing many individuals, is always at least as large as the normal range determined by observing one individual many times. It will, of course, be larger if in-

terperson variation is, on the average, greater than intraperson variation. We often use the rather arbitrarily chosen range 2½ to 97½%, the central 95% of a sample of values obtained from normal subjects, as the normal range of a measurement, including thereby both intra- and interperson variability.

3. Some patients seek their doctors' attention when their conditions seem to worsen. If the worsening simply represents temporal and not biologic variation in their illness, their illness is likely to improve irrespective of therapy. ("Most things, in fact, are better by morning"—Lewis Thomas.) The technical name in statistics for such changes is the "regression effect," meaning regression toward the mean. Thus a patient who feels spectacularly well today will probably not feel as well tomorrow.

4. The physician who observes a patient numerous times, or orders numerous laboratory studies, may observe "abnormalities" that do not reflect a biologic variation but are due to temporal and/or measurement variation. These, too, are likely to be "better" or changed soon.

Distributions

Frequency distributions, relative frequency distributions, and histograms are convenient (and equivalent) methods of summarizing collections of multiple observations. Typically, a frequency distribution is obtained by dividing observations into 10 to 20 classes such that each observation must fall into one and only one class. In a frequency distribution, the number of observations belonging to each class is recorded.

> The relative frequency distribution assigns to each class an estimated probability (observed relative frequency) that an observation will be in that class.

If conditions are constant, the larger the sample, the better the estimate. The estimated probability of each class is easily computed as the number of observations in that class divided by the total number of observations.

Measures of Location

Many questions in medicine hinge on determining whether one probability distribution differs from another. Such a difference may be difficult to determine because the distributions themselves are unknown and must be estimated with some degree of error, also unknown. Estimated distributions can be compared in terms of many different properties. Perhaps the most important is the "center" or location of the distribution, which may be defined as:

The **mean**: the ordinary average of the observations.
The **median**: defined earlier.

Another important property is the degree of "spread" or dispersion of observations about their center. Again, "spread" may be defined in several ways, such as:

The **range**: the difference between the largest and smallest observed values. This should not be confused with the "normal range" discussed earlier, although the ideas are similar.

The **interquartile range**: the range of values remaining when the largest 25% and smallest 25% are—temporarily—set aside.

The **standard deviation**: a frequently used measure of spread to be defined in Chapter 5. For some distributions met in practice, about two-thirds of the observations lie between 1 standard deviation above and 1 standard deviation below the mean. We treat spread in more detail in Chapter 5.

4-4. Changing Treatment

Clinical Problem 4-2. *Does the Patient Respond to Propranolol?*

Mr. Lewis returns for an unscheduled appointment and complains to his physician that he had 11 attacks of angina during the past week. His physician decides to try another drug, propranolol. The following week, Mr. Lewis reports that he has been feeling well, with only 2 attacks of angina. No symptoms of congestive heart failure or asthma (recognized complications of propranolol) have developed. Has propranolol proved effective for Mr. Lewis?

Variability or Treatment Effect? Implications of a Clinical Trial

Unfortunately, Mr. Lewis has stopped recording each of his angina attacks, so one can only look in the literature to see whether such a change in frequency of angina from week to week is compatible with the biologic effect of propranolol and/or temporal variation often observed in patients with angina. (Whether he stopped recording or not, do we need to know more than the weekly totals he reported: 11 and 2?) Rabkin *et al.* (1966) reported the results of an investigation into the effects of propranolol on angina: 16 patients with "a history of angina on effort which was virtually unchanged in severity and frequency over the previous six months" were treated in a crossover study comparing placebo and propranolol. Each patient was studied for a period of 9 weeks, divided into a pretreatment 1-week period and four treatment periods of 2 weeks each. Four different crossover

Table 4-4. Weekly glyceryl trinitrate consumption for each medication sequence in a double-blind crossover trial of propranolol and placebo[a]

Case	Pretrial week	Week		Week		Week		Week		Total for 4 weeks	
		1st	2nd	1st	2nd	1st	2nd	1st	2nd	A	B
		A		*B*		*B*		*A*			
1	10	0	0	4	6	8	5	0	0	0	23
2	3	2	0	0	0	1	0	0	0	2	1
3	8	3	1	6	4	1	6	4	0	8	17
4	37	49	29	43	37	61	35	43	41	162	176
		B		*A*		*B*		*A*			
5	10	6	6	2	2	8	5	2	4	10	25
6	3	3	7	0	2	4	2	1	1	4	16
7	43	62	49	33	21	57	43	30	15	99	211
8	5	4	6	4	2	1	1	1	3	10	12
		A		*B*		*A*		*B*			
9	5	1	1	1	2	1	1	1	1	4	5
10	8	0	0	3	7	2	2	7	8	4	25
11	15	4	6	13	17	4	6	13	17	20	60
12	16	6	4	5	6	5	5	5	3	20	19
		B		*A*		*A*		*B*			
13	2	13	10	2	1	3	1	2	8	7	33
14	8	2	4	0	0	0	0	1	3	0	10
15	8	5	6	6	1	12	11	13	11	30	35
16	6	3	6	1	1	0	0	0	1	2	10
Total	187									382	678
Average per week	11.5									5.9	10.6

[a] A, drug therapy; B, placebo.

Source: Reprinted by permission from R. Rabkin, D. P. Stables, N. W. Levin, and M. M. Suzman, *Am. J. Cardiol.*, **18**:370–80, 1966.

sequences were used. Let *A* be drug therapy and *B* be placebo. Then the four sequences were *ABBA*, *BABA*, *ABAB*, and *BAAB* and each was applied to four patients. These are all the possible "double" crossovers. That is, the first and last pair of the four have both treatments. So each pair is itself a crossover experiment. We see that *AABB* and *BBAA* are not double crossovers.

Although the number of angina attacks might be the most important variable, the authors state that the number of attacks did not differ materially from the number of TNG tablets consumed. Table 4-4 shows weekly TNG consumption.

Table 4-5 shows, for each patient, the maximum and minimum weekly usage of TNG tablets when only the pretreatment observation period and the weeks that the patient received placebo are considered. Each patient except numbers 4, 7, and 11 showed at least twofold variation, and several showed variation comparable to that reported by Mr. Lewis. Indeed, half showed multiples of 5 or more. For example, patient 13 reported just 2 TNG tablets used in the pretreatment week, immediately followed by 13 tablets in the first week of treatment (in this case, placebo). Therefore, it does not seem unusual that a patient would have 10 or 11 anginal attacks one week and 2 the next, merely because of temporal variation rather than a biologic or therapeutic variation. Mr. Lewis's visit to his physician was prompted by his 11 attacks of angina during the preceding week. If this represented a temporal variation in his angina pattern, one might expect the subsequent week to show the number of anginal attacks returning to the more usual 6 or 7 or even 2. A change from 11 attacks to 2 attacks seems compatible either with temporal variation or therapeutic effect. A longer period of observation would be needed to determine which.

Table 4-5. Maximum and minimum weekly use of TNG, excluding weeks on propranolol regimen

Patient	Minimum	Maximum
1	4	10
2	0	3
3	1	8
4	35	61
5	5	10
6	2	7
7	43	62
8	1	6
9	1	5
10	3	8
11	13	17
12	3	16
13	2	13
14	1	8
15	5	13
16	0	6

4-5. Summary

Repeated observations in medicine often yield results that vary. Variation may reflect biologic differences between patients, biologic or therapeutic changes in a patient, as well as temporal or measurement variability. The ease with which a therapeutic maneuver can be evaluated depends on whether the variation due to therapy is markedly different from (and greater than) the other sources of variation. Frequency distributions and histograms are useful to summarize data that vary. Demonstrating therapeutic or biologic differences depends on demonstrating differences between frequency distributions.

Problems

In evaluating their method for determining serum thyroxine (T_4) Murphy *et al.* (1966) measured T_4 in over 1000 patients. Table 4-6 shows the frequency distribution of serum T_4 in three series of patients, each series containing some patients judged to be hypothyroid, some euthyroid, and some hyperthyroid on clinical grounds.

4-1. Construct probability histograms for hypo-, eu-, and hyperthyroid patients using the data from series 2.

4-2. What would you choose as the normal range? Why?

4-3. Within the normal range there is considerable variation, both intraperson and interperson. Figure 4-3 shows the variation in plasma T_4 in an individual subject observed on repeated occasions. How much of the normal range is explained by the intrasubject variability as seen in this patient? (Note that the data from this

Table 4-6. Frequency distribution of T_4 in three patient series by type of thyroid disease (males and females combined)

	Series 1			Series 2			Series 3		
	HYPO	EU	HYPER	HYPO	EU	HYPER	HYPO	EU	HYPER
>16			2			6			1
15.1–16			3			2			0
14.1–15			0		1	1			1
13.1–14			1		1	2			2
12.1–13			2		0	5			1
11.1–12			2		3	2			0
10.1–11			1		6			3	0
9.1–10		10			29			9	1
8.1– 9		21			75			15	
7.1– 8		20			153			6	
6.1– 7		42			200			37	
5.1– 6		37			172			36	
4.1– 5		31			121			34	
3.1– 4	3	12		10	21			8	
2.1– 3	1	1		14	4		3		
1.1– 2	3	1		13	1		7		
0– 1				11			2		
Mean T_4	2.6	6.3	13.8	2.1	6.5	14.4	1.7	6.5	13.4

Source: Reprinted by permission from B. E. P. Murphy, C. J. Pattee, and A. Gold, *J. Clin. Endocrinol.* **26**:247–56, 1966.

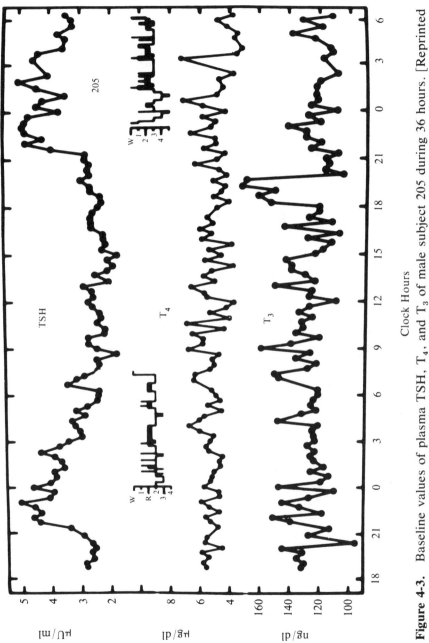

Figure 4-3. Baseline values of plasma TSH, T_4, and T_3 of male subject 205 during 36 hours. [Reprinted by permission from Azukizawa, M.; Perkary, A. E.; Hershman, J. M.; and Parker, D. C. (1970). *J. Clin. Endocrinol. Metab.*, **43:**533–42.]

figure represent a sampling over a 36-hour period. Seasonal variation in serum T_4 level has also been reported.)

4-4. You have a patient who is the sister of a woman with Graves' disease and hyperthyroidism. You suspect that your patient may be more likely than usual to develop a similar illness. Her serum T_4 determined one year ago was 5 μg per 100 ml. On a recent office visit it was 10 μg per 100 ml. Does this change indicate that this patient is developing hyperthyroidism? Why or why not?

4-5. The data in Table 4-4 record TNG consumption for 16 patients by week and treatment. Total TNG consumption over 4 weeks for each treatment is shown in the two rightmost columns.
   a. Compute the difference in 4-week total TNG consumption for each patient (placebo minus propranolol). Order the differences from smallest to largest. What is the median difference?
   b. How many patients consumed less TNG over 4 weeks when treated with propranolol? If propranolol is no more effective than the placebo, how many would you expect to consume less TNG? Why?

4-6. Table 4-7 shows plasma digoxin (ng/ml) for 10 patients on 6 consecutive days. Each patient received a daily digoxin dose of .25 mg. Compute the median of the 6 values for each patient.

Table 4-7. Digoxin concentration on 6 consecutive days in 10 patients on a continuous daily digoxin dose of .25 mg

Patient						
1	1.65	1.95	2.10	1.55	1.80	1.75
2	1.50	1.25	1.50	1.30	2.15	2.20
3	.80	.95	.95	.80	1.45	1.15
4	.70	.65	.80	1.10	1.05	1.00
5	.75	.75	1.00	.90	.80	.80
6	.55	.80	.80	.95	.80	.85
7	.70	.80	.80	.70	.80	.95
8	.75	.65	.75	.65	.50	.70
9	.55	.60	.75	.80	.55	.65
10	.60	.60	.75	.60	.60	.55

References

Azukizawa, M.; Pekary, A. E.; Hershman, J. M.; and Parker, D. C. (1976). Plasma thyrotropin, thyroxine, and triiodothyronine relationships in man. *J. Clin. Endocrinol. Metab.*, **43**(3):533–42.

Danahy, D. T.; Burwell, D. T.; Aronow, W. S.; and Prakash, R. (1977). Sustained hemodynamic and antianginal effect of high dose oral isosorbide dinitrate. *Circulation*, **55**:381–87.

Murphy, B. E. P.; Pattee, C. J.; and Gold, A. (1966). Clinical evaluation of a new method for the determination of serum thyroxine. *J. Clin. Endocrinol.*, **26**:247–56.

Rabkin, R.; Stables, D. P.; Levin, N. W.; and Suzman, M. M. (1966). The prophylactic value of propranolol in angina pectoris. *Am. J. Cardiol.*, **18**:370–80.

Additional Reading

Louis, T. A.; Lavori, P.; Bailar, J.; and Polansky, M. Crossover and self-controlled designs in clinical research. In Bailar, J. and Mosteller, F., eds. (1986), *Medical Uses of Statistics* (Chapter 4). Waltham, MA: New England Journal of Medicine.

CHAPTER 5

Blood Pressure and Hypertension: Distribution and Variability

The diagnosis and treatment of patients with high blood pressure forces us to consider the variability and distribution of blood pressure measurements in the individual patient. In this chapter we examine the blood pressure variation that might be observed in one office visit, and the variation from one visit to the next. The gaussian (or normal) distribution aids our thinking, as does the *t* distribution, when we try to estimate a patient's average blood pressure. Applying these ideas helps us to determine what we need to observe and why, and how to detect changes in blood pressure in response to antihypertensive therapy or other interventions.

5-1. Diagnosis of Hypertension

Clinical Problem 5-1. *Moderately Elevated Blood Pressure at a Routine Physical*

A company refers Mr. W.P., a 25-year-old computer programmer, to you for a preemployment physical. He has a family history of stroke, he smokes one package of cigarettes a day, and his blood pressure is 150/100 mm Hg.

Background

The following statements are excerpted from recommendations of the Joint National Committee (JNC) on the evaluation and treatment of high blood pressure (Joint National Committee on Detection, Evaluation, and Treatment of High Blood Pressure, 1984).

> The diagnosis of hypertension in adults is confirmed when the average . . . diastolic blood pressure . . . is 90 mm Hg or higher. . . .
>
> [The following] gives a recommended categorical scheme for arterial blood pressure for use in persons aged 18 years or older.

Diastolic Pressure Range, mm Hg	Category
less than 85	Normal BP
85–89	High normal BP
90–104	Mild hypertension
105–114	Moderate hypertension
greater than 115	Severe hypertension

Discussion: Variability of Blood Pressure in the Individual Patient

One's first impulse is to decide that Mr. W.P. has a diastolic blood pressure between 90 and 104 mm Hg, placing him in the mild hypertension category. This view may turn out to be correct, but before settling on it, let us review the variability of blood pressure measurements.

Figure 5-1 shows blood pressure measurements taken at 5-minute intervals over a 23-hour period from noon one day to late next morning for a single subject. We observe that the graph varies a great deal from time to time. The lower jagged curve gives the diastolic blood pressure. During the waking period of the first day, the diastolic pressure varies from about 50 to about 75, excluding the coital period. During the waking period the second day, the diastolic pressure varies from about 40 at waking time to about 95. Thus a waking period measurement the first day would have indicated that subject A.B. was well below the low cutoff of 90 given by the Joint National Committee, while on the second day, a measurement might have been over 90, thus throwing the patient into the mild hypertension category.

Armitage and Rose (1966) provide a second instructive set of data (Figure 5-2) showing how diastolic blood pressure varies in the individual. Even if we exclude the extreme right-hand measurements for subjects 8 and 10, some subjects have ranges of measurements (largest minus smallest) of more than 30 mm Hg. Therefore, Mr. W.P.'s measurement of 100 could possibly be a high measurement for him, and perhaps he averages 15 mm Hg lower,

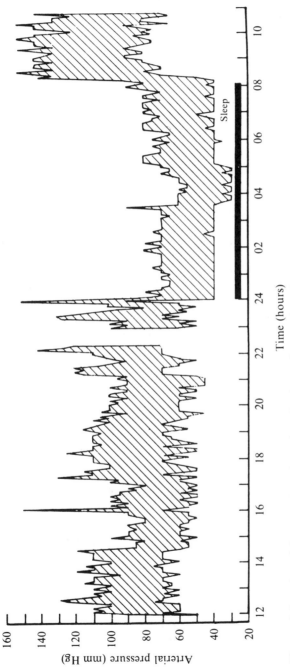

Figure 5-1. Arterial pressure, plotted at 5-minute intervals, of subject A.B. The period of sleep is shown by the horizontal bar. The high pressures shown at 16.00 and 24.00 hours are due to a painful stimulus and coitus, respectively. [Reprinted by permission from Bevan, A. T.; Honour, A. J.; and Scott, F. H. (1969). *Clin. Sci.*, **36**:329–44.]

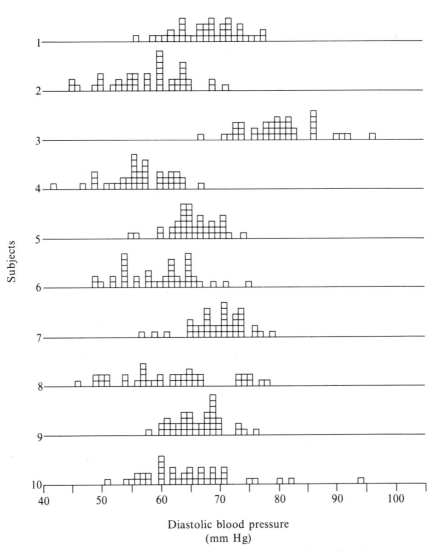

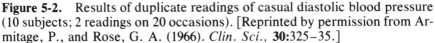

Figure 5-2. Results of duplicate readings of casual diastolic blood pressure (10 subjects; 2 readings on 20 occasions). [Reprinted by permission from Armitage, P., and Rose, G. A. (1966). *Clin. Sci.*, **30**:325–35.]

which would take him out of the hypertensive range. Or 100 might be a low measurement for him, and his average would be, say, 10 units higher, which would take him into the moderate category.

The message from Figures 5-1 and 5-2 is that Mr. W.P.'s diastolic blood pressure of 100 is ambiguous. We need to discuss distributions of blood pressures, probabilities, and their relation to means and standard deviations to get a better grip on the interpretation of blood pressure measurements.

The next section discusses some properties of one type of distribution, the gaussian (or normal) distribution. The distribution of an individual's blood pressure measurements often resembles the normal distribution. Therefore, assuming that our patient's blood pressure is normally distributed is reasonable. The relevance of the normal distribution to blood pressure measurements and to many other physiologic observations arises from a powerful and general rule that we discuss in this chapter: If a measurement arises from the summation of several small independent contributions, its distribution tends to assume a gaussian shape. This rule is known as the Central Limit Theorem.

Although it may seem impractical to try to force the empiric uneven histogram of actual observations into the smooth theoretical gaussian curve, subsequent sections will show that this gaussian model can provide us with very practical methods for interpreting measurements.

5-2. Gaussian or Normal Distribution

One distribution commonly used to describe measurement data is called the gaussian distribution after the great German mathematician Karl Friedrich Gauss, who did not invent it, and is also called the normal distribution, for no very good reason. It is the distribution that emerges when one sums many independent random effects. Figure 5-3 shows the general shape, often called bell-shaped.

> The gaussian distribution is one of the central concepts introduced in this book. It is fundamental to much of statistical inference.

The curve is symmetric, so the mean and the median (50% point) fall at the center of symmetry. The mean is labeled μ, lowercase Greek mu, pronounced "mew." The measure of spread for the normal distribution is the standard deviation, labeled σ, lowercase Greek sigma. In continuous distributions, probabilities are represented as areas. About 68% of the probability falls within 1 standard deviation of the mean, about 95% within 2 standard deviations, and practically all—99.7%—within 3 standard deviations.

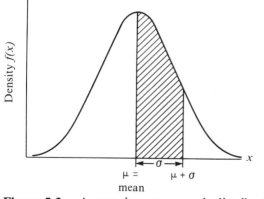

Figure 5-3. A gaussian, or normal, distribution with mean μ and standard deviation σ. Probabilities are represented as areas under this curve. For example, the shaded area from the mean to the mean plus 1 standard deviation is about .34.

Computing Normal Probabilities

Table 5-1 and Table III-1 in Appendix III give the probabilities for a gaussian distribution that has mean 0 and standard deviation 1.

EXAMPLE 1.   Find the probability between 0 and .5 standard deviations.

SOLUTION.   The table gives .1915.

EXAMPLE 2.   Find the probability beyond 1.96 standard deviations from the mean.

SOLUTION.   The table shows an area of .475 between 0 and 1.96 standard deviations, so $.5 - .475 = .025$ lies to the right of $+1.96$ and .025 lies to the left of -1.96.

COMMENT.   Although we said earlier that 95% of the probability lay within 2 standard deviations of the mean, the more precise result is that 95% lies within a distance of 1.96 standard deviations from the mean.

EXAMPLE 2A.   What is the probability that Mr. W.P.'s observed blood pressure of 100 mm Hg is within 1.96 standard deviations of his mean blood pressure (assuming that his blood pressure follows a normal distribution)?

SOLUTION.   We know that 95% of the probability lies less than 1.96 standard deviations away from the mean. Therefore, a blood pressure observation, such as Mr. W.P.'s, "chosen at random" has a 95% chance of being in this interval. More generally, the normal distribution allows us to estimate the probability or confidence that the average level falls in any given interval based on one measurement of blood pressure.

Although each mean and standard deviation pair produces a different gaussian distribution, provided that we know μ and σ, we can compute probabilities using the table of the standard gaussian distribution. We merely convert any measurement x into a standard measurement z, by measuring distance from the mean in standard deviation units, thus

$$z = \frac{x - \mu}{\sigma}$$

Then z has a gaussian distribution with a mean of 0 and a standard deviation of 1. We should speak of the family of gaussian distributions when we refer to the whole collection of curves, each with its own μ and σ.

Table 5-1. Probabilities, $A(z)$, for the standard gaussian or normal distribution

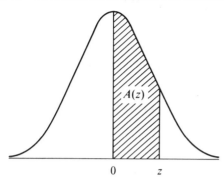

Area under the standard normal curve from 0 to z, shown shaded, is $A(z)$.

EXAMPLES. If Z is the standard normal random variable and $z = 1.54$, then

$A(z) = P(0 < Z < z) = .4382$

$P(Z > z) = .0618$

$P(Z < z) = .9382$

$P(|Z| < z) = .8764$

z	.00	.01	.02	.03	.04	.05	.06	.07	.08	.09
.0	.0000	.0040	.0080	.0120	.0160	.0199	.0239	.0279	.0319	.0359
.1	.0398	.0438	.0478	.0517	.0557	.0596	.0636	.0675	.0714	.0753
.2	.0793	.0832	.0871	.0910	.0948	.0987	.1026	.1064	.1103	.1141
.3	.1179	.1217	.1255	.1293	.1331	.1368	.1406	.1443	.1480	.1517
.4	.1554	.1591	.1628	.1664	.1700	.1736	.1772	.1808	.1844	.1879
.5	.1915	.1950	.1985	.2019	.2054	.2088	.2123	.2157	.2190	.2224
.6	.2257	.2291	.2324	.2357	.2389	.2422	.2454	.2486	.2517	.2549
.7	.2580	.2611	.2642	.2673	.2704	.2734	.2764	.2794	.2823	.2852
.8	.2881	.2910	.2939	.2967	.2995	.3023	.3051	.3078	.3106	.3133
.9	.3159	.3186	.3212	.3238	.3264	.3289	.3315	.3340	.3365	.3389
1.0	.3413	.3438	.3461	.3485	.3508	.3531	.3554	.3577	.3599	.3621
1.1	.3643	.3665	.3686	.3708	.3729	.3749	.3770	.3790	.3810	.3830
1.2	.3849	.3869	.3888	.3907	.3925	.3944	.3962	.3980	.3997	.4015
1.3	.4032	.4049	.4066	.4082	.4099	.4115	.4131	.4147	.4162	.4177
1.4	.4192	.4207	.4222	.4236	.4251	.4265	.4279	.4292	.4306	.4319
1.5	.4332	.4345	.4357	.4370	.4382	.4394	.4406	.4418	.4429	.4441
1.6	.4452	.4463	.4474	.4484	.4495	.4505	.4515	.4525	.4535	.4545
1.7	.4554	.4564	.4573	.4582	.4591	.4599	.4608	.4616	.4625	.4633
1.8	.4641	.4649	.4656	.4664	.4671	.4678	.4686	.4693	.4699	.4706
1.9	.4713	.4719	.4726	.4732	.4738	.4744	.4750	.4756	.4761	.4767
2.0	.4772	.4778	.4783	.4788	.4793	.4798	.4803	.4808	.4812	.4817
2.1	.4821	.4826	.4830	.4834	.4838	.4842	.4846	.4850	.4854	.4857
2.2	.4861	.4864	.4868	.4871	.4875	.4878	.4881	.4884	.4887	.4890
2.3	.4893	.4896	.4898	.4901	.4904	.4906	.4909	.4911	.4913	.4916
2.4	.4918	.4920	.4922	.4925	.4927	.4929	.4931	.4932	.4934	.4936
2.5	.4938	.4940	.4941	.4943	.4945	.4946	.4948	.4949	.4951	.4952
2.6	.4953	.4955	.4956	.4957	.4959	.4960	.4961	.4962	.4963	.4964
2.7	.4965	.4966	.4967	.4968	.4969	.4970	.4971	.4972	.4973	.4974
2.8	.4974	.4975	.4976	.4977	.4977	.4978	.4979	.4979	.4980	.4981
2.9	.4981	.4982	.4982	.4983	.4984	.4984	.4985	.4985	.4986	.4986
3.0	.4987	.4987	.4987	.4988	.4988	.4989	.4989	.4989	.4990	.4990

Source: F. Mosteller, R. E. K. Rourke, and G. B. Thomas, *Probability with Statistical Applications,* 2nd ed., © 1970, Addison-Wesley, Reading, Mass. (Table III); reprinted with permission.

Blood Pressure Distribution for a Population

The progress report of the Hypertension Detection and Follow-up Program Cooperative Group (1977) gives a figure showing the frequency distribution of diastolic blood pressures in a large sample. It is based on about 159,000 home screening measurements. Figure 5-4 shows this distribution, which although slightly asymmetrical, is approximately gaussian in shape, with mean about 85 and standard deviation about 13. Thus we expect about 95% of the population to measure between $85 - 2(13) = 59$ and $85 + 2(13) = 111$.

The important point that the Cooperative Group emphasized in presenting this figure is that the fraction of the population regarded as hypertensive varies considerably depending on the cutoff chosen. For example, about 25% of screenees are hypertensive if the cutoff is 90 mm Hg, whereas only 1.4% would be if the cutoff is 115.

The distribution shown in Figure 5-4 combines the variation among people with the variation within them. Figure 5-2 showed these variations separately for 10 subjects. Therefore, Figure 5-4 shows more variation than the individuals do in Figure 5-2.

Using the Normal Distribution in Clinical Problem 5-1

To think further about Mr. W.P.'s blood pressure, we might formulate the question: Suppose that his diastolic blood pressure is 110 on the average. What are the chances that on a single measurement we observe 105 or less? or 90 or less?

To use the gaussian distribution to answer these questions, we need a mean and a standard deviation. We have been given the hypothetical mean of 110, and we need a standard deviation for variation within an individual. Let us use Figure 5-2 to get a rough idea. Since about 1/6 of the observations for a gaussian distribution fall beyond 1 standard deviation at each end, we might count in 1/6 from each end, get the distance between these points, and divide by 2 as an estimate of the standard deviation. With 40 measurements, 7 is about 1/6.

Let us carry out this exercise for the first and last three subjects.

	Upper value	Lower value	Difference	Estimate of σ
Subject 1	73	64	9	4.5
Subject 2	64	51	13	6.5
Subject 3	87	75	12	6.0
$\vdots$				
Subject 8	75	51	24	12.0
Subject 9	70	62	8	4.0
Subject 10	71	57	14	7.0

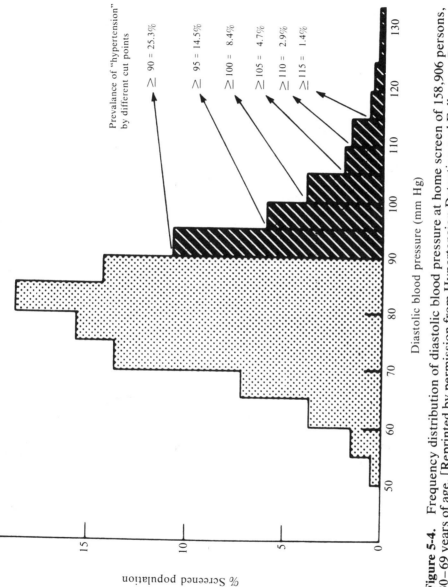

Figure 5-4. Frequency distribution of diastolic blood pressure at home screen of 158,906 persons, 30–69 years of age. [Reprinted by permission from Hypertension Detection and Follow-up Program Cooperative Group (1977). *Circ. Res.*, **40**:106–109.]

The average standard deviation estimated this way is about 7 mm Hg. Let us estimate W.P.'s standard deviation as 7. Now we are ready to tackle the two questions.

EXAMPLE 3.   If the true average is 110, we find the probability that the observation is less than 105 by computing the distance between 110 and 105 in standard deviation units. Calling this measure z,

$$z = \frac{105 - 110}{7} = \frac{-5}{7} = -.71$$

The probability of a smaller value, using Table 5-1, is about .24.

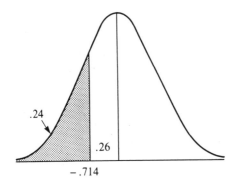

Thus even if Mr. W.P. is 5 points over the 105 level, he would have about a 24% chance of a measurement below the cutoff.

EXAMPLE 4.   We also want the probability that he produces a measurement less than 90, indicative of no hypertension. We compute similarly

$$z = \frac{90 - 110}{7} = \frac{-20}{7} = -2.86$$

The probability of a value more than 2.86 standard deviations below the true one is less than .01.

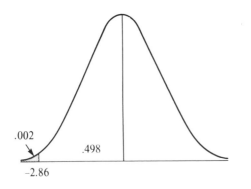

Thus the chance is slight that the value would fall into the normotensive region.

These illustrative calculations suggest that a single measurement of blood pressure can leave us rather uncertain about the average pressure and the treatment class to which the patient should be assigned.

5-3.   Confidence Limits

We know that if we have a gaussian distribution, the probability that an observation, x, falls between

$$\mu - 1.96\sigma \quad \text{and} \quad \mu + 1.96\sigma$$

is .95. The distance from μ to x is the same as the distance from x to μ. So if x falls within $\pm 1.96\sigma$ of μ with a probability of .95, μ must be within $\pm 1.96\sigma$ of x with a probability of .95.

> Thus, (given an observation x) we can compute "limits"
>
> $$x - 1.96\sigma \quad \text{and} \quad x + 1.96\sigma$$
>
> for the unknown mean μ. These are called **confidence limits.** Intervals so calculated include the value of μ in 95% of the calculations made.

The situation is shown graphically in Figure 5-5. In Figure 5-5A, the observation is within 1.96σ of μ and the 95% confidence interval contains μ. In Figure 5-5B, x is farther than 1.96σ from μ, so the 95% confidence interval does not contain μ. Since x falls within $\pm 1.96\sigma$ of μ 95% of the time, the interval contains μ 95% of the time. Thus we say that we are 95% **confident** that the interval from $x - 1.96\sigma$ to $x + 1.96\sigma$ contains μ.

Those who would like to see algebraically what is going on may want to run through the steps starting with

$$\mu - 1.96\sigma \leq x \leq \mu + 1.96\sigma \tag{5-1}$$

We want x on the outside and μ on the inside. Adding $-\mu - x$ to each of the three parts of the inequality yields

$$-x - 1.96\sigma \leq -\mu \leq -x + 1.96\sigma$$

Multiplying through by -1 changes all the signs and switches the order of the inequality to give

$$x + 1.96\sigma \geq \mu \geq x - 1.96\sigma \tag{5-2}$$

If the probability that inequality (5-1) was true was .95, say, then so is the probability that inequality (5-2) is true, as we have neither gained nor lost any information.

In inequality (5-1), the middle symbol is a random quantity and the outer

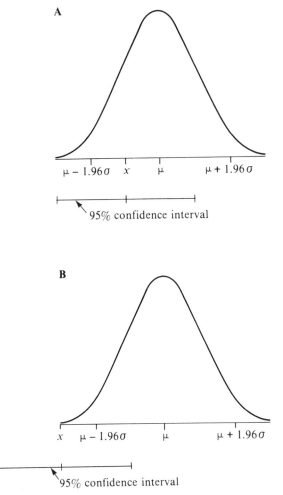

Figure 5-5. (*A*) When x is within 1.96σ of μ, the 95% confidence interval contains μ. (*B*) When x is farther than 1.96σ from μ, the 95% confidence interval fails to contain μ. Thus, the 95% confidence interval contains μ in A but does not in B.

limits are fixed. In inequality (5-2), it is the other way around. Since the limits fluctuate from sample to sample in inequality (5-2), we **cannot** say things like "the next observation from the same population has probability .95 of falling within the same numerical limits we got in this sample."

For instance, the first observation, x_1, might fall at a value equal to $\mu - \sigma$, while the second observation, x_2, might be equal to $\mu + 1.3\sigma$. Although x_2 did not fall within the limits established by x_1 ($x_1 \pm 1.96\sigma$), the "limits" estimated from each observation included μ. Indeed, this is the meaning of "confidence limits," namely that the next, or any sample's limits have a probability of .95 of including the true mean.

Essentially, each time we make a confidence statement using limits, we are right or wrong. Using 1.96σ, we are right 95% of the time when we have gaussian data.

EXAMPLE 5. In our original measurement, Mr. W.P. had observation $x = 100$ mm Hg. Using our estimated value $\sigma = 7$, the 95% confidence limits for μ are (rounding 1.96 to 2)

$$100 - 2(7) = 86 \quad \text{and} \quad 100 + 2(7) = 114$$

The interval from 86 to 114 leaves us with the reasonable possibility that Mr. W.P. falls into the high normal, the mild hypertensive, or the moderate hypertensive range.

The use of confidence limits can be carried one step further. Suppose that we wish to compute the confidence that Mr. W.P.'s true blood pressure is in the interval 90 to 104.

If μ belongs to the interval 90 to 104, then the observed value, 100, is not more than 10 mm Hg above the mean and not more than 4 mm Hg below the mean. We write this in an equation as

$$\mu - 4 \leq x \leq \mu + 10$$

Since $\sigma = 7$, we can express this as

$$\mu - .57\sigma \leq x \leq \mu + 1.43\sigma \tag{5-3}$$

As before, we add $-\mu - x$ to each of the parts to obtain

$$-x - .57\sigma \leq -\mu \leq -x + 1.43\sigma$$

Multiplying through by -1 gives

$$x + .57\sigma \geq \mu \geq x - 1.43\sigma \tag{5-4}$$

The confidence level for inequality (5-4) is the probability for inequality (5-3). We can use Table 5-1 to compute this probability if we note that inequality (5-3) can occur if either

$$\mu - .57 \leq x \leq 0$$

or

$$0 \leq x \leq \mu + 1.43\sigma$$

These probabilities can be obtained from Table 5-1 as .216 and .424, respectively. Thus the probability of occurrence of inequality (5-3) is the sum of these, .64. Inserting $x = 100$ and $\sigma = 7$ into (5-4) gives

$$104 \geq \mu \geq 90$$

as intended, and our level of confidence in these inequalities is .64.

One important asymmetric confidence interval is known as the **one-sided confidence** interval. Suppose that we wish to compute the confidence that Mr. W.P.'s true blood pressure is 105 mm Hg or above. If μ is 105 or above, then the observed value, 100, is at least 5 mm Hg below the mean, or

$$x \leq \mu - 5$$

Repeating the argument applied to inequalities (5-3) and (5-4), we infer that this condition will be satisfied if

$$x \leq \mu - .71\sigma$$

From Table 5-1, we compute this probability as .26. Thus our confidence is .26 that Mr. W.P.'s diastolic blood pressure is 105 mm Hg or above.

Characterizing the Patient's Blood Pressure Distribution

Before going far with the treatment program, we may want to establish Mr. W.P.'s average pressure more firmly. We know from the law of averages that as we take more measurements, we can make the confidence limits shorter. If we had more measurements, we could also find out more about this patient's own variability instead of guessing it on the basis of the variation of others.

One snag emerges. If we take many measurements in a short interval of time, their variation is smaller than when we allow substantial time—several days—between measurements. So taking several measurements in a short time, although useful, can be misleading. We need a few measurements at several different times.

5-4. Using Averages for Tighter Diagnostics. Known Standard Deviation

If we have an observation on the patient at each of several times, we can use the average measurement, which we will call $\bar{x}$ (read: x bar), to make inferences about the long-run average blood pressure for the patient. We will be helped because averages are ordinarily less variable than individual measurements. In fact, if one increases the number of measurements by a factor of 4, this gives a mean whose standard deviation and length of confidence interval are half that of the single measurement. More generally:

> If a population has standard deviation σ, then averages of samples of size n drawn from that population have mean μ and standard deviation $\sigma_{\bar{x}} = \sigma/\sqrt{n}$. The square, σ^2, of a standard deviation is called a **variance**, and $\sigma_{\bar{x}}^2$ is called the variance of the sample mean.

Furthermore, averages tend to be more nearly gaussianly distributed than do single measurements if they were nongaussian, so our use of the gaussian distribution usually produces a closer approximation to the probabilities than we would have gotten from single measurements. This property of averages rests on a powerful and general result known as the **Central Limit Theorem**. Leaving aside the mild conditions that are typically satisfied in practice, the Central Limit Theorem states that:

When n measurements are independently drawn from a population with mean μ and standard deviation σ, as the sample size, n, increases, the distribution of $\bar{x}$ tends to the gaussian (normal) distribution with mean μ and variance σ^2/n.

EXAMPLE 6.   If we have six measurements, each taken on a different occasion for Mr. W.P., and their average is 113, the new confidence limits for his mean are

$$113 \pm 2 \left(\frac{7}{\sqrt{6}}\right) = 113 \pm 2(2.86) = 113 \pm 5.72 = 107.3 \text{ to } 118.7$$

Thus these limits put his mean rather firmly above 105—into the moderate hypertensive range where diagnostic studies are indicated.

5-5.   Graphical Representation of Confidence Limits

Some readers will benefit from a graphical understanding of confidence limits. Figure 5-6 shows the plan. The horizontal axis gives values of $\bar{x}$, the observed mean, and the vertical axis gives values of μ, the true mean, which is ordinarily unknown. The general plan is to start with probability limits and generate confidence limits from them.

Given μ, we can set upper and lower probability limits, $\mu - z_{1-\alpha}\sigma_{\bar{x}}$ and $\mu + z_{1-\alpha}\sigma_{\bar{x}}$. We can set $z_{1-\alpha}$ so that we are $100(1 - \alpha)\%$ sure that an observed $\bar{x}$ will fall in the probability interval. If we choose $z_{1-\alpha} = 1.96$, we are 95% sure that any given $\bar{x}$ sampled from the population with a mean of μ will lie in the interval determined by the probability limits. For any value of $1 - \alpha$, and thus $z_{1-\alpha}$, we say that our confidence level is $100(1 - \alpha)\%$.

We can now generate the upper and lower 45° lines shown in the figure to construct a probability belt by setting probability limits for every μ. The figure illustrates the situation where we have set $z_{1-\alpha} = 2$ (rounding from 1.96 for convenience). There is a 95% probability that any pair of observed $\bar{x}$ and the μ from which it was sampled will lie within the belt.

We are now going to use this horizontally constructed probability belt to fashion a vertical confidence interval. Suppose that we observe a specific $\bar{x}$, call it $\bar{x}_a$, sampled from a population with unknown μ. We know that 95% of all pairs of observed mean and the true mean from which it was sampled lie within the belt.

Thus, given an observed mean $\bar{x}_a$, we are 95% confident that the μ from which it was sampled lies on the vertical interval $\bar{x}_a \pm 2\sigma_{\bar{x}}$.

Constructing the confidence limits by using probability limits horizontally and then inverting the process to get confidence limits is the general plan for other problems as well. The upper and lower limits may be curved or discrete, but the idea is to construct a belt that is $1 - \alpha$ sure to contain the

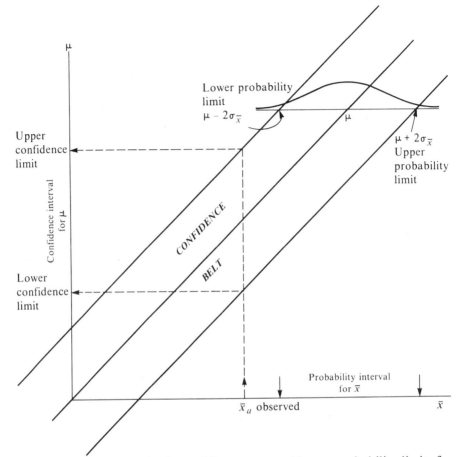

Figure 5-6. On every horizontal line, upper and lower probability limits for $\bar{x}$ appear, centered on the true μ. When $\bar{x}_a$ is observed, the vertical line to the upper and lower bounds of the confidence belt gives the upper and lower confidence limits for μ. The confidence level is about .95 when 2 $\sigma_{\bar{x}}$ limits are used.

statistic, here $\bar{x}$, and then invert the argument when the statistic is observed to get lower and upper confidence limits on the true value of the parameter, here μ.

5-6. Tests of Hypotheses and Tests of Significance

Although we use confidence limits here to emphasize that an observed mean is an uncertain measure of the true mean, an alternative method in use is called significance or hypothesis testing. We illustrate in the context of a sample mean.

If the true mean of a distribution is μ, and if a sample mean $\bar{x}$ is observed, then, as discussed above, the standard score

$$z = \frac{\bar{x} - \mu}{\sigma_{\bar{x}}}$$

has mean 0 and standard deviation 1. Thus we expect z to fall into the in-

102

terval -1.96 to $+1.96$ about 95% of the time. If, on the other hand, the true mean differs from the μ that we have hypothesized, z will be more likely to lie outside these limits.

We might form up a test using the following language:

One hypothesis is that the true mean is μ_0.

Alternative hypotheses are that the true mean differs from μ_0. (Each possible value for the true mean is a different hypothesis.)

We choose a significance level which is the probability that z will fall outside corresponding limits when the true mean is μ_0. In the present instance, we employ a 5% significance level corresponding to the probability that z, computed using μ_0, will fall outside the interval -1.96 to $+1.96$ if the true mean is μ_0.

The Test

> When the experiment is performed, a value of $\bar{x}$, and therefore of z, is observed. If z falls inside the limits, we say that we accept the hypothesis that the true mean is μ_0; and if it falls outside the limits, we say that we reject the hypothesis that the true mean is μ_0. Usually, we also quote the significance level we have chosen as a P value. In our example, if we reject, we would say that $P < .05$, as the 5% significance level has been used; if we accept, we would say that $P \geq .05$.

Sometimes we are more precise about the value of P instead of giving a limit like .05. For example, if we observed $z = -1.28$, then using a normal table, we would find that the probability of a value outside the interval -1.28 to $+1.28$ is .2, and we would report the probability of a more extreme value as $P = .2$.

The authors' view is that this testing approach is an alternative way of reporting results, and not quite as informative as the confidence limit approach, but the journal reader needs to appreciate both these methods.

In many problems a close connection between confidence limits and tests of significance applies. If the confidence level is chosen as $1 - \alpha$, the corresponding significance level is α. In our illustration a confidence interval of 95% is an interval of $(100 - 5)\%$, and this corresponds to a 5% significance level. The confidence interval either contains a hypothetical value of μ, or it does not. If it does, the corresponding significance test will accept the hypothesis, otherwise reject.

EXAMPLE 7. The observed mean blood pressure for a patient is $\bar{x} = 97$, and the standard deviation of the mean is $\sigma_{\bar{x}} = 3.4$. We ask whether $\mu = 90$. Use both the method of confidence limits and of significance tests to examine this question.

SOLUTION. Confidence limits: The 95% confidence limits on μ are (rounding the multiplier 1.96 to 2)

$$97 \pm 2(3.4) = 97 \pm 6.8 = 90.2 \text{ to } 103.8$$

The limits do not include 90, although it is just outside. We would reject 90.

Significance test: Compute

$$z = \frac{97 - 90}{3.4} = 2.06$$

Since 2.06 falls outside -1.96 to $+1.96$, we reject the hypothesis that $\mu = 90$ at the 5% significance level. If we report more carefully, we find from the gaussian table that the probability of a more extreme value of z is about .04. This number is the P value.

We shall not always run through both the confidence limit method and the test of significance method, even though both may apply for the statistical techniques presented in this and subsequent chapters.

5-7. Unknown Standard Deviation

In the discussion thus far, we have assumed that we have a good approximation for the value of σ for Mr. W.P. We got it by looking at the variability of other patients. Work by Rosner and Polk (1979) suggests that people vary considerably not only in their average blood pressure, but also in their variability. This suggests that, where possible, we should base our estimate of variability on data obtained directly from the patient.

We have seen that when $\bar{x}$ is normally distributed with mean μ and standard deviation $\sigma_{\bar{x}}$, where

$$\sigma_{\bar{x}} = \frac{\sigma}{\sqrt{n}}$$

then the quantity

$$z = \frac{\bar{x} - \mu}{\sigma_{\bar{x}}} = \frac{\bar{x} - \mu}{\sigma/\sqrt{n}}$$

is distributed according to the standard normal distribution. Therefore, the 95% confidence limits on μ are

$$\bar{x} \pm 1.96\sigma_{\bar{x}} \quad \text{or} \quad \bar{x} \pm \frac{1.96\sigma}{\sqrt{n}}$$

When we want to estimate σ from our sample of observations (as is usually the case), we can no longer use the normal distribution to compute confidence limits. In this case we no longer know σ (the standard deviation of the distribution), but we have to rely on s, the standard deviation of our sample (s can differ from σ just as $\bar{x}$ differs from μ).

The t *Distribution*

When σ is unknown, we use the t statistic instead of the z statistic, where

$$t = \frac{\bar{x} - \mu}{\sqrt{\dfrac{\Sigma (x - \bar{x})^2}{n(n - 1)}}} = \frac{\bar{x} - \mu}{s/\sqrt{n}}$$

Here s, the sample standard deviation, is

$$s = \sqrt{\frac{\Sigma (x - \bar{x})^2}{n - 1}} \qquad \text{definitional form}$$

$$= \sqrt{\frac{\Sigma x^2 - n\bar{x}^2}{n - 1}} \qquad \text{computational form}$$

The symbol Σ, capital Greek sigma, represents summation. Hence Σx^2 is the sum of the squares of the observations, or

$$x_1^2 + x_2^2 + \cdots + x_n^2$$

Similarly, $\Sigma(x - \bar{x})^2$ is the sum of the squares of the differences $(x_i - \bar{x})$ for each of the n observations.

When n is large, s is close to σ, so t is distributed very much like z (as a gaussian distribution with $\mu = 0$ and $\sigma = 1$). For small values of n, however, t is much more variable than z because, of course, we have to use s to estimate σ instead of knowing it.

From mathematical considerations not given here, the distribution of t belongs to a family of distributions called the t distributions. These distributions depend on the sample size in different ways in different problems. Here, in the one-sample problem, where one mean is computed, they depend on $n - 1$, and this number is called the degrees of freedom. Essentially, "degrees of freedom" is the number of independent measurements available for estimating σ. (We lose a degree of freedom here because we use $\bar{x}$ to estimate μ. This means that the sum of $x - \bar{x}$ over the values of x in the sample is zero. If we knew what $n - 1$ of the residuals were, then the other one would be determined and there would be only $n - 1$ "independent" values used to compute s.) As a practical matter, the table of the t distribution is entered with the number of degrees of freedom and also the confidence level desired. We abbreviate degrees of freedom by "df."

Table 5-2 and Appendix Table III-2 are such t-distribution tables. We note in Table 5-2 that the bottom line for infinitely many degrees of freedom (∞) gives values identical with those computed from the gaussian distribution. For example, the 95% level has both z and t equal to 1.96. But for small numbers of degrees of freedom, the penalty for having to use the observations to estimate σ is severe. For example, with only 1 degree of freedom the 95% multiplier is 12.71 instead of 1.96, a factor of more than 6. By the time we get to 10 degrees of freedom, the loss is not great, considering the dis-

Table 5-2.   Values of $t_{df,1-\alpha}$ for Student
t distributions

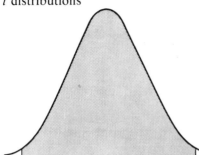

Degrees of freedom (df)	(Two-sided) probability level $(1 - \alpha)$				
	.50	.80	.90	.95	.98
1	1.00	3.08	6.31	12.71	31.82
2	.82	1.89	2.92	4.30	6.96
3	.76	1.64	2.35	3.18	4.54
4	.74	1.53	2.13	2.78	3.75
5	.73	1.48	2.02	2.57	3.36
6	.72	1.44	1.94	2.45	3.14
7	.71	1.41	1.89	2.36	3.00
8	.71	1.40	1.86	2.31	2.90
9	.70	1.38	1.83	2.26	2.82
10	.70	1.37	1.81	2.23	2.76
15	.69	1.34	1.75	2.13	2.60
30	.68	1.31	1.70	2.04	2.46
50	.68	1.30	1.68	2.01	2.40
100	.68	1.29	1.66	1.98	2.36
1000	.67	1.28	1.65	1.96	2.33
∞^a	.67	1.28	1.64	1.96	2.33

[a] Standard normal distribution.

Source: Adapted from D. B. Owen, *Handbook of Statistical Tables,*
© 1962, U.S. Department of Energy, Table 2.1; published by Addison-Wesley, Reading, Mass.; reprinted with the permission of the
publisher.

tance from 10 to infinity. The loss is only about 14% (2.23 compared with
1.96).

The t distribution is also used to compute confidence limits for μ based
on s computed from our sample.

The 95% confidence limits on μ are

$$\bar{x} \pm t_{df,.95} \frac{s}{\sqrt{n}}$$

Here $t_{df,.95}$ is read from the .95 two-sided column and the
appropriate degrees of freedom row, here $n - 1$, of Table 5-2.

EXAMPLE 8.   On six occasions, Mr. W.P.'s diastolic blood pressures are:
100, 125, 110, 112, 113, 118
We want the 95% confidence limits on μ.

We compute s as

$$s = \sqrt{\frac{\Sigma (x - \bar{x})^2}{n - 1}}$$

Here $\bar{x} = 113$. The sum of squares then is

$$(100 - 113)^2 + (125 - 113)^2 + \cdot \cdot \cdot + (118 - 113)^2 = 348$$

Thus

$$s = 8.34$$

The 95% confidence limits are

$$\bar{x} \pm t_{5,.95} \frac{s}{\sqrt{n}} = 113 \pm 2.57 \left(\frac{8.34}{\sqrt{6}}\right) = 113 \pm 8.75 = 104.2 \text{ to } 121.8$$

Thus μ seems fairly firmly established as larger than 105 and therefore in the moderate hypertensive range.

Multiple Observations on Single Occasions

As long as the patient is available for a blood pressure measurement, why not take several measurements a few minutes apart and thus cut down on the variability? This is a useful device, but it does not rid us of the between-occasion variability.

Armitage and Rose have estimated that (for some individuals) the standard deviation of blood pressure measurements during one office visit, σ_w, is as follows:

$$\sigma_w = \sqrt{12} = 3.5$$

When the same individuals were measured during different visits, they found the following variability, σ_t:

$$\sigma_t = \sqrt{42} = 6.5$$

If we had many measurements on each occasion, we could nearly eliminate the 12 by taking an average, but there still would be variation from time to time. To reduce that, we need measurements on repeated occasions.

Because of the rules for combining variability due to "independent" effects, the variability of single measurements made at each of several visits is considered to be composed of the variability of measurements within a single visit and the variability between visits according to the formula

$$\sigma_t^2 = \sigma_w^2 + \sigma_b^2$$

where σ_b^2 is the variability between visits. To distinguish σ_b^2 from σ_t^2, imagine the situation in which a great many measurements were obtained at each visit. Averages computed at each visit would still vary between visits. The variability of these averages is called the variability between visits, σ_b^2.

Within- and Between-Visit Blood Pressure Variability

Rosner and Polk (1979) have reported measurements on the two kinds of variability on hundreds of individuals. In Table 5-3 we give the sample deciles of the distributions of σ_b^2 and σ_w^2 they have obtained for their sample of

Table 5-3. Sample percentiles of the components of variability (σ_b^2, σ_w^2) for systolic and diastolic BP ($N = 326$)

BP	Component	N	Percentile									Range	
			10	20	30	40	50	60	70	80	90	MINIMUM	MAXIMUM
Systolic	σ_b^2	326	1.3	4.3	7.0	11.3	18.5	23.2	32.3	41.9	67.9	.0	171.9
Systolic	σ_w^2	326	4.3	6.2	8.0	9.3	11.0	13.3	16.0	18.3	25.0	1.3	93.0
Diastolic	σ_b^2	326	.0	3.1	5.9	9.0	12.3	16.7	23.0	32.3	48.6	.0	104.6
Diastolic	σ_w^2	326	2.7	4.0	5.0	7.0	8.7	10.0	12.0	14.2	18.3	1.0	46.7

Source: Reprinted by permission from B. Rosner and B. F. Polk, *J. Chronic Dis.*, **32**:451–61, © 1979, Pergamon Press, Ltd.

individuals. For instance, 40% of the estimated values of σ_b^2 for systolic blood pressure were 11.3 or less.

Some of the findings on diastolic blood pressure from their paper are:

1. No association was found between the within-visit variability and the between-visit variability.
2. Between-visit variability was not associated with either age or mean level of blood pressure.
3. Within-visit variability was negatively associated with both age and mean level; that is, variability was smaller on the average for persons who were older or had larger mean levels.

From Table 5-3, a patient with middle-sized (50th percentile) values of σ_b^2 and σ_w^2 would have $\sigma_t^2 = 21$; this person's standard deviation of single measurements made on different visits is $\sigma_t = \sqrt{21} = 4.6$. It may be worth remembering that an average standard deviation of diastolic blood pressure is about 5. However, looking at the 90th percentile in Table 5-3, we see that an extremely variable person may have

$$\sigma_t^2 = 48.6 + 18.3 = 66.9$$

$$\sigma_t = 8.2$$

Thus most people have a standard deviation of less than 10.

Multiple Measurements on Several Occasions

If we want several measurements on each occasion, the patient has to be detained for considerable time as does the person taking the measurements. A compromise is to take two (or three) measurements on each occasion, the same number each time. We let $\bar{y}_i$ be the average of the measurements on occasion i. Then we can treat $\bar{y}_i$ as if it were a single measurement, and the degrees of freedom will be one less than the number of visits. We shall not here try to separate the components σ_b^2 and σ_w^2 because for our purposes the variability of the $\bar{y}_i$'s tells us about the confidence intervals.

EXAMPLE 9.   Let us illustrate with two measurements on each of four occasions for Mr. W.P.

Visit	Measurement on visit		$\bar{y}_i$
	FIRST	SECOND	
i			
1	100	88	94.0
2	80	84	82.0
3	75	83	79.0
4	85	81	83.0
			338.0
			$\bar{x} = 84.5$

$$s^2 = \frac{(94.0 - 84.5)^2 + (82.0 - 84.5)^2 + (79.0 - 84.5)^2 + (83.0 - 84.5)^2}{4 - 1}$$

$$= 43$$

$$\frac{s^2}{4} = 10.75$$

$$\frac{s}{\sqrt{4}} = 3.28$$

The 95% confidence interval is (where $\bar{x}$ is the average of the $\bar{y}_i$'s)

$$\bar{x} \pm t_{3,.95} \frac{s}{\sqrt{4}} = 84.5 \pm 3.18(3.28) = 84.5 \pm 10.4 = 74.1 \text{ to } 94.9$$

The confidence interval straddles the normotensive and mild hypertensive ranges, although leaning toward normotensive. Sometimes the first blood pressure measurement in a sequence is unusually high, so it seems reasonable to continue observing this patient a bit longer without special treatment.

5-8.   Evaluating the Response to Antihypertensive Therapy

Clinical Problem 5-2.   *Did Diuretic Therapy Lower the Blood Pressure?*

A 50-year-old asymptomatic woman, Mrs. O.M., comes for a routine physical examination and you discover a blood pressure of 150 mm Hg systolic and 105 mm Hg diastolic. You start her on 50 mg of hydrochlorothiazide daily (a frequently used diuretic, antihypertensive drug), and one month later her blood pressure is 140 mm Hg systolic and 95 mm Hg diastolic. She complains that she thinks the new medicine has made her slightly weak and she wants to stop taking it. Before you urge her to continue with the hydrochlorothiazide, you should be sure that it has lowered her blood pressure. Do the blood pressure measurements noted above convince you that the medicine has in fact lowered her blood pressure, or is there a reasonable chance that the observed difference in blood pressure might have occurred without therapy?

As we have seen, blood pressures are variable and we may well need measurements from more than one occasion to get a solid basis for decisions. In the case of Mrs. O.M., we have two diastolic measurements from each of four pretreatment visits with average values

    102, 105, 110, 103

On three recent visits since beginning the treatments, her averages have been

    95, 93, 97

We want to use these two sets of measurements to assess the improvement.

Assessing Differences

The general statistical idea is that:

> If $\bar{x}$ and $\bar{y}$ are (independent) means of samples of size n_x and n_y from populations with means μ_x and μ_y and standard deviations for individual measurements of σ_x and σ_y, then
>
> $$\bar{x} - \bar{y}$$
>
> is approximately normally distributed with mean
>
> $$\mu_x - \mu_y$$
>
> and standard deviation
>
> $$\sigma_{\bar{x}-\bar{y}} = \sqrt{\frac{\sigma_x^2}{n_x} + \frac{\sigma_y^2}{n_y}}$$

In the special situation where $\sigma_x^2 = \sigma_y^2 = \sigma^2$, this reduces to

$$\sigma_{\bar{x}-\bar{y}} = \sqrt{\sigma^2\left(\frac{1}{n_x} + \frac{1}{n_y}\right)}$$

In our problem, the x's correspond to the first set of measurements and the y's to the second. We may be willing to assume that $\sigma_x = \sigma_y = \sigma$. Our problem is to estimate the common σ^2. We do this by getting a sum of squares for each sample separately, then pooling these and dividing by the total degrees of freedom. This is the same result as if we had estimated σ_x^2 and σ_y^2 separately and then got the weighted average of the estimates, weighting by their degrees of freedom. In a formula:

> $$\text{Estimate of } \sigma^2 = \frac{\Sigma\,(x - \bar{x})^2 + \Sigma\,(y - \bar{y})^2}{n_x + n_y - 2} = s_p^2$$

where the subscript p indicates a pooled variance estimate.

> Then 95% confidence limits on
>
> $$\mu_x - \mu_y$$
>
> are given by
>
> $$\bar{x} - \bar{y} \pm t_{n_x+n_y-2,.95}\sqrt{s_p^2\left(\frac{1}{n_x} + \frac{1}{n_y}\right)}$$
>
> where the coefficient comes from the t distribution with $n_x + n_y - 2$ degrees of freedom.

EXAMPLE 10. For the two sets of measurements, we have

$$\bar{x} = \frac{102 + 105 + 110 + 103}{4} = 105$$

$$\bar{y} = \frac{95 + 93 + 97}{3} = 95$$

$$\bar{x} - \bar{y} = 105 - 95 = 10$$

$$\Sigma (x - \bar{x})^2 = 9 + 0 + 25 + 4 = 38$$

$$\Sigma (y - \bar{y})^2 = 0 + 4 + 4 = 8$$

$$s_p^2 = \frac{38 + 8}{4 + 3 - 2} = \frac{46}{5} = 9.2$$

$$\sqrt{s_p^2 \left(\frac{1}{n_x} + \frac{1}{n_y} \right)} = \sqrt{9.2 \left(\frac{1}{4} + \frac{1}{3} \right)} = \sqrt{5.36} = 2.32.$$

Then the 95% confidence limits on the difference are

$$\bar{x} - \bar{y} \pm t_{5,.95} \sqrt{s_p^2 \left(\frac{1}{n_x} + \frac{1}{n_y} \right)} = 10 \pm 2.57(2.32) = 10 \pm 6.0$$

$$= 4 \text{ to } 16$$

Thus the interval runs from 4 to 16. We notice that 0 is not in the interval. This means that we are very confident that Mrs. O.M. has reduced her diastolic blood pressure. The change is not readily accounted for by sampling variation. Because we are reasonably confident that the hydrochlorothiazide has reduced her pressure, we might urge her to continue it. Her weakness may be unrelated to the drug and it may disappear.

5-9. Summary

The initial evaluation of a patient is often complicated by the fact that observations vary from minute to minute and day to day. The greater the potential for variation, the more the need for performing many observations over time to establish the patient's average condition.

Computing the mean and standard deviation of the observations enables one to make statements about our confidence that the true mean falls within a specified interval. The Central Limit Theorem ensures that $\bar{x}$, the mean of n observations, is approximately normally distributed. Thus, when the variance of the observations, σ^2, is known, the interval

$$\bar{x} - z_{1-\alpha} \frac{\sigma}{\sqrt{n}}, \qquad \bar{x} + z_{1-\alpha} \frac{\sigma}{\sqrt{n}}$$

is a $100(1 - \alpha)\%$ confidence interval for μ, the true mean. When σ^2 is unknown, we compute the sample variance

$$s^2 = \frac{\Sigma (x - \bar{x})^2}{n - 1}$$

and the $100(1 - \alpha)\%$ confidence interval has the form

$$\bar{x} - t_{n-1,1-\alpha} \frac{s}{\sqrt{n}}, \qquad \bar{x} + t_{n-1,1-\alpha} \frac{s}{\sqrt{n}}$$

where $t_{n-1,1-\alpha}$ is a critical value read from the $1 - \alpha$ column of Table 5-2 or Appendix Table III-2.

We also use the normal distribution and the t distribution to test the hypothesis that the true mean is μ_0. When the variance is unknown, we compute the t statistic

$$t = \frac{\bar{x} - \mu_0}{s/\sqrt{n}}$$

and compare its value to critical values of the t distribution with $n - 1$ degrees of freedom. When the variance is known, we use instead the critical ratio

$$z = \frac{\bar{x} - \mu_0}{\sigma/\sqrt{n}}$$

and compare its value to critical values of the normal distribution.

Given two sample means, $\bar{x}$ and $\bar{y}$, and their true means, μ_x and μ_y, confidence intervals and tests for $\mu_x - \mu_y$ are based on the critical ratio

$$z = \frac{(\bar{x} - \bar{y}) - (\mu_x - \mu_y)}{\sqrt{\dfrac{\sigma_x^2}{n_x} + \dfrac{\sigma_y^2}{n_y}}}$$

when the variances, σ_x^2 and σ_y^2, are known and the sample sizes are n_x and n_y. When the variances are unknown but assumed to be equal, tests and confidence intervals are based on the critical ratio

$$t = \frac{(\bar{x} - \bar{y}) - (\mu_x - \mu_y)}{s_p \sqrt{\dfrac{1}{n_x} + \dfrac{1}{n_y}}}$$

where

$$s_p^2 = \frac{(n_x - 1)s_x^2 + (n_y - 1)s_y^2}{n_x + n_y - 2}$$

If we plan to measure a patient several times to improve our estimate of the true mean of a characteristic, we need to consider both within-occasion and between-occasion variability. Usually, it will be important to see the patient more than once to reduce uncertainty due to long-term variability.

Problems

The report of the JNC included the following recommendations concerning the goals of antihypertensive therapy:

The initial goal of antihypertensive therapy is to achieve and maintain diastolic pressures at below 90 mm Hg, if feasible.

5-1. Mr. W.P. is started on treatment. He has the following blood pressures at his next 4 visits:

86, 92, 82, 84

a. Assuming that the standard deviation of his blood pressure is 5, about average, compute the 80% and 95% confidence intervals for his mean blood pressure. What is your confidence that his mean blood pressure is below 90 mm Hg?
b. Use the measurements to estimate Mr. W.P.'s standard deviation (compute s).
c. Compute the 80% and 95% confidence limits for his mean blood pressure using s, n, and the t distribution.

5-2. Mr. W.P. is followed and his average blood pressure over many visits is 85 mm Hg. Suppose that his true standard deviation for individual measurements is 6 mm Hg.
a. How often would you expect a reading of 95 or higher? 100 or higher?
b. On the next visit, his blood pressure is 95. How could you settle whether his average pressure is no longer below the goal of 90 mm Hg?

5-3. (continuation) After measuring Mr. W.P.'s blood pressure on several visits, you find his new average to be 95.
a. How many measurements must you have made to be 90% confident that his new mean is 90 mm Hg or greater?
b. How many observations would be required if the new observed mean were 91?

5-4. You follow Mr. W.P. and his blood pressure is consistently above 90 mm Hg. His pulse on 3 visits is 80, 85, and 75. You prescribe propranolol (an antihypertensive agent which also slows the pulse). On the next 5 visits, his blood pressure is unchanged but his pulse is 70, 65, 75, 60, and 65.
a. Compute the 95% confidence limits for the change in Mr. W.P.'s pulse.
b. Do you think the reason his blood pressure has not responded is that he has not taken the propranolol, or that the dose prescribed was not effective? Why?

5-5. Patients are often followed by monitoring a particular variable such as pulse, respiration, blood urea nitrogen (BUN), creatinine, serum drug concentration, pulmonary function, and blood count. Because there is always some sampling variability, part of the trick to interpreting the results is to know how much variability to expect. Pick one or two stable patients and collect at least 5 measurements (either through chart review or through visits to hospitalized patients) and estimate the standard deviation of some of these or other measurements.

5-6. Suppose we have decided that, given a single diastolic blood pressure measurement, we will classify a person as hypertensive if it is at least 90 mm Hg.
a. Using the data given in Figure 5-2, estimate the probability of misclassifying a normal person. (Note: We are assuming our patient is from the same population as the persons used.)
b. Compute the probability of misclassifying the following patients when a single measurement is used.

Patient 1 has a mean of 85 and a standard deviation of 5.
Patient 2 has a mean of 85 and a standard deviation of 10.
Patient 3 has a mean of 92 and a standard deviation of 5.

 c. What is the probability of misclassifying the above patients if the average of 4 measurements is used?

 d. Patients often have unusually high blood pressures when they visit a physician for the first time. How would this affect the chance of misclassification if a single observation on the initial office visit is used to diagnose hypertension?

5-7. Varady and Maxwell (1972) reported that there was no evidence in their data to suggest that the measurement variability of diastolic blood pressure changed with the mean value. Are the data in Figure 5-2 consistent with Varady and Maxwell's observation? Hint: Plot the median against the range for each patient.

5-8. A patient with a sphygmomanometer at home reports to you that she measured her diastolic blood pressure once a day for the last 8 days, that the pressure varied and the average was 85 mm Hg. You check her diastolic blood pressure and observe a value of 95 mm Hg. Compute the probability of observing such a large difference given no true difference. Assume a gaussian distribution with $\sigma = 6$ mm Hg. What do you suspect?

5-9. The coordinator of the hypertension treatment program described in Problem 5-1 wants to change the criteria for success in treating hypertension so that no more than 5% of persons with no change in diastolic blood pressure level will be declared successfully treated. Assuming 2 observations before and after treatment, $\sigma = 5$ mm Hg, and gaussian distributions, how large a difference between the mean diastolic blood pressures before and after treatment will be needed to classify a person as having a reduced diastolic blood pressure level after treatment while ensuring no more than a 5% error rate for persons whose blood pressure has not declined?

References

Armitage, P., and Rose, G. A. (1966). The variability of measurements of casual blood pressure. *Clin. Sci.*, **30**:325–35.

Bevan, A. T.; Honour, A. J.; and Scott, F. H. (1969). Direct arterial pressure recording in unrestricted man. *Clin. Sci.*, **36**:329–44.

Hypertension Detection and Follow-up Program Cooperative Group (1977). The hypertension detection and follow-up program: a progress report. *Circ. Res.*, **40**(Suppl. 1):106–9.

Joint National Committee on Detection, Evaluation, and Treatment of High Blood Pressure (1984). The 1984 Report of the Joint National Committee on Detection, Evaluation, and Treatment of High Blood Pressure. *Arch. Intern. Med.*, **144**: 1045–57.

Rosner, B., and Polk, B. F. (1979). The implications of blood pressure variability for clinical and screening purposes. *J. Chronic Dis.*, **32**:451–61.

Varady, P. D., and Maxwell, M. H. (1972). Assessment of statistically significant changes in diastolic blood pressures, *JAMA*, **221**(4):365–67.

Additional Reading

All readings are from Bailar, J. and Mosteller, F., eds. (1986) *Medical Uses of Statistics*. Waltham, MA: New England Journal of Medicine.

Bailar, J.; Louis, T. A., Lavori, P.; and Polansky, M. Studies without internal controls. Chapter 5.

Godfrey, K. Comparing the means of several groups. Chapter 10.

Lavori, P.; Louis, T. A.; Bailar, J.; and Polansky, M. Designs for experiments—parallel comparisons of treatment. Chapter 3.

5A

Ideas of Exploratory Data Analysis:
Reanalyzing an Exercise Data Set

OBJECTIVES

Exploratory data analysis
Relation of variability and size
Assessing outliers
Reanalysis with outliers set aside

5A-1. An Addendum on Variability

The exercise data in Table 4-2 provide six control measurements for each patient. These motivate us to explore variability within patients and across patients. Let us, for the moment, regard all six measurements from each patient as repeated observations under the same conditions for these patients. Although the patients may themselves differ, we pretend that the measurements for a given patient all come from a single distribution describing the responses of that patient. We will see the patients varying more among themselves than within themselves. We find how to control for outliers and get an appreciation of the main body of the data.

Useful Summary Statistics

For each patient, using Table 4-2, we have computed several statistics for measuring location and variability. Some are defined below. The statistics presented are the average (mean), median, range, interquartile range, standard deviation, and the minimum and the maximum values. For example, patient 1 has 6 control observations, ordered from least to greatest as follows:

117

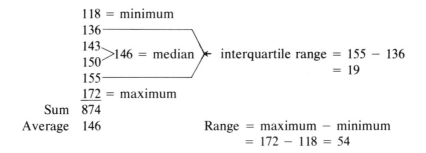

$$\begin{array}{ll} 118 = \text{minimum} \\ 136 \\ 143 \\ 150 \\ 155 \\ 172 = \text{maximum} \end{array}$$

Sum   874

Average   146   Range = maximum − minimum

                                         = 172 − 118 = 54

We can arrange the data in a similar manner for the other patients, and these data together with their summary statistics appear in Table 5A-1. Notice that we round the summary statistics to integer values to avoid clutter.

Definitions

The **average** is, as usual, the sum of the measurements divided by the number of measurements.

The **median** is the middle measurement for an odd number in the sample and the average of the two middle measurements for an even number in the sample. In the example, the number of measurements is 6. If n is the number of measurements, to get the median we count in from either end of the ordered measurements the number $(n + 1)/2$, here $3\frac{1}{2}$, which is midway between the third and fourth ordered measurements; and rounding 146.5 to the nearest even number gives 146.

Just as the median is the halfway or middle observation, the **fourths** are the first and last 25% points of the ordered sample. We find upper and lower fourths as follows: If the depth we count in for the median is a whole number, add 1 to the depth and divide by 2 to get the depth for a fourth. So for $n = 5$, the depth for the median is $(5 + 1)/2 = 3$. The depth for a fourth is $(3 + 1)/2 = 2$. So to get the fourths we count in 2 from each end for $n = 5$. When $n = 6$, the depth for the median is $(6 + 1)/2 = 3\frac{1}{2}$. To get the fourths for even sample sizes we drop the $\frac{1}{2}$ that goes with the median, and again add 1 and divide by 2. Thus, for $n = 6$, the depth for a fourth is $(3 + 1)/2 = 2$, the same as for $n = 5$. In the example, the fourths are 136 and 155.

The difference of the fourths gives the **interquartile range**, I, here 155 − 136 = 19. It is one measure of variability.

The **range**, R, is:

$$R = \text{range} = \text{maximum} - \text{minimum} = 172 - 118 = 54$$

The standard deviation, s, was defined in Chapter 5.

The distributions in Table 5A-2 show that, in these patients, the maximum is more broadly distributed than the minimum or the median and that the minimum and the median have similar spreads for these data.

From everyday experience with nature, we expect larger things to vary more than smaller things of the same kind. For example, we expect weights

Table 5A-1. Basic data for 21 patients, control observations only: I, interquartile range; R, range; s, standard deviation[a]

Patient	1.	118	2.	205	3.	147	4.	150		
		136		207		215		205		
		143		211		221		210		
		150		250		244		235		
		155		269		250		290		
		172		287		408*		308		
Average		146		238		248		233		
Median		146		230		232		222		
I		19		62		35		85		
R		54		82		261		158		
s		18		35		87		58		
	5.	87	6.	301	7.	342	8.	182		
		105		357		390		188		
		129		388		441		189		
		135		388		441		208		
		135		409		455		210		
		157		425		468		215		
Average		125		378		423		199		
Median		132		388		441		198		
I		30		52		65		22		
R		70		124		126		33		
s		25		44		48		14		
	9.	99	10.	89	11.	206	12.	147		
		115		104		207		190		
		125		108		224		198		
		131		111		228		199		
		141		114		250		222		
		154		136		251		243		
Average		128		110		228		200		
Median		128		110		226		198		
I		26		10		43		32		
R		55		47		45		96		
s		19		15		20		32		
	13.	231	14.	191	15.	152	16.	328		
		234		194		156		406		
		241		197		176		417		
		249		218		197		431		
		267		223		199		448		
		274		224		329*		458		
Average		249		208		202		415		
Median		245		208		186		424		
I		33		29		43		42		
R		43		33		177		130		
s		18		15		66		47		
	17.	159	18.	251	19.	406	20.	115*	21.	199
		168		258		424		229		200
		168		264		437		232		223
		188		276		478		234		227
		191		277		512		237		227
		213		490*		560		281		265
Average		181		303		470		221		224
Median		178		270		458		233		225
I		23		19		88		8		27
R		54		239		154		166		66
s		20		92		59		56		24

[a] The asterisk (*) denotes outliers (see text).

Table 5A-2.   Distribution of statistics across
21 patients grouped by class intervals of
size 50

Class interval	Median	Minimum	Maximum
50–100		3	
101–150	4	5	1
151–200	4	5	3
201–250	8	3	4
251–300	1	1	5
301–350		3	2
351–400	1		
401–450	2	1	2
451–500	1		3
501–550			
551–600			1

of 1-month-old babies to vary less than weights of 40-year-old men. In our exercise data, let us see this visually by relating the interquartile range, I, to the median, Med. We do this graphically in Figure 5A-1.

By eye or otherwise, we might pass a line through the cloud of points as a device for summarizing the relation between I and Med. Approximately, we get

$$\hat{I} = 6 + \frac{1}{8} \text{Med}$$

where $\hat{I}$ is the fitted value given by the line. The line uses the pooled variability of all the samples to estimate a better value of I for each Med than the original sample can provide. By introducing the line, we can get a better idea of the variation that a single individual might produce if we had more measurements. Furthermore, we can get an idea about outlying or wild measurements, which is a major feature of our inquiry.

Identifying Outliers

Almost any collection of data will be infested with **outliers**—observations that do not seem to belong with the rest. For example, there may be laboratory errors; some patients may not have met the criteria for entering the study; and there may be recording errors. The data themselves often can be made to give some indication that a particular observation may be an outlier.

When we look at the measurements in Table 5A-1, a few stand out like sore thumbs. For example, patient 18 has

251, 258, 264, 276, 277, 490

The thought is that patients have a natural variation but that occasionally this variation is contaminated by a wild measurement. We assume that the wild ones are relatively few.

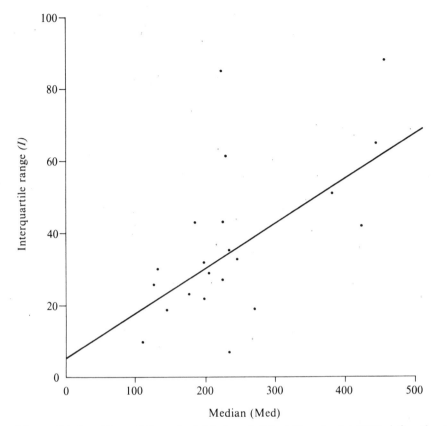

Figure 5A-1. Plot of I against Med and fitted line $I = 6 + \frac{1}{8}$Med for the 6 control exercise observations for the 21 patients.

If we base our estimate of variability on the central part of the distribution, as we have here by using the interquartile range, we tend to diminish the effect of the wild observations on the measurement of variation.

Theory not given here says that when we have the interquartile range for a sample based on a large number of measurements, we can identify probable outliers by a very simple procedure. Construct **upper** and **lower cutoff points** as

$$\frac{\text{Upper}}{\text{cutoff}} = \frac{\text{Upper}}{\text{fourth}} + 2I \quad \text{and} \quad \frac{\text{Lower}}{\text{cutoff}} = \frac{\text{Lower}}{\text{fourth}} - 2I$$

Diagrammatically, these cutoffs look as follows:

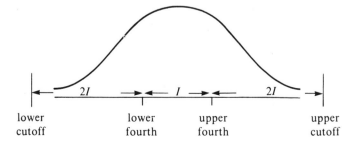

If there are no outliers, practically all measurements would lie inside the interval between the two cutoffs. Although examples can be contrived to violate this principle, many sets of measurement data obey this rule. The rule provides us with an objective way of finding and thinking about outliers.

Here, we have only six observations for each patient. That is not enough for comfortable use of the rule of thumb given above because, with such a small sample, I varies a great deal from one set of measurements to another because of random variation. However, Figure 5A-1 can be used to estimate the value of I we would have gotten if we had more observations per patient. We use the whole set of data to reinforce our conclusions about individuals. This procedure is an example of what is called borrowing strength.

When we use the value of $\hat{I}$, the fitted I for each patient, we find the following measurements lying beyond the cutoffs:

Patient 3	408	(too large)
Patient 15	329	(too large)
Patient 18	490	(too large) [upper cutoff: $277 + 2(40) = 357$]
Patient 20	115	(too small)

In the example of patient 18, the median is 270 and $\hat{I} = 6 + 270/8 \doteq 40$.

Thus we identified only four outliers out of $6 \times 21 = 126$ measurements. As sets of measurements on the performance of people go, these are rather well behaved.

Had we used the I from the sample instead of the fitted I, patient 20 would also have had a high observation, 281.

In plotting I against the "center" of the sample, we have used the median instead of the average because the median is not much affected by a single outlier. For each patient, we computed the difference

median − mean

A stem-and-leaf diagram for these differences follows. The line 1|082* means differences of 10, 18, and 12, and the line −1|6*16*2 means −16, −11, −16, −12.

```
  3 |
  2 |
  1 | 082*
  0 | 791
  0 | 0000
 −0 | 812243
 −1 | 6*16*2
 −2 |
 −3 | 3*
```

The starred differences come from samples with outliers as defined by our cutoffs. Thus we see that once the outliers are removed, the mean and the median are rarely as much as 20 seconds apart in these data.

Following Up on Outliers

Depending on the problem, many further actions may be appropriate when an outlier is identified—some obvious but important and some not so obvious. A brief list of possibilities may be helpful.

TYPOGRAPHICAL ERROR. Whenever numbers are transcribed, reading, writing, and typing errors occur. Sometimes the entry can be checked back a step or two, and if an error is found, corrected. Permutations such as 431 instead of 341 are common. Reading errors from meters and dials are common, and except for impossible values, cannot usually be checked.

ARITHMETIC ERRORS. Many laboratory calculations offer the possibility of arithmetic errors, misplaced decimal points being among the most common errors. Again, often a suspected observation can be checked for its arithmetic and measurement errors.

LABORATORY TESTS. Occasionally, something goes wrong with a laboratory test.

REANALYSIS. By statistical methods not discussed here, we can reanalyze the data with the suspected measurements set aside to see what their effect may be. Some people think that this is cheating. We are just beginning to have objective systematic ways to analyze measurements, ways that pay suitable attention to outliers. When we set aside outliers, we are not suppressing them. They are separately reported and discussed—indeed, heavily emphasized.

When large changes in a small part of the data make a big difference in the results, it seems best to ask how the data can be analyzed so as to emphasize the indications of the main body of the data. The methods that do this are called "resistant" because they are not much affected by large changes in small parts of the data. Medians illustrate resistant methods.

If, in addition, the methods are especially good at summarizing the data when the data are well behaved—outlier free—we say that the methods are robust. For example, in many problems the average of the middle 60% of the data is resistant to outliers, and more stable than the median, and so we say it is more robust than the median.

MEDICAL FOLLOW-UP. Where possible, one explores the clinical implications and related data for an unusual measurement. Sometimes, one can find corroborative evidence that the outlier represents a substantial change in the patient, and sometimes the absence of confirming evidence casts additional doubt on the validity of the outlier. Outliers often pose the clinical dilemma: early warning or false alarm.

UNUSUAL PATIENTS. Biological variability of people is a medical key-note, and some otherwise normal people have remarkably different values for some important measurements, such as very low heart rates.

Rounding and Cutting

In these analyses, we have rounded to the nearest second because a second seemed already a finer unit than we needed. When individuals vary by hundreds in six comparable measurements, carrying seconds themselves could be wasteful.

In many kinds of work, cutting the last few digits from a lengthy number may be much cheaper and lose little more information than rounding. A stem-and-leaf diagram is shown below for all 126 measurements, in which we cut the last digit.

```
Cum.
   1     5 | 6
   2     5 | 1
   7     4 | 56697
  19     4 | 002440134023
  23     3 | 5889
  28     3 | 00422
  44     2 | 5686955675567786
 (34)    2 | 0011240130110022243344122123330222
  48     1 | 5575588859999995579566899
  22     1 | 13440233123400 11341
   3     0 | 898
 126 = 48 + 34 + 44
```

Med depth = $63^1/_2$ Med 220
Fourth depth = $^{64}/_2 = 32$ Upper fourth 270
                              Lower fourth 160
                              $I = 110$

The result is that when we went back to the full three digits, we probably should have added 5 to the numbers. For example, instead of median depth being 220, we should probably have taken 225 to recover the average of the missing digits. Recall that 220 is standing for any of the numbers from 220 through 229.99. On the other hand, we are not making comparisons where a change of 5 in the last digit makes much difference.

Variability Related to Size

We see in our analysis of Figure 5A-1 of I plotted against Med, and in our scatterplot of the exercise measurements against their median, Figure 5A-2, that the larger medians go with larger variability. If the variability is

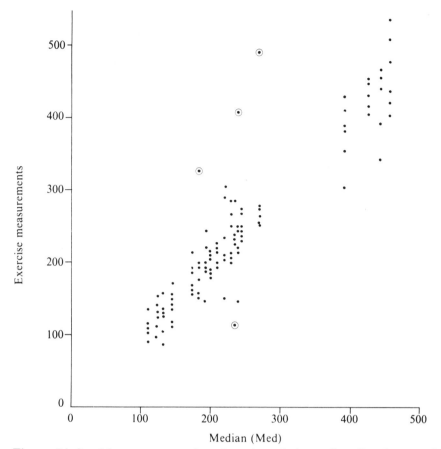

Figure 5A-2. Measurements plotted against their median for the exercise tests. Patients 11 and 21 were omitted because their points would crowd the region with medians between 225 and 235. Note the distinct break in the medians between 270 and 370. Circled points were identified in the text as outliers.

directly proportional to size, we would have

$$\hat{I} = k\text{Med} \qquad \text{or} \qquad \hat{I} = c\bar{x}$$

where k and c are constants and $\bar{x}$ is the average. When this simple proportionality occurs, if we take the logarithms of the original measurements, the variability of the logarithms for a given average will be approximately constant. This is one of the reasons we often analyze the logarithms of data rather than the raw data. We do not discuss this further here.

Sometimes, data are more variable as they grow larger, but not directly proportional to size. For example, by methods not shown here, we know that when n coins are flipped, the standard deviation of the number of heads is measured by

$$\frac{\sqrt{n}}{2}$$

so this measure of variability increases with the number of coins being flipped, but not proportional to their number. Instead, it is proportional to the square root.

5A-2.   Reanalyzing the Exercise Data Without the Outliers

As mentioned above, we may wish to reanalyze our data after the outliers are removed. In the exercise data we have four outliers, one in each of four patients' control data. If we were to do a reanalysis, we would remove the offenders, recompute the medians, and recompute I for the set.

The new median would be based in each instance on a sample of size 5, so the median would be found by counting 3 in from either end to give the middle observation. Further, we would have to compute an I based on 5 measurements rather than 6. We have already noted that this would mean that we would count in 2 to get the fourths. When the outlier is high, this means that we would lower the median and we would lower both the upper and lower fourths. And if the outlier is low, the three measures would all increase.

In our data, these changes would all be very slight, with the result that the plot of I against Med in Figure 5A-1 would be scarcely changed, and the fitted line would also be practically the same. Thus when we use a fitted I to construct a new set of cutoffs, we will not change them by much. So in this problem, no further outliers will be found. In other data, this might not be true. The reason for the modest change is that we have used a method of analysis that is highly resistant to the effects of outliers, and the slight changes now found reflect this resistance.

5A-3.   Variability Between Patients and Within Patients

If we pool all 126 measurements and form the stem-and-leaf diagram to facilitate computing medians and fourths, we find for all measurements (see page 124) that

$$\text{Median} = 220$$
$$\text{Upper fourth} = 270$$
$$\text{Lower fourth} = 160$$
$$I = 110$$

If we entered Figure 5A-1 with Med = 220, we would read a fitted I of 32, whereas the I for the 126 measurements is 110. What does this mean? It means that the individual variation among the patients is so large that the interquartile range among the measurements is 110 instead of only about 32. Indeed, the variation between patient levels is swamping the variation within patients. Figure 5A-2 illustrates this large effect of variation in median.

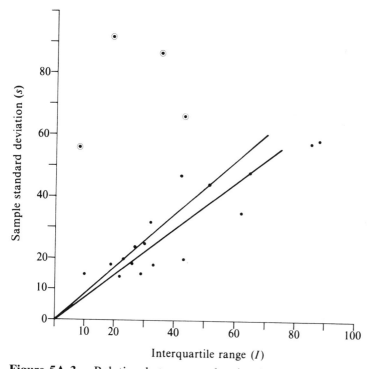

Figure 5A-3. Relation between s, the simple standard deviation, and I, the interquartile range. The circled points are associated with outliers.

5A-4. The Standard Deviation

One might expect that the sample standard deviation, which we label s, would be closely related to I, the interquartile range. In Figure 5A-3, we plot the relation between s and I for the 21 patients. Two lines through the origin have been plotted. The upper line has 10 points above and 10 below and one on the line. It is, therefore, the median line through the origin. Four points look far above the line. These are the points associated with samples that have outliers. The effect of an outlier on an individual's set of measurements is to make the standard deviation rather large. For this reason, s is a poor measure of variability in the presence of wild observations, unless it is the wildness one wants to measure. Here we want to know about baseline variation, so we prefer I. The observed median line relating s and I after we remove the four outliers is

$$\hat{s} = .76I$$

One might think of this as the estimated sample standard deviation in these samples of size 6, when they are cleared of outliers.

5A-5. Summary

Variability is an important characteristic of clinical data. Both variability within patients (over different occasions) and variability between patients can be important. This chapter introduces some fundamental techniques of exploratory data analysis and uses them to analyze variability in the exercise data introduced in Chapter 4.

We define the median, fourths, and the interquartile range. We also emphasize the importance of outliers (observations that are atypical) and introduce techniques for detecting and removing them from an analysis. Removing the outliers from the exercise data has relatively little effect on the analysis based on the interquartile range because this statistic is resistant to the effects of outliers. However, the standard deviation is sensitive to outliers and is highly inflated for the four samples, including observations identified as outliers.

Problems

5A-1. Use the data in Table 5A-1 to construct a stem-and-leaf diagram of the interquartile ranges for the 21 patients.

5A-2. Make a stem-and-leaf diagram for the maximum using hundreds as the stem and the tens digit as the leaf. Thus 451 is represented as 4|5.

5A-3. Why do we not use the range instead of the interquartile range in looking for outliers?

5A-4. Plot the 1-hour oral ISDN exercise time against the control value (Table 4-2) and discuss the resulting diagram. Do all patients show similar response, or do they seem to break up into sets?

5A-5. Do the four outliers we detected have a pattern with respect to the six kinds (columns) of control observations?

CHAPTER 6

Assessing Treatment Efficacy by Analyzing Counts

OBJECTIVES

Recognizing counts in clinical information

Using the binomial distribution to estimate the probability of disease occurrence

Using the Poisson distribution to estimate an attack rate

Computing confidence intervals for probability of disease and attack rate

Using the normal approximations to the binomial and Poisson distributions

Testing differences in binomial probabilities and Poisson rates

Many measurements and patient responses encountered in clinical medicine can be called **counts.** In this chapter we describe and analyze two important types of counted data. The first type includes clinical signs and symptoms that are recorded as present or absent in a single patient or on a single occasion. For example, migraine headache, chest pain, and manic-depressive illness are sometimes recorded in this way. The second type includes responses recorded as the number of occurrences of an event in a specified time interval: for example, the number of angina attacks experienced in the past week. Here, it is natural to say that the possible responses consist of zero and every positive integer, even though very large values surely have small probability.

We discuss both types of responses in this chapter. In each situation we have special techniques for describing a patient's status, for setting confidence limits, and for making tests of significance. We describe those techniques and apply them in several examples to assess the effect of treatment. Two important probability distributions, the binomial, which we met in Chapter 2, and the Poisson, which we introduce here, play an important role in the discussion, and we describe some of their properties. We also show how the Central Limit Theorem, the fundamental idea from probability theory introduced in Chapter 5, sometimes enables us to apply statistical techniques used for normally distributed responses to data consisting of counts.

6-1.   Preventing Migraine Headache

Clinical Problem 6-1.   *Has Propranolol Been Effective?*

Mr. I. has a history of migraine headaches. During the past year, he experienced attacks on 30 different weekends. Since then, he has started taking propranolol, 40 mg four times a day. In the 10 weeks since beginning treatment, he has had attacks on 2 weekends. Is propranolol having an effect?

Comparing Observed and Expected Results

By history, the probability that Mr. I. would have at least one attack in a weekend was about .6 (= 30/50) before treatment began. What do we expect on the next weekend? If the probability is still .6, even with propranolol therapy, then the probability of having no attacks on a single weekend, specifically the next, is .4. We can view a single weekend as a trial. The results of a trial are either attack or no attack. In the present instance, we have observed 10 trials (weekends) and wish to assess whether a change in the attack rate has occurred.

What do we expect to occur in 10 weekends if the probability of an attack is .6? Of course, we could observe any number of weekends with attacks from 0 to 10, and we expect 6 (= 10 $\times$ .6). Since 2 is considerably less than 6, the result looks promising, but we have learned that the binomial distribution has considerable variation. Let us investigate the new value of the probability of attacks.

Constructing a Confidence Interval for the Probability of an Attack-Free Weekend

In Chapter 5, we described a method for computing a confidence interval for the mean, μ, of a normal distribution. Here we want to construct or describe confidence intervals for the true binomial probability, π, the probability of at least one migraine attack during a weekend. It will speed our understanding if we describe our solution to this problem first and then explain where the results come from.

For a change, we will compute 80% confidence limits instead of 95%. We make this change for two reasons: first, because in dealing with single patients, as opposed to many, data come slowly, and one may wish to ease up on the confidence required; and second, we want to emphasize that there is nothing sacred about the 95% level. We discuss both points in more detail in Chapter 7.

Confidence intervals for π can be constructed either from the exact probabilities given by the binomial distribution or from an approximation to these probabilities based on the normal distribution. When the exact probabilities are used, special tables are required as we explain in the next section.

Using Tables and Charts to Obtain Confidence Intervals

> The principal methods for obtaining confidence intervals are through tables, charts, formulas, or computer programs. We describe the first three of these.

Tables and charts make confidence limits for the binomial probability π readily available for some levels of confidence and some sample sizes. Table 6-1 shows for each possible outcome in a sample of size $n = 10$ what the lower and upper 80% confidence limits are. Mr. I. has had 2 weekends with attacks in 10 trials (weekends). Assuming that the binomial distribution applies to his attacks, Table 6-1 tells us that the true value of π lies between .08 and .45 and that our confidence in this statement is at least 80%. Note that .6, Mr. I.'s previous rate, is not in the interval, so we can reject the hypothesis of $\pi = .6$ at the 20% level of significance. Thus we can reject the hypothesis of no propranolol effect at the 20% level of significance. The significance level is the complement of the confidence level.

Why do we say at least and not exactly 80%? Unlike the normal distribution, where the probability is continuously distributed, the binomial distribution's probability comes in discrete amounts. We can arrange to have the confidence be at least 80% but not exactly 80%.

To help us understand the tabular approach to confidence limits and also to prepare us for using charts, we give in Figure 6-1 the same information contained in Table 6-1. In that figure, for each possible $\bar{p}$ on the horizontal axis, we have plotted a lower and an upper point. These are the lower and upper 80% confidence limits for π. On the figure, we show with arrows the

Table 6-1. 80% confidence limits for the binomial probability, π, for an investigation consisting of $n = 10$ independent trials

Number of occurrences	Confidence limit	
	LOWER	UPPER
0	.00	.20
1	.02	.31
2	.08	.45
3	.15	.55
4	.20	.62
5	.31	.69
6	.38	.80
7	.45	.85
8	.55	.92
9	.69	.98
10	.80	1.00

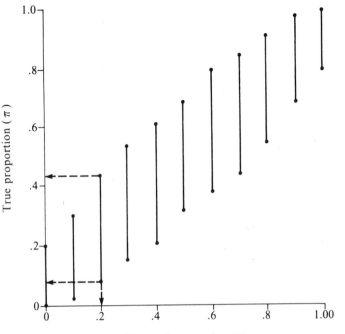

Observed proportion ($\bar{p}$)

Figure 6-1.   An 80% confidence belt for the binomial proportion π for samples of size $n = 10$ and each possible observed proportion. Compute $\bar{p}$ and read lower and upper limits vertically, for example, 2 successes in 10 trials: $\bar{p} = .2$, the lower 80% confidence limit is .08, the upper limit is .45.

confidence limits for our problem, 2 attacks in 10 trials. Again, the limits are .08 and .45.

To **use the chart,** we start with $\bar{p}$ and read vertically to get limits for π, as described above. The idea is very similar to that we saw in Chapter 5 for the normal distribution, but we have to deal with the discreteness of the distribution in the binomial problem. If you want only to understand how to use charts and tables, you may skip on to "The Confidence Statement." If you want the details of the reasoning, continue reading.

To **understand what is going on,** we start with π and read horizontally. On any horizontal line, the possible counts 0, 1, 2, . . . , 10 can occur and each count has a probability of occurring. We want a set of these counts whose probabilities add to at least .8. The set should include as few counts as possible to keep the limits narrow. When π is near 0, sets of small counts such as "0" or "0 and 1" will include at least a probability of .8. When π is near 1, sets of large counts such as "10" or "9 and 10" will include at least a probability of .8. Near the middle, when π is near .5, a set of middle counts such as 3, 4, 5, 6, 7 will be required.

By making the probability at least .8 that $\bar{p}$ is in the set of counts chosen for the π line, we make sure that our confidence is at least .8 that the confidence limits include π. The configuration of dots in Figure 6-1 forms a confidence belt. When we draw a sample from a binomial population, the population has an associated value of π, and the sample produces $\bar{p}$, the observed proportion of successes (here migraine attacks). These coordinates $(\bar{p}, \pi)$ designate a point, and the probability is at least .8 that the point is inside the belt. Since we know the value of $\bar{p}$ for the sample, we can read upper and lower values of π from the chart or table.

One way to understand Figure 6-1 is to imagine a horizontal line near the bottom of Figure 6-1. If we start with very small values of π, the probability of 0 successes is very high, higher than .8, the required confidence level. When π rises to .02 (more precisely .022), the binomial probability of 0 successes, $(.978)^{10}$, is about .8, the required confidence level. When π increases a little beyond .02, we will need to include both the count of 0 and the count of 1 in our set to have a probability of at least .8. From $\pi = .02$ to $\pi = .08$, the probability is at least .8 that 0 or 1 success occurs. We are arranging that for every π, we have a set of consecutive counts with probability totaling at least .8. Working up from the bottom, we have, using extensive tables of the binomial distribution (or a computer), the results shown in Table 6-2. It gives sets of integers that carry probabilities totaling at least .8. Thus for π between .20 and .31, the **probability limits** for $\bar{p}$ are

$$.1 \le \bar{p} \le .4$$

whereas Table 6-1 gives us the **confidence limits** for π. When we draw a horizontal line through Figure 6-1, at level π, it intersects the verticals associated with the integers in Table 6-2.

Our situation now is that on every horizontal line, we have a consecutive set of points with probability totaling at least .8. We have chosen the smallest set that guarantees this, so that we have short probability sets, the horizontal sets. In turn this shortness helps make the vertical confidence sets short.

Table 6-2. Sets of counts that achieve probabilities of at least .8 for binomial samples of size $n = 10$

Interval for π	Counts required for .8 probability
.00–.02	0
.02–.08	0, 1
.08–.15	0, 1, 2
.15–.20	0, 1, 2, 3
.20–.31	1, 2, 3, 4
.31–.38	2, 3, 4, 5
.38–.45	2, 3, 4, 5, 6
.45–.55	3, 4, 5, 6, 7
and so on	

> The Confidence Statement: When we observe the proportion, $\bar{p}$, of successes in n binomial trials and use a chart like Figure 6-1 or a table like Table 6-1 to set the lower and upper $100(1 - \alpha)\%$ confidence limits, then if we state that
>
> "π is between the lower limit and the upper limit"
>
> the statement will be true or false. If we make such statements for every binomial sample drawn with limits set in this manner, then at least $100(1 - \alpha)\%$ of the statements will be true.

This confidence limit method gives us one way of reporting the outcome of a binomial investigation and making an inference about π.

Appendix Table III-4 gives 80% and 95% confidence limits for the binomial probability for n up to 20. Appendix Charts III-1 and III-2 also give 80% and 95% confidence limits for π for several larger sample sizes.

Assuming Independence

The assumption of independence is not necessarily reasonable. Suppose that we looked over the dates of Mr. I.'s 30 attacks and found that he typically had attacks for three or four weekends in a row, then enjoyed a several-week period free of attacks. This would suggest that the outcomes on successive weekends are dependent and would invalidate the probability calculation just performed. Then we would be less encouraged by his 8 weekends free of migraines. Recall that the outcomes on successive weekends are independent if the probability of an attack on any weekend is not affected by the outcomes on other weekends.

6-2.   Manic-Depressive Illness

Clinical Problem 6-2.   *What Is the Risk of Discontinuing Lithium Therapy?*

Lithium has been used for several years to prevent the occurrence of episodes of mania and/or depression in patients with a history of manic-depressive illness. R.K. has had two prior episodes of depression (one suicidal) and one episode of mania. The patient has been taking lithium for one year without a recurrence. R.K. dislikes taking any medicine, and lithium in particular, because the drug has caused some polyuria and intermittent diarrhea. It is important, therefore, to estimate the magnitude of the benefit that this patient can expect from continued treatment.

Background

Baastrup *et al.* (1970) report a trial of discontinuing lithium therapy in female patients who had been taking lithium for a year without relapse. They studied 56 patients with a history of manic depression prior to starting lithium therapy. These patients were paired according to the total number of prior episodes to form 28 pairs.

One patient in each pair was randomly assigned to lithium and the other to placebo. During the follow-up observation of these patients, in 12 pairs, at least one relapse occurred, and in 10 pairs, no relapses occurred. The remaining 6 pairs failed to complete the experiment. In each of the 12 pairs with at least one relapse, the placebo patient was the first or the only patient to relapse.

Discussion: Comparing the Probability of Relapse with Lithium and Placebo

We want to compare the performances of lithium and placebo. Although the pairs with no relapses tell us something about the probability of a relapse, they appear to tell us nothing about the comparative performance of the two therapies. We will set them aside. For the remaining 12 pairs, let π be the probability that a lithium patient will relapse first. If lithium does not matter, then $\pi = .5$, but if it is effective, $\pi < .5$.

The responses in the 12 pairs will be assumed to be independent, since they apply to different pairs of randomly assigned patients. If lithium has no effect, the probability that the placebo patient will relapse first in every pair if the experiment is repeated is

$$(1 - .5)^{12} = (.5)^{12} = .000244$$

Clearly, the result suggests that lithium defers relapse, since a result as extreme as the one observed favoring lithium would occur only 2 times in 10,000 if lithium and placebo were equally effective.

In the study of Baastrup *et al.*, the probability of at least 1 relapse and the conditional probability that the lithium patient relapsed first could depend on the length of follow-up. If pairs were followed for varying intervals, the assumption that π is the same for all pairs could be incorrect. Still, it is very likely that R.K. will avoid relapse for a longer period if she continues on lithium therapy than if she discontinues therapy, assuming that these results apply to her circumstances. We say more about applying the results of a study to a single patient in Chapter 11.

More About Confidence Intervals

Using Appendix Table III-4, we can set 80% confidence limits on π, the probability of lithium relapsing first when a relapse occurs, using the results from our 12 pairs. For $n = 12$, we read for 12 successes, $0 \leq \pi \leq .17$. Thus the probability favoring lithium is very high.

6-3. Using a Normal Approximation to the Binomial Distribution

Clinical Problem 6-3. *Estimating the Probability of Relapse*

R.K. is still concerned about the side effects of lithium. She con-
cedes that lithium decreases the probability of relapse, but is still
interested in the probability of relapse in the absence of treatment.
What can you tell her?

Discussion: Applying the Normal Approximation

Recall that Baastrup *et al.* followed 22 pairs of patients for an average of
20 weeks. During this time, 12 of the 22 placebo patients relapsed. Let π_P be
the probability that a placebo patient will relapse during 20 weeks of
follow-up and π_L be the corresponding probability for lithium-treated patients.

Because the patients respond independently, the number of relapses
among placebo patients has a binomial distribution with parameters $n(=22)$
and π_P. The **observed proportion** of relapses among placebo patients is

$$\bar{p} = \frac{\text{number of placebo relapses}}{\text{number of placebo patients}}$$

In this case,

$$\bar{p} = \frac{12}{22}$$

$$= .55$$

Mean and Variance of $\bar{p}$: For a binomial distribution based on
a sample of n independent trials with π the probability of success on each trial, the observed proportion, $\bar{p}$, has

$$\text{Mean of } \bar{p} = \mu_{\bar{p}} = \pi$$

$$\text{Variance of } \bar{p} = \sigma_{\bar{p}}^2 = \frac{\pi(1 - \pi)}{n}$$

Furthermore, $\bar{p}$ is approximately normally distributed for
large n and π not near 0 or 1.

You should memorize these facts because they are repeatedly useful.

We use this information in two ways, to set approximate confidence
limits, by using $\bar{p}$ to estimate π in the variance, and to make a test of significance.

> Setting $100(1 - \alpha)\%$ Confidence Limits: For a binomial π and a sample of size n, the $100(1 - \alpha)\%$ two-sided confidence limits on π are approximately
>
> $$\bar{p} \pm z_{1-\alpha} \sqrt{\frac{\bar{p}(1 - \bar{p})}{n}}$$
>
> where $z_{1-\alpha}$ is obtained from the normal Table III-1.

For 80% confidence limits,

$$z_{.8} = 1.28$$

and for 95% confidence limits

$$z_{.95} = 1.96$$

In general, $z_{1-\alpha}$ is chosen so that the standard normal distribution has probability $1 - \alpha$ in the interval $(- z_{1-\alpha}, z_{1-\alpha})$.

In our example, the 80% confidence interval for π is given by

$$\bar{p} \pm 1.28 \sqrt{\frac{\bar{p}(1 - \bar{p})}{n}}$$

$$.55 \pm 1.28 \sqrt{\frac{.55(.45)}{22}}$$

$$.55 \pm .14$$

$$.41 \text{ to } .69$$

Thus, the probability of a relapse in 22 weeks without lithium therapy is near .5. From the paper of Baastrup *et al.*, we learn that no relapses occurred among the 22 lithium-treated patients. Consequently, π_L is near zero. This is strong evidence that π_P is much larger than π_L. The increased risk of discontinuing treatment may be sufficient to convince R.K. to accept the side effects of lithium therapy.

The Continuity Correction

The binomial distribution is discrete and the normal distribution is continuous. Consequently, the normal approximation to the binomial distribution is improved by introducing a continuity correction. When we use the continuity correction, the 80% confidence interval for π becomes

$$\bar{p} \pm \left(1.28 \sqrt{\frac{\bar{p}(1 - \bar{p})}{n}} + \frac{1}{2n} \right)$$

and the 95% confidence interval becomes

$$\bar{p} \pm \left(1.96 \sqrt{\frac{\bar{p}(1 - \bar{p})}{n}} + \frac{1}{2n} \right)$$

In this example, the 80% confidence interval, since $1/2n = 1/44 = .02$, becomes .39 to .71 when the continuity correction is applied.

The continuity correction can be used whenever the normal approximation is employed. Since the correction factor is $1/2n$, it is less important when n is large. However, it can make a substantial difference when the sample size is small. As a final cautionary note, the normal approximation can be poor when the sample size is small.

Test of Significance

Instead of reporting confidence intervals, sometimes investigators report the probability of a result as extreme as or more extreme than the one they have observed, as we saw in the discussion of tests of significance in Chapter 5. For example, in the 12 out of 12 successes for lithium paired with no lithium, we have seen that the probability of 12 successes when the probability of success is $\pi = .5$ is $(.5)^{12} \approx .000244$. Reports of significance are often two-sided, allowing for a result as far away or further in the opposite direction, here 12 failures. So the significance level reported would be $2(.000244) \approx .00049$.

Let us review the features of significance tests in this binomial setting:

The **null hypothesis**: a value of the parameter set up for comparison.

In the example, the natural null hypothesis is $\pi = .5$—the two treatments are equally likely to relapse.

The **alternative hypotheses**: alternative values of parameters of interest.

In the example, we want to know whether either treatment is preferable to the other, so all the other values of π that differ from .5 are the alternative hypotheses. We would be interested in small values of π because they mean that lithium is preferable, or large values of π because they mean that lithium is not preferable.

In reporting the outcome, we may use

The **significance level**: the probability chosen as a criterion.

The significance level most widely used is $\alpha = .05$, and the letter α is widely used for the significance level. The corresponding **confidence level** is labeled $1 - \alpha = .95$. In our example with confidence level .80, the significance level is $\alpha = .20$. Ordinarily, the significance level is thought of as prechosen. Sometimes it is chosen as the level associated with the departure observed and then it is called

The descriptive level of significance, or
The nominal level of significance, or
The P value

In the example, the investigator might choose the $\alpha = .05$ level of significance. Then when the outcome favored lithium 12 to 0, the investigator might say that $P < .05$ to indicate significance at the .05 level. The actual value of P, as we have seen, is .00049 (two-sided). So P, the descriptive or nominal level of significance, is .00049.

> The general idea is that some value of π is a possible neutral or standard value to be tested, in our problem .5. In probability problems, we usually cannot prove as in mathematical logic that a statement is false. Instead, we set a significance level, α, and if the P value is less than α, we say that we reject the null hypothesis in favor of the alternative hypotheses.

Because the choice of α is rather arbitrary, this method does not provide a solid decision-making mechanism, although it does provide a way of reaching intellectual conclusions. When P is very small, we are confident that the departure from the null value is real. We discuss this matter in more detail in Chapter 7.

In the confidence limit approach to testing hypotheses, we say that we accept the null hypothesis if the null hypothesis value of π falls inside the confidence limits, and reject it if the null hypothesis value falls outside. We introduced this idea earlier in discussing the migraine problem. There the null hypothesis was $\pi = .6$, based on past history, $\alpha = .2$, and the confidence limits were .08 to .45. Since .6 was not in the interval, we rejected the null hypothesis at the .2 level.

6-4. Does the Chest Pain Respond to Nitroglycerin?

Background

One feature of classic angina pectoris is that the chest pain responds to sublingual nitroglycerine (TNG). It has been suggested that TNG should be tried as a method of diagnosing patients who have chest pain of unclear etiology. The following hypothetical doctor (Dr.), patient (P) interview is taken from Harrison and Reeves' *Principles and Problems of Ischemic Heart Disease.**

Dr.: The physical examination did not tell us anything. And the resting electrocardiogram as well as the ones with mild and moderate exercise were all normal. But the other one after that last and very strenuous exercise is very suggestive of early and absolutely minimal heart trouble. When I gave you one of these pills, the ones in the box labeled A, to put under your tongue you noticed a little headache. But pill B didn't bother you at all. There is a good chance that one or even both of these pills will prevent that full feeling if you take it just before sexual intercourse. Now the first time, I want you to take pill A immediately before and the second time pill B. Then continue to alternate until you have taken 3 of each. Each time, you are to keep a record of whether you have that full feeling or not. And

* Reprinted by permission from T. R. Harrison and T. J. Reeves. *Principles and Problems of Ischemic Heart Disease.* Copyright © 1968 by Year Book Medical Publishers, Chicago.

if you do have it, you are to take the other pill the moment it starts. Thus, if you take pill A before and the discomfort develops during intercourse, you stop and take pill B. But if the discomfort comes after you have taken pill B, you are to take pill A. And when you have the discomfort and take the pill you must keep a record of the exact number of seconds before the pain disappears. Is that clear?

P.: Yes.

Dr.: Now I want to see you and your wife just as soon as you have finished the test. How soon do you think that will be?

P.: You mean six times? Three after taking each pill before?

Dr.: Yes

P.: Certainly no more than 2 weeks.

Dr.: Then I not only congratulate you but shall make an appointment for 2 weeks from today. And you can tell your wife that these pill tests are an essential part of my examination. And the better she co-operates, the more accurate the test will be.

P.: Thank you, doctor. I should have seen you sooner. My wife was right. She usually is.

It is unnecessary to add that in the great majority of patients like this one, the pill A (nitroglycerin) vs. pill B (saccharin) procedure will provide the clinching evidence. There will be either no discomfort (usually) or less discomfort (occasionally) when intercourse is preceded by nitroglycerin. The chest fullness will be unaffected by saccharin but will vanish sooner than usual when nitroglycerin is taken the moment it starts. The consequent normalization, at both physical and psychic levels, of the sex relationship will dispel the slight cloud that was beginning to threaten the marriage.

Two weeks later the patient returns and says that he had no fullness the 3 times he used pill A first, but every time he used pill B he experienced the discomfort, which subsided before he could use A. Is this "clinching evidence" favoring TNG?

Discussion: Comparing Two Probabilities

This is a question about comparing two probabilities. Let π_A be the probability that fullness occurs after using pill A (nitroglycerin), and π_B the corresponding probability for pill B (saccharin). Our null hypothesis is

$$\pi_A = \pi_B$$

and the set of alternatives, corresponding to improvement with nitroglycerine, is all values of π_A and π_B satisfying

$$\pi_A < \pi_B$$

Our data consist of two sample proportions

$\bar{p}_A$ = proportion of times fullness occurs with pill A

$$= \frac{0}{3}$$

$$= 0$$

and

$$\bar{p}_B = \frac{3}{3}$$

$$= 1$$

The **critical ratio** for $\pi_A - \pi_B$ is

$$z = \frac{\bar{p}_A - \bar{p}_B - (\pi_A - \pi_B)}{\sqrt{\bar{p}(1 - \bar{p})(1/n_A + 1/n_B)}}$$

where $\bar{p}$ is the proportion of events in the combined sample and n_A and n_B are the two sample sizes.

In this instance, $\bar{p}_A - \bar{p}_B = -1$. In our problem, $\bar{p} = 3/6 = .5$, and if the null hypothesis is true, $\pi_A - \pi_B = 0$; so

$$z = \frac{0 - 1}{\sqrt{.5(.5)(^1/_3 + {}^1/_3)}} = -2.45$$

This result could be looked up in Appendix Table III-1, and the two-sided P value is .0142. Thus we reject the null hypothesis at the .05 level and at other smaller levels down to .0142.

6-5. The Poisson Distribution

As you may have noticed, one general strategy in statistical analysis is to assume that our data represent a sample from an unknown probability distribution, and then find a basis for approximating that distribution by one with known and convenient mathematical properties. In Chapter 5 we investigated averages of blood pressure distributions. The Central Limit Theorem made it reasonable to assume that these averages had a normal distribution. In this chapter we have been discussing trials with two possible outcomes, and have used the binomial distribution as the underlying probability distribution. When we can think of the data as coming from a binomial distribution with large n and small π, we can approximate the binomial distribution with a Poisson distribution. The advantage of the Poisson distribution is its simplicity and flexibility. Once we know its true mean, we can obtain all its properties. The **Poisson distribution** gives the probability distribution of the number of events that occur in an interval of time or in a region of space.

When the number of events has a Poisson distribution, the probability that x events will occur is given by the expression

$$P(x|m) = \frac{e^{-m}m^x}{x!}$$

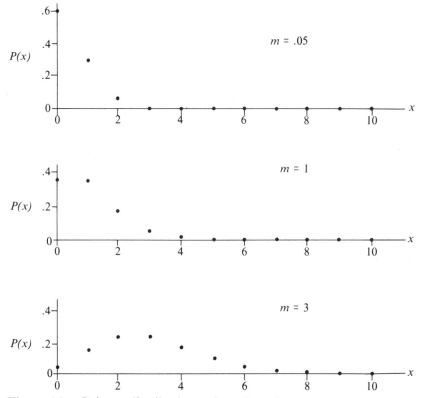

Figure 6-2.   Poisson distributions of number of occurrences x when the average is m, for $m = .5$, 1, and 3.

where m is the mean or expected number of events. To illustrate the shapes assumed by Poisson distributions as their mean m increases, Figure 6-2 shows the probability distributions for $m = .5$, 1, and 3. For $m < 1$, a count of 0 has the highest probability and the distribution descends as shown in the top graph. When $m > 1$, the distribution has the rounded shape associated with the bottom graph, and the shape looks more nearly symmetrical and more nearly like a normal distribution when m becomes large.

An additional attractive feature of Poisson distributions is:

> The sum of independent Poisson variables is again Poisson with mean equal to the sum of the means. Thus we can add results for two time periods, say, to get a single Poisson variable as we illustrate below.

Even though the Poisson distribution may be an approximation in many applications, it is a powerful tool for analyzing counts. We illustrate first with angina attacks.

Clinical Problem 6-4. *Is the Patient's Angina Becoming More Severe?*

Mr. W.K. has had an average of 3 angina attacks a week for many weeks. This week he has had 6. Does this mean that he is getting worse?

Discussion: Comparing Attack Rates

Each week, Mr. W.K. could have many angina attacks because they are short-lived compared with the amount of time in a week. We could think of the week as broken up into many small intervals of time during each of which an attack could start. The intervals could be short enough that more than one attack would not be likely to start. Let us think of a week as 120 active hours. We could imagine breaking it up into 60 two-hour periods or, alternatively, 100 or even 300 periods. We will discuss all three possibilities.

We could think of applying the binomial distribution to each of these situations. We do this to show one way the Poisson arises and how it is related to the binomial distribution. We have sample size n (periods) and probability π of an angina attack in a chosen period. A period either has an attack or it does not. We know that the average number of attacks is $n\pi$, equal to 3 for Mr. W.K. We could see what the binomial distribution produces as the probability of 0, 1, 2, 3, . . . attacks for each choice of π.

To get a mean of $n\pi$ equal to 3, we have chosen three pairs:

$$n = 60, \quad \pi = .05$$

$$n = 100, \quad \pi = .03$$

$$n = 300, \quad \pi = .01$$

Table 6-3. Probabilities for three binomial distributions with $n\pi = 3$: $n = 60, \pi = .05$; $n = 100, \pi = .03$; and $n = 300, \pi = .01$; and for the Poisson distribution with mean 3

Number of events, x	$\pi = .05$, $n = 60$	$\pi = .03$, $n = 100$	$\pi = .01$, $n = 300$	Poisson mean = 3	
0	.046	.048	.049	.050	
1	.145	.147	.149	.149	$\leftarrow m - \sqrt{m}$
2	.226	.225	.224	.224	
3	.230	.227	.225	.224	$\leftarrow m$
4	.172	.171	.169	.168	$\leftarrow m + \sqrt{m}$
5	.102	.101	.101	.101	
6	.049	.050	.050	.050	
7	.020	.021	.021	.022	
8	.007	.007	.008	.008	
9	.002	.002	.003	.003	
10	.001	.001	.001	.001	
.	.	.	.	.	
.	.	.	.	.	
.	.	.	.	.	

From large tables of the binomial distribution, we can read the probabilities we have entered in Table 6-3. To this set of three binomial tables, we have adjoined the probabilities for a Poisson distribution with a mean of 3. This Poisson distribution is the distribution the others are leading up to as we increase n and decrease π so that $n\pi$ is always 3. All these distributions are close to one another, so it does not matter much which we choose, but the Poisson is especially convenient because we need only recall the mean count of 3, not the n and the π. Of course, this trick works only for very large n and very small π.

Assuming, then, that the active week can be broken up into intervals in which attacks occur independently with small but not necessarily equal probabilities, we see by summing the probabilities in Table 6-3 that the chance of 6 or more attacks is about .08 or close to 1 in 12. Mr. W.K. might have this number of attacks 1 week in 12 without having changed his average. On the other hand, 8, 9, or 10 or more attacks would very strongly suggest that something had changed.

> One important feature of the Poisson distribution is that when its mean is m, its standard deviation is $\sqrt{m}$.

The arrows at the right on Table 6-3 indicate values corresponding to the mean and the mean plus or minus 1 standard deviation. Since $\sqrt{3} \doteq 1.7$, the points are

1.3, 3, 3.7

The interval $m \pm \sqrt{m}$ includes a probability of about .616. When m is large, the interval $m \pm \sqrt{m}$ contains about 68% of the probability and the interval $m \pm 2\sqrt{m}$ contains about 95% of the probability. This remark is based on the normal approximation to the Poisson distribution discussed below.

Clinical Problem 6-5. *Observing the Attack Rate over Several Weeks*

After averaging 3 attacks per week for many weeks, Mr. W.K. brings in a 4-week record totaling 22. Is this evidence for change?

Discussion: The Normal Approximation to the Poisson Distribution

Of course it is some evidence. To make it more quantitative, we could look up the probability of 22 or more in a large table of Poisson probabilities. Here the null hypothesis value of m is 12 instead of 3 because the count is for 4 weeks. Alternatively, we can apply the Central Limit Theorem. When the mean is large, the Poisson distribution can be approximated by a normal probability distribution with mean m and variance also m.

The critical ratio, defined for x successes from the Poisson distribution and using a **continuity correction** of $^1/_2$, is.

$$z = \frac{|x - m| - ^1/_2}{\sqrt{m}}$$

Then for our example, we compute

$$z = \frac{|22 - 12| - ^1/_2}{3.46}$$

$$= \frac{9.5}{3.46} = 2.74$$

By finding the probability of this value or any larger value of z from the normal table, we learn that the probability of 22 or more successes is only .003 if m equals 12. The evidence is strong that Mr. W.K. has changed for the worse.

WARNING.   In the calculation just made, we computed the critical ratio as if this were the only 4-week period under consideration. In an earlier calculation we used the time interval starting with a change in medication. Here the time interval was chosen by the size of the observation. Let us look at the matter a bit more carefully.

Suppose that Mr. W.K. comes in every 4 weeks, and that he has been doing this for 12 such periods. If there were enough periods, even if he were stabilized at 3 per week, sometimes he would have 22 or more attacks in 4 weeks. What is the chance of 22 or more in at least one of 12 periods?

This is a binomial problem. The normal distribution probability to the right is 2.74 is about .003. The probability of not getting 22 or more in any of the 12 periods is

$$(1 - .003)^{12} \approx 1 - 12(.003) = .964$$

Thus the probability of getting at least one such period is about .036, since the first equality is approximate. This result is still small. Even after making allowance for multiplicity, we have reason for concern at the high count.

6-6.   Treating Paroxysmal Cardiac Arrhythmias

Background

Patients with heart disease often have paroxysmal cardiac arrhythmias which may be of ventricular or supraventricular origin. Either type of arrhythmia may cause fainting. Ventricular tachycardia is thought to predispose to sudden death. Often it is felt that medical treatment to suppress these arrhythmias is important to decrease symptoms and lessen the risk of sudden death. As different patients respond to different therapeutic agents, patients frequently undergo 6- to 24-hour continuous electrocardiographic monitoring while taking different treatments to determine which therapy is effective. In some patients it is possible to provoke paroxysmal arrythmias with exercise and/or emotional stress. In many patients the arrythmia seems to occur unpredictably.

Clinical Problem 6-6. *Evaluating the Response to Antiarrhythmic Therapy*

Mr. A. has a history of several episodes of fainting and he does not think exercise, stress, or any other situation induces his attacks. He is started on the antiarrhythmic medication procainamide. Subsequent 6-hour monitoring reveals no episodes of ventricular tachycardia. How would you evaluate this response to therapy?

Discussion: Computing Confidence Limits for the Poisson Mean

Let us set confidence limits on the expected number of episodes per hour, m, for Mr. A. We assume that the number follows a Poisson distribution with mean m per hour, so that in 6 hours the mean is $6m$.

We can use Appendix Table III-7, which gives confidence limits for the Poisson mean. Let us compute 95% confidence limits. The table gives limits 0 and 3.69. Thus we are 95% confident that the 6-hour mean is between 0 and about 3.7. Dividing by 6 gives the 1-hour confidence limits 0 and .62 episodes.

When x is large, the normal approximation can be used. The 95% confidence interval for m is (using the continuity correction of $1/2$)

$$(x - 1.96\sqrt{x} - 1/2, x + 1.96\sqrt{x} + 1/2)$$

Because the variance of the Poisson distribution is equal to the mean value, exact confidence limits for m will be asymmetric about an observed x, even when x is very large. The normal confidence intervals are symmetric by construction and will fail to capture this feature of the exact interval.

EXAMPLE.   A man has 16 angina attacks in the first week that he counts them. Set 95% confidence limits on the true rate m using the Poisson formula and the Poisson table (Table III-7).

SOLUTION 1.   Formula: If we round 1.96 to 2, we get limits

$$(16 - 2\sqrt{16} - 1/2,   16 + 2\sqrt{16} + 1/2)$$

or

7.5 to 24.5

SOLUTION 2.   Poisson confidence limit table: Using Appendix Table III-7 and the .95 level, we read 9.15 and 25.98 as the lower and upper limits, respectively.

DISCUSSION.   Note that the length of this confidence interval from the table is $25.98 - 9.15 = 16.83$, or about 17. That is the length we got from the formula. Had we not rounded the 1.96, we would have been even closer because the limits would have been .32 closer together and the formula would have given length 16.68. More to the point, the table deals with the asymmetry of the distribution and the formula does not, as we mentioned earlier.

Difference of Two Poisson Counts

EXAMPLE.   Mr. R. had 24 angina attacks in a 2-week period before taking his new medication. In the 2 weeks since then, he has had 12 angina attacks. Test for improvement.

Let m_B be the rate before and m_A the rate after medication, and x_B and x_A the corresponding observed counts. The difference, $x_B - x_A$, is approximately normally distributed, with mean $m_B - m_A$ and variance $m_B + m_A$. We estimate $m_B + m_A$ by $x_B + x_A$. Thus:

> To test the hypothesis $m_B = m_A$, we compute the critical ratio
>
> $$z = \frac{x_B - x_A}{\sqrt{x_B + x_A}}$$
>
> and refer the result to a normal table.

For our example,

$$z = \frac{24 - 12}{\sqrt{24 + 12}} = \frac{12}{6} = 2$$

Thus the result is just beyond the 2½% (one-sided) level of significance and we conclude that Mr. R.'s angina has improved.

6-7.   Summary

Clinical data frequently consist of independent observations, each having two possible outcomes, which we arbitrarily label success and failure. If π is the probability of success on each trial, n is the number of trials, and d is the number of successes observed, then d follows a binomial distribution with parameters n and π.

If $\bar{p} = d/n$ is the observed proportion, we can set confidence limits on π using Appendix Table III-4 or Appendix Charts III-1 and III-2, or an approximate formula based on the gaussian distribution

$$\bar{p} - z_{1-\alpha}\sqrt{\frac{\bar{p}(1-\bar{p})}{n}} - \frac{1}{2n} < \pi < \bar{p} + z_{1-\alpha}\sqrt{\frac{\bar{p}(1-\bar{p})}{n}} + \frac{1}{2n}$$

where $z_{.8} = 1.28$, $z_{.95} = 1.96$ from normal tables.

To test whether a null value of π is consistent with the data, we use the approximation

$$z = \frac{\bar{p} - \pi}{\sqrt{\pi(1 - \pi)/n}}$$

and enter the normal table, Table III-1, with this value of z.

To compare two observed proportions and thus test whether their π's are equal, we use the normal critical ratio

$$z = \frac{\bar{p}_A - \bar{p}_B}{\sqrt{\bar{p}(1 - \bar{p})(1/n_A + 1/n_B)}}$$

where $\bar{p}_A$ and $\bar{p}_B$ are the observed proportions in samples A and B, and n_A and n_B are the corresponding sample sizes, and $\bar{p}$ is the pooled observed proportion of successes:

$$\bar{p} = \frac{n_A \bar{p}_A + n_B \bar{p}_B}{n_A + n_B}$$

If the observed situation is Poisson, then m, the mean count, determines the distribution. To set confidence limits on m when x events are observed, we can use Appendix Table III-7, or when x is large (with the ½ correction), we use

$$x \pm (z_{1-\alpha} \sqrt{x} + \tfrac{1}{2})$$

where $z_{1-\alpha}$ comes from a normal table as above. Or equivalently, to test whether a null value of m is correct, we can use the critical ratio (with the ½ correction)

$$z = \frac{|x - m| - \tfrac{1}{2}}{\sqrt{m}}$$

and refer the resulting value of z to a normal table.

To test whether two independent Poisson samples have equal m's, we compare the counts x_A and x_B using the critical ratio

$$z = \frac{x_A - x_B}{\sqrt{x_A + x_B}}$$

and refer z to a normal table.

Problems

6-1. Suppose that the lithium trial had reported that of 12 pairs with a relapse, 4 were a placebo success.
   a. Compute the P value corresponding to the null hypothesis of no effect for lithium, $\pi = .5$.
   b. Compute the P value for the null hypothesis that lithium is 90% effective, $\pi = .10$.

6-2. The World Health Organization reports that in 1965, the number of suicides in Scotland averaged 8.0 per week. Assuming that the number of suicides is a

Poisson random variable, find how frequently, on the average, 12 or more suicides might occur in a given week.

6-3.   Baastrup *et al.* found 0 relapses among 22 patients treated with lithium and 12 relapses among the 22 patients treated with placebo. Use the normal approximation formula to test the null hypothesis of the equality of the relapse rates.

6-4.   In the previous 2 weeks, Mr. M. had 20 attacks of angina, and under his new medication he had 5 in 2 weeks. Assuming a Poisson distribution, test whether it is reasonable that the rates are unchanged.

6-5.   Before his new medication, Mr. J. had migraine headaches on 8 out of 10 weekends. Since then he has had 3 in 10 weekends. Test for a change in rate.

6-6.   In Problem 6-5, set 80% confidence limits on the difference in rates.

6-7.   Use Table 6-1 to set 80% confidence limits on Mr. J.'s rate of headaches before he got the new medication (see Problem 6-5).

6-8.   Use the normal approximation with the correction term to set 80% confidence limits on Mr. J.'s headache rate before his new medication (see Problem 6-5). Compare the result from the formula with that from Problem 6-7.

Background for Problems 6-9 through 6-12: Your patient has a complete blood count with 7200 white blood cells (WBC). The differential count of 100 WBC shows 60 polymorphonuclear leukocytes (polys), 38 lymphocytes, and 2 eosinophiles. His asthma is under good control and the patient is on 15 mg of prednisone a day.

6-9.   What is the 90% confidence interval for the poly count (using the binomial distribution)?

6-10.   Compute the 80% confidence interval for the eosinophile count.
   a.   State why the binomial model might be applicable and use a binomial calculation based on the differential count for the eosinophiles.
   b.   State why the Poisson model might be applicable and use a Poisson calculation to compute the interval for the eosinophiles.

6-11.   What is the 80% confidence interval for the lymphocyte count (using the binomial distribution)?

6-12.   The patient returns and says that he may have had more wheezing. His WBC count is still 7200, but he has 10 eosinophiles. What is the P value and the 80% confidence interval for this change from 2 eosinophiles, based on the Poisson analysis?

6-13.   In a certain factory, the probability of an accident occurring on any given day is .002. Use the Poisson approximation to find the probability that 1000 days go by without an accident.

6-14.   If x is a Poisson random variable with $m = 16$, use the normal approximation to find (a) $P(x \le 20)$; (b) $P(x \ge 18)$.

6-15.   If the number of clicks on a Geiger counter is, on the average, 6 per second, find the probability of getting 330 clicks or fewer in 1 minute. Assume a Poisson distribution, and use the normal approximation.

References

Baastrup, P. C.; Poulsen, J. C.; Schou, M.; Thomsen, K.; and Amdisen, A. (1970). Prophylactic lithium: double blind discontinuation in manic-depressive and recurrent-depressive disorders. *Lancet,* **2:**326–30.

Harrison, T. R., and Reeves, T. J. (1968). *Principles and Problems of Ischemic Heart Disease.* Chicago: Year Book Medical Publishers.

CHAPTER 7

What Are *P* Values?

P values are a standard device for reporting quantitative results in research where variability plays a large role. They scatter through articles in medical research, clinical trials, epidemiologic studies, sample surveys, and laboratory work. Statements like ". . . bile saturation increases significantly ($P < .05$) in both sexes during pubertal growth" (Bennion *et al.*, 1979) appear in many scientific papers. The *P* value measures the dissimilarity between two or more sets of measurements, or between one set of measurements and a standard. In the quotation cited, it measures the dissimilarity between values of bile saturation obtained before and during pubertal growth.

> The *P* value is actually a **probability,** usually the probability of getting a result as extreme as or more extreme than the one observed if the dissimilarity is entirely due to variability of measurements or patient response or, to sum up, due to chance alone.

In this chapter we show how to compute the *P* value in some situations and discuss its value in scientific reporting. We also explain the relationship between *P* values and tests of significance.

151

We turn first to some instances where *P* values help communicate the results of investigations. The first example shows how *P* values are computed.

7-1.   Computing the *P* Value

Clinical Problem 7-1.   *Is There an Epidemic of Hepatitis?*

In his office practice, Dr. Walker sees, on the average, 1 new patient a week with viral hepatitis. Yet, during the past week, he saw 3 patients with this disease. He is surprised and wonders whether something unusual may be going on—perhaps a small epidemic that should be referred to the health department. But just how surprising is 3 cases in a single week? How often should he expect to see 3 or even more cases in one week even though the average rate is only 1 per week? A *P* value answers this question.

Discussion

To start, we need a measure. The natural choice is the number of cases of hepatitis in 1 week. We also have a natural view of direction—the more cases, the more worrisome the situation. We want to know the probability that 3 or more patients would acquire hepatitis in a given week if chance variation alone were at work with an average rate of 1 case per week. This is our *P* value for this investigation.

To perform this calculation, we need a probability model that explains how the caseload varies from week to week. Suppose we assume that the case number per week will vary according to a Poisson distribution with an average rate of 1 case per week (see Chapter 6). Referring to Table III-5, a table of the cumulative Poisson distribution, we see in the row corresponding to a rate of 1 that 3 or more cases in a week occur with probability .08, that is, 8% of the time. Thus, if we ask how unusual 3 cases is if no cause but chance is at work, we find that this outcome, or one even more extreme, occurs about 8 times in every 100 weeks, or about 4 times a year in the absence of an epidemic. Although our decisions about notifying the health department may depend on many factors, this result is not a remarkably rare occurrence.

The calculation just performed required a model, in this problem the Poisson distribution, for describing the influence of chance on the measure chosen. In some problems the *P* value may be sensitive to the choice between several plausible models.

Notice that the *P* value is the probability of the result observed or one more extreme given the hypothesized rate, not the probability that the rate is 1 given the data. Confusion about this point often leads to misinterpretation of *P* values.

Warning: A common blunder begins by saying that the *P* value is the probability that the data arose by chance, or that the outcome came about by chance, or that the null hypothesis is true. The implication in Clinical Problem 7-1 would be that when we find 3 cases of hepatitis in a week, we are .080 sure that the hepatitis rate is 1 per week. This conclusion is nonsense. The *P* value is the probability of observing 3 or more cases in a week if the assumed rate of 1 per week is correct, not the probability that the assumed rate of 1 per week is correct if 3 cases occur. In fact, the chance that the true hepatitis rate is 1.000 . . . per week is zero. We know that there are so many possible numbers near 1 that we cannot expect the rate to be correct even to two decimals.

Clinical Problem 7-2. *Computing the P Value When the Number of Cases of Hepatitis Is Small*

Suppose in the same setting that Dr. Walker observes 0 cases of hepatitis in a week. Should he believe that the disease is on the wane? Possibly, but what is the corresponding *P* value?

Discussion

We now ask whether it is rare for the outcome to be as good as or better than the one we have observed. It cannot be better here if we study only the number of new cases in one week. Again using the Poisson model with mean 1, we find that the probability of 0 cases in 1 week is .368, which is the *P* value. Thus, if chance alone is at work, we expect the week to be hepatitis-free about one-third of the time. About two-thirds of the weeks (.632) would have one or more cases of hepatitis. The observation of 0 cases is encouraging but scarcely gives us reason to think the long-term incidence of hepatitis has decreased.

7-2.   Elements of a *P*-Value Calculation

In our discussion thus far, we have given few details on how to compute the *P* value. Our primary concern is understanding what the *P* value is. The foregoing examples emphasized that the *P* value reports the rarity of events as extreme as or more extreme than the one observed.

We now review the four concepts we need before making a *P*-value calculation. They are: measure, direction, null model, and alternative model.

Measure

First, we need a measure of the quantity we care about—the number of deaths, the rate of disease, or some quantitative measure for a physiological process. For example, a study that compared treatments for angina measured four variables: deaths, myocardial infarctions, exercise tolerance, and myocardial lactate extraction (Kloster *et al.,* 1979). Each of these measures was used to compute a *P* value. Sometimes the measure may require complicated calculations. A study of *Herpes simplex* labialis virus used, among other measures, the "virus titer-hour area." This measure was computed as "the area under the curve in which the $\log_{10}$ lesion virus titer is on the ordinate and the age of the lesion in hours is on the abscissa" (Spruance *et al.,* 1979).

Direction

Second, we need to know for what regions—ranges of measurement— we are to compute the *P* value. We might compute the probability of doing this well or better, the probability of doing this poorly or worse, or we might just ask for the probability of being at least this far from a given value. In Clinical Problem 7-1, we computed the *P* value for the region corresponding to "this poorly or worse," namely 3, 4, 5, or more cases of hepatitis in a week; in Clinical Problem 7-2, we focused on the region corresponding to "this well or better," namely exactly 0 cases in a week.

Chance or Null Model

Third, we need some way of generating the probabilities so that we can compute the actual *P* values. Ideas from probability and statistics together with the theories and data of medical science usually provide the information needed to make such calculations. In Clinical Problems 7-1 and 7-2, we supposed that the Poisson distribution was appropriate and that the average case rate was 1 per week. This is the null model. We then supposed that chance alone was operating and computed the probability of getting 3 or more in one week. Small values of the probability cast doubt on the chance hypothesis. Large values give it some support.

For many, it is a puzzle why small or large *P* values give us any feeling about the truth of the chance model. After all, they might say, what is the use of computing these probabilities? If chance alone is at work, these outcomes will occur with about their calculated probabilities, and when we get a small *P* value, we merely know that a rare event has occurred, just as a large *P* value goes with a likely event. This argument calls attention to another concept we need to help us understand *P* values.

The Alternative or Nonnull Models

Fourth, when *P* values are reported, an underlying but usually unstated notion is that the chance model may be mistaken and that some other model may be true. In Clinical Problem 7-1, we had such a nonnull model, namely that there might be an epidemic of hepatitis with a higher attack rate, say, 3 or 4 cases per week. Furthermore, if a nonnull model holds, then the probability of an outcome at least as extreme as the one observed may be high or at any rate higher than when the null model is true. Thus it is not only that the probability of an event is small when the chance model holds, but also that the probability is larger when calculated under other plausible models. These two ideas together encourage us to believe that an alternative model may be true when the null model yields a small *P* value for the observations.

Clinical Problem 7-1 Continued: *An Alternative Model of an Increased Hepatitis Rate*

Dr. Walker observed 3 hepatitis cases in 1 week. The chance model was that the average rate was 1 case per week. Suppose, as an alternative, that the weekly rate is 4 cases per week. Then the probability that 3 or more cases will occur in a week is .762 rather than .080. Thus, we know that 3 or more cases are quite likely to occur if the rate has increased to 4 per week, perhaps because of an epidemic.

In interpreting an observation of 3 cases in a week when the usual rate is 1 per week, but an alternative is 4 per week, we have two main choices:

1. We can conclude that the situation is as usual, but an event having probability .080 has occurred, or
2. We can conclude that the hepatitis rate has somehow changed for the worse this week and that an event having probability .762 has occurred in these new circumstances.

Deciding between these two possibilities can be difficult. Inevitably, a number of issues arise. Most revolve around the question of how small the probability should be before we can conclude that the null model is mistaken, others about the chance that an alternative model is true.

Computing **P** *Values in Other Settings*

In Clinical Problem 7-1, we employed the number of hepatitis cases per week as our measure. This measure has a relatively simple distribution, namely the Poisson, and the *P* value could be computed easily. In other problems, the computation of the *P* value will be different and sometimes more difficult if the distribution of the measure is more complex. Nevertheless, the four ideas stated above remain the same.

Clinical Problem 7-3.   *Has Diuretic Therapy Been Effective?*

At his annual physical examination, Mr. S.M. had a diastolic blood pressure of 97. You encouraged him to reduce the salt in his diet, lose weight, and exercise regularly. Although he made modest gains toward these goals, his diastolic blood pressure at three successive monthly visits was 92, 96, and 93. At the last of these visits, you began a program of diuretic therapy, and continued to see Mr. S.M. monthly. The next three diastolic blood pressure measurements were 88, 91, and 86. Has diuretic therapy been effective?

Discussion

The measure of interest is the change in Mr. S.M.'s diastolic blood pressure level once therapy has begun. The observed change in mean diastolic blood pressure level between the pretreatment and posttreatment periods is $93.7 - 88.3 = 5.4$. We want to compute the probability of a change this large or greater under the null model that no true change has occurred and only chance is at work. Of course, the alternative model is that Mr. S.M.'s diastolic blood pressure has been reduced by treatment.

In Chapter 5 we learned that the *t* statistic is used when comparing two means. We use the pooled variance because the variability of Mr. S.M.'s measurements seems to be about the same before and after treatment. Using formulas introduced in Chapter 5, the pooled variance estimate is

$$s^2 = 5.33$$

and the *t* statistic is

$$t = \frac{93.7 - 88.3}{\sqrt{5.33(1/3 + 1/3)}}$$
$$= 2.86$$

Referring to the critical values of the *t* distribution with 4 degrees of freedom (Table III-2), we find that a positive value of *t* as large as or larger than 2.78 occurs with probability .025. Thus, a value as large as or larger than 2.86 occurs with probability less than .025 and we say that the *P* value for the observed reduction in diastolic blood pressure is less than .025 ($P < .025$).

7-3.   The .05 Level — Pros and Cons

Instead of reporting the actual *P* value, researchers often merely report whether the *P* value is greater or less than .05. Sometimes this is called the 5% level. In classroom experiments with dice or coins, most people begin to disbelieve the chance model at about the .05 level. For example, when one tosses a coin repeatedly and gets the same side every time, most of the audience is comfortable for the first 4 heads (or tails) in a row; but when 5 in a row occur, most no longer believe that pure chance is operating. The chance of getting heads on the first five tosses is 1/32 with a fair coin, and the chance of five tails in a row is the same, so the chance of one or the other occurring with a fair coin is 1/16, which is about .05. Various experiments like this have suggested that many people find the .05 level to be one where their tolerance wears thin.

Another reason why statisticians may have adopted the .05 level early is that it is a round number. Moreover, when dealing with a normal distribution it corresponds (roughly) to a round number of standard deviations from the mean, namely 2 (more precisely 1.96).

Whatever the history, the .05 and .01 levels have been customarily used in reporting results. This standardization has both advantages and disadvantages, and it is certainly not compulsory.

Advantages

1. It gives a specific level to keep in mind, objectively chosen.
2. It may be easier to say whether a *P* value is smaller than or larger than .05 than to compute the exact probability.

Disadvantages

1. It suggests a rather mindless cutoff point having nothing to do with the importance of the decision or the costs and losses associated with the outcomes.
2. Reporting of "greater than" or "less than" .05 is not as informative as reporting the actual level.

To Illustrate Disadvantage 2. In Clinical Problem 7-1 the *P* value would be less than .05 only if 4 or more patients with hepatitis were seen in 1 week. If we see exactly 4, the *P* value is .019. This is much less than .05, and merely reporting a result of less than .05 might not do justice to the finding. Disadvantage 1 may be especially important in clinical situations where only a few observations are available and apparently striking effects fail to achieve statistical significance at the .05 level. We will return to this problem when we discuss the power of a test.

7-4. *P* Values and Hypothesis Tests

P values are closely related to a statistical procedure introduced in Chapters 5 and 6, hypothesis testing or significance testing. Hypothesis testing is a method for choosing between null and alternative models.

> To emphasize the relationship between *P* values and hypothesis testing, we can describe hypothesis testing as consisting of the following steps:
>
> 1. Choose the significance level, α, of the test.
> 2. Compute the *P* value as described above.
> 3. If the *P* value is smaller than α, reject the null model in favor of the alternative model or models; otherwise, accept the null model.

We see this as closely related to the idea of reporting *P* values as larger or smaller than .05. The significance level or α level is often taken to be .05. This leads to the familiar strategy of rejecting the null hypothesis if the *P* value is less than .05. Given the measure used to compute the *P* value, the significance level, and the number of observations, one can determine in advance the set of outcomes resulting in rejection of the null hypothesis. We could call this set of outcomes the **critical region.** Thus another description of the testing rule is:

3'. If the measure falls into the critical region, reject the null model. Otherwise, accept the null model.

We illustrate these ideas with an example based on the binomial distribution.

EXAMPLE 1.   Mr. R.K. has angina pectoris which is exacerbated by exercise. He has kept a diary and for the last few months he has noted angina 80% of the times that he climbed the two flights of stairs to his attic. He is started on an experimental drug for angina and returns 2 weeks later having climbed to the attic 11 times. The null hypothesis is that the drug will not change the .2 probability of climbing the stairs without angina. How can we test the null hypothesis at the .05 level of significance?

SOLUTION.   In this experiment the measure is the number of "successes" or the number of times out of 11 that Mr. R.K. climbed the stairs without angina. Under the null hypothesis, the number of successes has the binomial distribution where the sample size is 11 and the probability of success is .2.

The alternative hypothesis here is that the experimental drug is effective, implying that the probability of success (climbing the stairs without angina) is greater than .2. This is sometimes called a composite hypothesis, since it includes all values of the probability between .2 and 1.0. The test statistic or measure is the number of successes, and large values suggest that the null hypothesis may be incorrect.

Table 7-1 shows the *P* values associated with all possible outcomes of the trial. These probabilities can be obtained from Table III-3 by adding the probabilities for outcomes included in the critical region. If we have chosen .05 as the significance level of the test (more precisely, .051), the critical region includes all outcomes yielding 5 or more successes. That is, if Mr.

Table 7-1.   Probability of *d* or more successes in a binomial sample of size 11 when the success probability equals .2

d	Probability	d	Probability
0	1.000	6	.012
1	.914	7	.002
2	.678	8	.000
3	.383	9	.000
4	.162	10	.000
5	.051	11	.000

R.K. climbed the stairs 5 or more times without angina, we will reject the null hypothesis of ineffectiveness at the .05 level of significance.

Note that the words "accept" and "reject" are merely formal labels like "success" and "failure"; they do not correspond to any particular action. The implication of "reject" is that substantial evidence against the null model has been supplied. The words "accept" and "reject" are a holdover from sampling methods for quality control where the outcome of the sampling was to decide what to do with a lot of manufactured materials.

EXAMPLE 1 CONTINUED. In the trial just described, Mr. R.K. climbed the stairs 7 of the 11 times without angina. What do we conclude?

SOLUTION. We reject the null hypothesis of no treatment effect at the .05 level of significance. We have explained above that we prefer to report the actual P value of the result, in this case .002, rather than acceptance or rejection at a prespecified level. In this case, exact reporting makes a substantial difference in the apparent strength of the evidence.

7-5. Additional Considerations in Computing and Interpreting P Values

One-Sided Versus Two-Sided Tests

In Example 1 we proposed a statistical test with significance level .05. The critical region is 5 or more successes; thus it consists of the numbers 5, 6, 7, 8, 9, 10, and 11. This is a **one-sided test** because we ask only whether 5 or more successes occur.

In some situations we ask whether the result is either too high or too low. Then we have a two-sided test, and the critical region consists of two separated pieces.

EXAMPLE 2. Two randomized studies have reported that 60% of patients with manic-depressive disease who were treated with lithium remained free of relapse over two years. The corresponding success rate for placebo-treated patients was about .20 (Fleiss *et al.*, 1978). The drug imipramine is also of possible value in the prevention of relapse. Consider an investigation that follows 13 patients treated with imipramine. What should the critical region be for a significance level of .05 to test whether imipramine therapy has a different, that is, a higher or lower, success rate than lithium?

SOLUTION. In this case the null hypothesis is that each imipramine-treated patient has a probability of success (remaining relapse-free) of .6, equal to that of lithium. The probability of observing exactly d relapse-free patients among 13 is given by a binomial distribution, with sample size 13 and probability .6. The alternative hypothesis is two-sided, including probabilities of success for imipramine larger and smaller than .6. Thus we should reject the null hypothesis if the number of successes is too large or too small.

Table 7-2. Probability of exactly
d successes in a binomial sample
of size 13 when the success
probability equals .6

d	Probability	d	Probability
0	.000	7	.197
1	.000	8	.221
2	.001	9	.184
3	.006	10	.111
4	.024	11	.045
5	.066	12	.011
6	.131	13	.001

Table 7-2 gives the probability distribution for the number of successes when the probability of success is .6. One convention for choosing a critical region is to assign approximately half the significance probability to the large values in the critical region and the other half to small values. If the critical region includes 4 or fewer successes, that goal is nearly achieved for the small values. For large values, only the outcomes 12 and 13 can be included. The critical region for rejection, including 0, 1, 2, 3, 4, 12, and 13, yields a significance level of .043. Adding another point to the critical region would push the significance level over .05.

> Often a great deal of fuss is made over the issue of one-sidedness versus two-sidedness, but the intellectual issue here is minimal. If we are told what the test is about, what measure is being used, and whether one-sided or two-sided results are being reported, we usually can take the data and recompute what we think is the appropriate *P* value.

In this example, and again in Example 5, we describe therapeutic trials having no concurrent comparison group. Although this type of investigation is a convenient setting for introducing the ideas related to *P* values, it can have serious deficiencies as a strategy for evaluating therapies. We discuss some of the potential problems with this design in Chapter 10.

Problems of Multiplicity

Although we make light of the intellectual issue in the one-sided versus two-sided problem, situations often arise with many sides or at least many opportunities for a result to be statistically significant, and we then have to reconsider the meaning of the *P* value.

EXAMPLE 3.   In one investigation of imipramine and lithium in manic-depressive illness (Veterans Administration and National Institute of Mental Health Collaborative Study Group, 1973), three different treatments were given: placebo, lithium, and imipramine. Moreover, patients were divided into two separate groups with different types of manic-depressive illness (unipolar and bipolar). Thus there are 4 possible opportunities to test active treatments against placebo. These are imipramine versus placebo and lithium versus placebo in each type of illness. Finding at least one comparison significant at the .05 level is more likely than the P value may lead us to believe. Under the null hypothesis, each comparison has a .95 chance of **not** being significant at the .05 level. If we assume for a moment that the 4 tests are independent, there is a $(.95)^4$ or .81 chance of observing no comparison significant at the .05 level.

Here, the four tests are unlikely to be independent, since the experience of each placebo group contributes to 2 different tests. However, an inequality due to Bonferroni implies that the probability that **none** of the 4 P values falls below .05 is at least $1 - 4(.05) = .80$ when the null hypothesis of equal efficacy holds. Thus when we compute 4 separate P values simultaneously, there is a probability of up to **.20** that at least one will be smaller than .05 even when the treatments do not differ. In such situations, a significance level of .01 is sometimes chosen as the criterion for "statistical significance." If an individual P value of .01 is used, the significance level of the 4 tests together is less than .04.

> In general, Bonferroni's inequality implies that when k tests are performed, each with significance level P, the probability of one or more significant tests by chance alone is at most kP. Thus we sometimes say that the significance level should be multiplied by the number of tests.

EXAMPLE 4.   In an investigation with 50 variables, we compute correlation coefficients for all pairs of variables, and 65 pairs turn out to have correlation coefficients with P values less than .05, based on the null hypothesis that the true correlations are zero. What shall we make of this?

SOLUTION.   The number of pairs of variables is $50 \times 49/2$, or 1225, and 5% of this number is about 61. Thus 65 pairs that turn out to be statistically significant give us the queasy feeling that not much is doing among all these variables, since the correlations achieve statistical significance at about the rate expected under the null hypothesis. We might reasonably conclude that these variables are showing approximately chance variation. We could look to see whether a very few correlations are highly significant, say at the .001 level or smaller. Such strong correlations are less likely to have arisen by chance.

Many of the problems of multiplicity are unsolved, partly because it is difficult to know exactly how many independent comparisons were made by the investigators, and partly because it is difficult to figure out how to "adjust" the P value of single comparisons given the comparisons that were made.

7-6.   The Essential Role of Power in the Interpretation of *P* Values

Sample Size and Power

One popular question asked of statisticians is how large a sample is needed for an investigation. This question is closely related to how big the gain or difference is that we are trying to detect, and to the method we have for detecting it. It can be answered by examining the power of the statistical procedure. Once the test and the significance level are chosen, we ask what the chance is of rejecting the null hypothesis if a specific alternative hypothesis is true. This probability is the power of the test for that alternative hypothesis.

> The **power** of a hypothesis test is the probability of rejecting the null hypothesis when the alternative hypothesis is true.

In a study of 71 randomized trials, Freiman *et al.* (1978) found that 94% (67) of the trials that reported "no effect" had insufficient power to detect a reduction in morbidity or mortality of 25%, and 70%(50) of the trials could not detect a 50% reduction, using a one-tailed significance level of 5%. Thus power matters as a practical clinical issue, not just as a statistical nicety.

Note that power is a technical term. Like the *P* value, power is a probability, but power involves two hypotheses and a significance test, while *P* values require only one hypothesis.

EXAMPLE 5.   Consider evaluating a new drug for the prevention of relapse of manic-depressive illness. The spontaneous relapse rate for untreated patients is known to be .80 in two years. Suppose also that it is hoped that the new drug will reduce this rate to a two-year relapse rate of .50. The plan of the investigation is to treat a sample of patients with the drug and observe the number who do not relapse over two years. What power would various sample sizes give us?

SOLUTION.   To determine power, we first specify a hypothesis test. Suppose that we agree to choose a significance level near .05 ("near" to allow for discreteness of the data) and to reject the null hypothesis of no effect if large numbers of successes occur. Table 7-3, computed from the cumulative binomial probabilities given in Table III-3, and more extensively elsewhere, gives the critical region and significance level of the hypothesis test for sample sizes ranging from 5 to 26. It also gives the probability of obtaining an observed value in the critical region when the true success probability is .5. This probability is the power of the test for that sample size. We see that as the sample size grows, we could approximately maintain the level of statistical significance near .05, and at the same time, increase the power; that is, increase the probability that the outcome will fall into the critical region when the alternative holds rather than the null hypothesis.

Table 7-3. Critical region, significance level, and power of a test of the null hypothesis: "the success probability is .2" against the alternative hypothesis "the success probability is .5" for binomial samples of various sizes

Sample size	Successes required	Significance level[a]	Power[b]
5	≥3	.057	.499
8	≥4	.056	.636
11	≥5	.051	.726
15	≥6	.061	.849
18	≥7	.051	.881
22	≥8	.056	.933
26	≥9	.059	.962

[a] Assuming a null success rate of .2.
[b] Assuming an actual success rate of .5.

Although Table 7-3 tells us how the critical region changes with the sample size and what the power would be for an alternative probability of success, $\pi = .5$, that specific value is drawn from the air. On the other hand, the null value, $\pi = .2$, has some empirical backing. Thus it is reasonable to ask what the probability of rejection would be for other values of π in addition to .2 and .5. Figure 7-1 shows curves that give the power for all values of π for

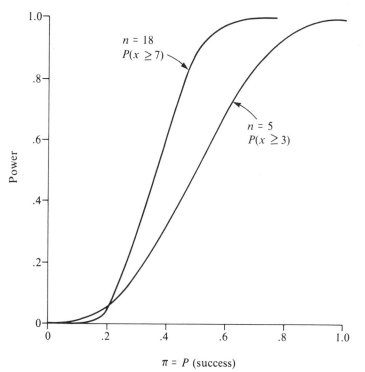

$\pi = P \text{ (success)}$

Figure 7-1. Power functions for binomial samples of size $n = 5$ and 18 with significance levels between .05 and .06 at $\pi = .2$.

$n = 5$ (when we reject the null hypothesis for $x \geq 3$) and for $n = 18$ (when we reject for $x \geq 7$).

The graph shows, as would be expected, that as π increases beyond .2, the test using a sample of size 18 has a better chance of detecting a gain. For $\pi = .4$, the test based on 18 has twice the power of that based on the sample of 5. For $\pi = .6$, the test based on $n = 18$ has power .98 to find a difference, while not until $\pi = .95$ does the test based on $n = 5$ have power .99. It is easy to see from these curves that small pilot studies, such as those based on samples of 5, have not much chance of detecting a doubling of the cure rate when we use a significance level of .05.

Relaxing the Significance Level When Treating the Individual Patient

Many clinical decisions concerning an individual patient's response to therapy are based on small data sets. We have seen that tests based on a significance level of .05 can have very poor power in that situation, even when medically substantial effects are observed.

> When using significance tests in the clinical setting, it makes sense to relax the significance level, for instance, to .10 or .20.

In this way, large effects can be recognized as possibly real, allowing the possibility for further inquiry to separate real effects from false positives. The alternative strategy, routine use of the .05 criterion, essentially ignores the costs of insensitive testing.

Power in Detecting an Epidemic of Hepatitis

Let us return to Clinical Problem 7-1 concerning hepatitis. Suppose that the attack rate has been 1 case per week for several years. The health department is concerned that there might be an epidemic with a rate of 2 cases per week (or more). The department plans surveillance. From Poisson tables, we can determine the number of cases that have to be observed in any period to get a significance level of between .02 and .05 under the null hypothesis. We can also compute the power—namely the probability that an epidemic rate of 2 per week will produce such a number of cases. We can then vary the observation period and determine the power of the test at each length of observation as shown in Table 7-4. Again we see that increasing the sample size, in this situation corresponding to duration, increases the power of the investigation.

The studies of power illustrated in these situations can be most instructive in more general problems. Power tells us about the size of sample required in order to have a given chance of detecting an effect. When the

Table 7-4. Critical region, significance level, and power of a test of the null hypothesis "the hepatatis case rate is 1 per week" against the alternative hypothesis "the rate is 2 per week" for various observation periods

Weeks of observation	Cases required	Significance level assuming 1 per week (null hypothesis)	Power assuming 2 per week (alternative hypothesis)
1	≥4	.02	.14
2	≥5	.05	.37
3	≥7	.03	.39
4	≥8	.05	.55

sample size is small, failure to reject the null hypothesis can often be attributed to low power rather than the correctness of the null hypothesis. Our examples treat the simplest situations, but the methods can be generalized to many other problems.

Clinical Investigations Often Have Poor Power

One common, though often disappointing, outcome of an examination of the power of a statistical investigation is the discovery that the investigation is too small or that the scientific tools are too insensitive for a study of feasible size to give us the assurance we desire. It also may show that there is a great difference between the size of an effect that can be desirable and one that can be detected. For example, a reduction in death rate of even 1% is desirable, but it may take thousands of patients under the most carefully controlled conditions to verify such an improvement. Investigations of power tell us what is required to detect an advance and what sizes of advances have a reasonable chance of being detected.

Considerations of power have great importance during the design of an investigation. Some feel that, after the investigation, power holds no role—that a report on the confidence interval or some other way of reporting the reliability of the investigation is adequate. Although this position has some merit, the question is partly one of adequacy of reporting. In an investigation of size $n = 5$, if 2 successes occurred, we might report the 95% confidence interval of .08 to .81. But this is rather different from telling the reader that the investigation as designed had only 1 chance in 6 of detecting an improvement in success rate from 20% to 30%. Thus the reader may find a somewhat different impact from the two statements. We believe that power has a role in discussions after as well as before an investigation.

Sometimes power cannot be adequately estimated before an investigation, and the information for evaluating it may come from the study itself. Thus lack of planning may not always be the cause of a study too small to detect an important effect.

Statistical Significance and Practical Significance

It is unfortunate that the statistical profession introduced the word "significance" in association with statistical testing of hypotheses. It muddies the waters. We have seen precisely what statistical significance means. It is a technical term. Substantive significance is more vague because it usually means "important." Four possibilities exist. Findings can be both statistically significant and practically significant or neither one. The more troublesome outcomes occur when one or the other holds. When we have very large samples, we may find small differences statistically significant even though they have no clinical importance. At the other extreme, when we have small samples, as in Clinical Problem 7-1, even large differences in rates will often not be significant at the levels usually required to recognize the difference as real. Although one needs to keep an eye out for the distinction between statistical and practical significance, the clinician is not likely to be misled by this language.

Reporting and Decisions

We should not equate *P* values or the results of hypothesis testing with decisions. *P* values are a way of reporting the results of a statistical analysis. Similarly, the result of a hypothesis test might be best described as a conclusion, rather than a decision, to emphasize that the results of hypothesis tests are another way of reporting data. Decisions depend on costs, risks, consequences, and policy considerations. For example, treatment decisions for individual patients may depend on the costs and potential for adverse and beneficial effects of alternative therapies, as well as the importance of these effects. Issues of institutional decision making and many other factors can affect the decision. What hypothesis tests and *P* values can do is give us a form of reporting that has value because it can be standardized. The practical decision is a separate matter that uses *P* values and related information, but not the *P* value alone.

7-7.   Summary

P values are quoted widely in the medical literature, yet their meaning is not always understood. The *P* value is actually a probability—specifically, the probability of getting a result as extreme as or more extreme than the one observed if the proposed null model is correct. We have described the elements of the computation and interpretation of *P* values, and the relationship between *P* values and hypothesis tests. We have also reviewed the basis for the widely used .05 level, described one- and two-sided tests, and emphasized the need for an understanding of power to appreciate the results of tests of hypotheses. Problems of multiple tests and multiple endpoints decrease considerably the *P* values appropriate for quotation. A bound for the overall significance level is kP, where k is the number of tests and P the significance level for individual tests. *P* values and hypothesis tests are useful tools for reporting scientific evidence, in part because they are computed in a standardized way. However, the results of hypothesis testing should be distinguished from medical decisions, for example, regarding patient management. Decisions require the consideration of costs, risks, benefits, and other factors that play little or no role in hypothesis testing.

Problems

7-1. Explain to a physician who has not studied statistics what is meant by a *P* value.

7-2. What is the power of a statistical test, and why is it important?

7-3. Why can't we speak of the *P* value as giving us the probability that the null hypothesis is true?

7-4. Can we have more than one confidence level at the same time? If so, how many?

7-5. Why does the *P* value not settle medical decisions?

7-6. What are problems of multiplicity?

7-7. When are *P* values especially useful?

References

Bennion, L. J.; Knowles, W. C.; Mott, D. M.; Spagnola, A. M.; and Bennett, P. A. (1979). Development of lithogenic bile during puberty in Pima indians. *N. Engl. J. Med.*, **300**:873–76.

Fleiss, J. L.; Prien, R. F.; Dunner, D. L.; and Fieve, R. R. (1978). Actuarial studies of the course of manic-depressive illness, *Compr. Psychiatry*, **19**:355–62.

Freiman, J. A.; Chalmers, T. C.; Smith, H., Jr.; and Keuhler, R. R. (1978). The importance of beta, the type II error and the sample size in the design and interpretation of the randomized control trial. *N. Engl. J. Med.*, **299**:690–94.

Kloster, F. E.; Kremkau, E. L.; Ritzmann, L. W.; Rahimtoola, S. H.; Roosch, R.; and Kanarek, P. H. (1979). Coronary bypass for stable angina. *N. Engl. J. Med.*, **300**:149–57.

Spruance, S. L.; Crumpacker, C. S.; Haines, H.; et al. (1979). Ineffectiveness of topical adenine oral unoside 5'-monophosphate in the treatment of recurrent herpes simplex labialis. *N. Engl. J. Med.*, **300**:1180–84.

Veterans Administration and National Institute of Mental Health Collaborative Study Group: Prien, P. F.; Kleh, C. J.; and Caffey, E. M., Jr. (1973). Lithium carbonade and impramine in prevention of affective episodes. *Arch. Gen. Psychiatry*, **29**:420–25.

CHAPTER **8**

What Is Chi-Square?

The chi-square statistic, χ^2, helps us analyze data that come in the form of counts. Very often, chi-square is used to compare two proportions, for example, the proportions of successes in two groups exposed to different treatments. The chi-square statistic usually requires a substantial number of observations to be very informative, so we only occasionally apply it directly to data from single patients. Nevertheless, the chi-square statistic is widely used in reports of clinical trials, observational studies, case-control studies, sample surveys, and laboratory studies. In interpreting these studies for clinical practice, the physician will find an understanding of chi-square very helpful.

In this chapter we apply the chi-square statistic to observations on a series of hypertensive patients in a physician's practice and use the test to decide whether the outcome of therapy is different from expected based on results of the Hypertension Detection and Follow-up Program (1979). We then apply the chi-square statistic to test whether two physicians have equal success in blood pressure control.

We relate the chi-square statistic to the normal distribution. We also give the basic meaning of degrees of freedom in chi-square problems and explain how to count degrees of freedom in many contingency table problems.

168

8-1. Chi-Square for Counts

Clinical Problem 8-1. *Comparing Success Rates in Treating Hypertension*

The Hypertension Detection and Follow-up Program (HDFP) reported that four years after the program began, 62% of hypertensive patients participating in a stepped care treatment program at antihypertensive clinics had diastolic blood pressure levels at or below treatment goals. The goal diastolic blood pressure (DBP) was defined as 90 mm Hg for those entering with a DBP of 100 mm Hg or greater and a 10 mm Hg decrease for those entering with a DBP of 90 to 99 mm Hg. You have conducted a review of your practice and find 20 hypertensive patients whom you have followed for four years. Only 7, or 35%, of these are at goal blood pressure. How likely is a finding this different from 62% if your treatment is as effective as the HDFP in lowering blood pressure to goal (assuming that your patients are similar)?

Solution: Chi-Square for Contingency Tables

Although we could treat this as a binomial problem, we use it here to introduce chi-square for contingency tables. Let us make a table of observed and expected counts (Table 8-1).

Here the observed counts are the numbers of patients you observe among your 20 who were or were not at goal DBP. The expected counts are computed from the total number of patients observed (20) multiplied by the probability of observing a patient at goal or not at goal based on the findings of the HDFP. These probabilities are estimated as .62 and .38, respectively. We compute the expected number at or below goal based on the HDFP program as $.62 \times 20 = 12.4$.

Consider for a moment only the observed counts. We present them as the top line of Table 8-1 in the form of a contingency table. In the present instance, it is a 1×2 (read: one by two) contingency table because it has one row and two columns of **observed** cells. It is customary and often helpful to adjoin the total to the table. In some computations we use the total; in others we focus on the outcome cells. This table has two outcome cells—one containing a count of 7 and the other of 13.

Table 8-1. Observed counts for 20 patients in a physician's practice and expected counts for 20 patients treated according to the HDFP protocol

	At or below goal DBP	Not at goal DBP	Total
Observed count	7	13	20
Expected count	12.4	7.6	20

In contingency table problems, we create an index that computes for each **outcome** cell

$$\frac{(\text{observed count} - \text{expected count})^2}{\text{expected count}}$$

and then we sum this index over all cells. If O stands for observed count and E for expected count, the index is written

$$\chi^2 = \sum \frac{(O - E)^2}{E} \tag{8-1}$$

where Σ means "sum over all cells." This is a chi-square statistic.

It applies to contingency tables generally, as we discuss below.

When, for a term in the sum, O and E are close together, the numerator is small. The denominator implies that when an $O - E$ difference is associated with a large E, it counts less than with a small E. For our problem, this statistic is

$$\chi^2 = \sum \frac{(O - E)^2}{E} = \frac{(7 - 12.4)^2}{12.4} + \frac{(13 - 7.6)^2}{7.6}$$
$$= 6.19$$

This chi-square has 1 degree of freedom.

In contingency table problems, we can get the number of degrees of freedom by fixing the totals and seeing how many cells we can fill in freely.

In this problem we can fill in one cell, and then the other is determined because the outcome cells must add to 20. The degree of freedom is like a dimension in physics. If an object can move in a line, it has 1 degree of freedom; in a plane, 2; in space, 3. Having only one number we can fill in freely is like being constrained to a line.

Recall that we are trying to determine whether finding only 7 of 20 treated patients at goal pressure is probable assuming a treatment such as that in the HDFP. In other words, what is the P value of our observation, assuming as a null hypothesis that the probability of being at goal is .62? If we had the distribution of the chi-square statistic under the null hypothesis, we could compute the P value by noting the probability of observing a

Table 8-2.   Values of chi-square for various degrees of freedom

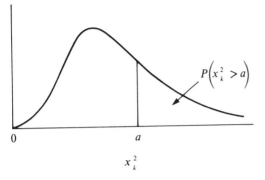

$$P\left(x_k^2 > a\right)$$

$$x_k^2$$

EXAMPLES.   (a) The probability that a χ^2 random variable with 1 degree of freedom exceeds 3.84 is $P(\chi_1^2 \geq 3.84) = .05$. (b) The table shows that $P(\chi_{10}^2 \geq 15)$ lies between .10 and .20. Interpolation gives $P(\chi_{10}^2 \geq 15) = .14$, approximately.

Degrees of freedom	Probability levels						
	.99	.95	.50	.20	.10	.05	.01
1	.00	.00	.45	1.64	2.71	3.84	6.63
2	.02	.10	1.39	3.22	4.61	5.99	9.21
3	.11	.35	2.37	4.64	6.25	7.81	11.34
4	.30	.71	3.36	5.99	7.78	9.49	13.28
5	.55	1.15	4.35	7.29	9.24	11.07	15.09
6	.87	1.64	5.35	8.56	10.64	12.59	16.81
7	1.24	2.17	6.35	9.80	12.02	14.07	18.48
8	1.65	2.73	7.34	11.03	13.36	15.51	20.09
9	2.09	3.33	8.34	12.24	14.68	16.92	21.67
10	2.56	3.94	9.34	13.44	15.99	18.31	23.21

chi-square greater than or equal to 6.19. The precise distribution of the chi-square statistic in our problem is discrete. An approximation is given by the continuous chi-square distribution with 1 degree of freedom. From a chi-square table such as Table 8-2 or Table III-8 one can read the probability of a value exceeding 6.19 as less than .05 but greater than .01.

This chi-square statistic suggests that it is unlikely that your 20 patients were sampled from a population whose patients had a .62 probability of being at goal blood pressure. You might consider whether either your patient population or your treatment strategy differs from that of the HDFP.

8-2.   Chi-Square Distribution

Learning to compute a chi-square statistic is trivial. Understanding where it comes from and how it relates to a chi-square distribution is not. Chi-square, written χ^2 (some say chi-squared), refers to several things, including:

1. A family of continuous probability distributions
2. Some statistics associated with tests of independence in contingency

tables and having distributions approximated by the continuous ones of item 1

Chi-Square with 1 df Is the Square of a Normally Distributed Variable

If we have a variable distributed according to a gaussian (normal) distribution with mean 0 and variance 1, as shown in Figure 8-1, its square is distributed according to the chi-square distribution with 1 degree of freedom. The shape is shown in Figure 8-2.

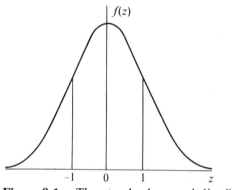

Figure 8-1.   The standard normal distribution.

Let us review how we get from Figure 8-1 to Figure 8-2. Recall that for the standard normal (gaussian) distribution, 95% of the probability is within the interval from -1.96 to 1.96 and 5% is outside with 2.5% in each tail, as shown in Figure 8-3.

When we square the z of the standard normal distribution to get the chi-square distribution with 1 degree of freedom, negative numbers become

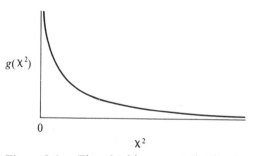

Figure 8-2.   The χ^2 (chi-square) distribution with 1 degree of freedom. It has mean 1 and variance 2.

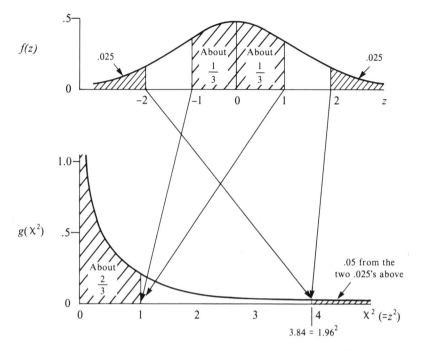

Figure 8-3. Relations between the standard normal distribution, with ordinates $f(z)$, and the χ^2 distribution with 1 degree of freedom, with ordinates $g(\chi^2)$.

positive, so we are folding the normal over about 0 and doing some stretching and compression. Figure 8-3 shows the relationship.

Thus we could get the P value of a chi-square with 1 degree of freedom by taking the square root of the chi-square statistic, entering a normal table, and reading out the two-sided probability level.

Chi-Square with More than 1 Degree of Freedom

> If we have several independent gaussian variables, z_i, $i = 1$, $2, \ldots, k$, all from the standard distribution ($\mu = 0, \sigma^2 = 1$), and we square each variable and add them, their sum
>
> $$\chi^2 = z_1^2 + z_2^2 + \cdots + z_k^2$$
>
> has the chi-square distribution with k degrees of freedom.

For $k \geq 3$, the distribution is shown in Figure 8-4. It has mean k and variance $2k$.

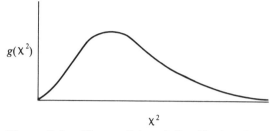

Figure 8-4.   Shape of the χ^2 distribution for 3 or more degrees of freedom.

Chi-Square and the Binomial Distribution

Next we relate this discussion of the chi-square distribution as the sum of squares of gaussian variables to our problem in the 1×2 contingency table. We illustrate that computing chi-square as

$$\chi^2 = \sum \frac{(O - E)^2}{E}$$

yields the square of a variable of mean 0 and variance 1.

As we mentioned above, one could view the data in Clinical Problem 8-1 as a binomial sample. The null hypothesis is that, in your practice, the probability of a hypertensive attaining goal blood pressure is the same as the probability in the HDFP, namely .62. Your review of 20 patients is then a sample of 20 binomial trials, so the outcome should have a binomial distribution with $n = 20$ and $\pi = .62$. In a binomial sample when π is not too small or too large and n is large as it is in this problem, the distribution of the number of patients at goal, say d, can be approximated by a normal distribution with mean $n \pi$ and variance $n\pi(1 - \pi)$ (see Chapter 6). Thus $d - n\pi$ is the difference between our sample value and the mean value and

$$z = \frac{d - n\pi}{\sqrt{n\pi(1 - \pi)}}$$

is approximately normally distributed with mean 0 and variance 1. Then z^2 is approximately the square of a standard gaussian variable and ought to have approximately a chi-square distribution with 1 degree of freedom. Substituting from our problem $n = 20$ and $\pi = .62$, we find that

$$z^2 = \chi^2 = \frac{[7 - (20)(.62)]^2}{(20)(.62)(.38)} = 6.19$$

which is the same as the value 6.19 which we computed using formula (8-1).

It is possible to show algebraically that this chi-square, which we computed as

$$\chi^2 = \frac{(d - n\pi)^2}{n\pi(1 - \pi)} \tag{8-2}$$

is equivalent to our sum formula

$$\chi^2 = \Sigma \frac{(O - E)^2}{E} \tag{8-1}$$

although we do not present the proof.

In larger contingency tables, it is also possible to demonstrate that the chi-square can be represented as the sum of the squares of a number of approximately normally distributed random variables that are nearly independent of one another, each standardized to have 0 mean and variance 1.

Just as in the normal approximation to the binomial, there is also a $1/2$ correction for chi-square with 1 degree of freedom. We describe this improvement in Section 8-5.

8-3.   2 × 2 Tables

Clinical Problem 8-2.   *Comparing Success Rates from Two Practices*

Because of your concern that your therapy for hypertension may not be as effective as it could be, you continued your investigations to determine whether your results are different from those of your colleague, Dr. M.W., who practices near you. You have now reviewed a total of 40 patients from your practice and 30 patients from Dr. M.W.'s. In all, 18 of your patients are at goal blood pressure, while 20 of his are at or below goal. This information is summarized in the 2 × 2 contingency table of counts, Table 8-3.

Discussion: Chi-Square for Association

What is being studied is whether the probability of being at goal DBP is the same for both populations (your patient's and Dr. M.W.'s). This is another way of saying that we want to know whether being at goal is independent of physician. If not, we say that there is an **association** between physician and success rate.

Table 8-3.   Number of hypertensive patients at goal DBP and number not at goal in two practices

	At goal DBP	Not at goal DBP	Total
Your patients	18	22	40
Dr. M.W.'s patients	20	10	30
Total	38	32	70

To relate this table to chi-square, we see that there are four cells which are observed counts—the O's. We need the expected counts—the E's.

If the proportions at goal are the same in each practice, then the conditional probabilities equal the unconditional probability:

$$P(\text{at goal}|\text{your practice}) = P(\text{at goal}|\text{M.W.'s practice})$$
$$= P(\text{at goal})$$

Under the "null hypothesis" that your practice is like Dr. M.W.'s, this is what we expect, and this is what we mean technically by independence. These P's are for the populations, but we can observe only the proportions in our samples.

We want to know what the expected number in each cell would be if we had independence. To compute this it is convenient to lay out the table of margins as shown in Table 8-4. What is given or "fixed" in the table of margins is the total number of patients (n), the total in each practice (r and $n - r$), and the total at and not at goal (t and $n - t$).

For independence in the table, we want the proportions at goal to be the same in both patient samples. Thus we want

$$\frac{y}{r} \quad \text{to equal} \quad \frac{t - y}{n - r}$$

and we ask what value of y makes them equal. That value is what we call the expected value, E, for the cell. We need to solve the following equation for y:

$$\frac{y}{r} = \frac{t - y}{n - r}$$

Carrying out the algebra gives

$$E = \text{expected count} = y = \frac{rt}{n} = \frac{(\text{row total})(\text{column total})}{\text{grand total}}$$

> More generally, for any cell, the **expected value** when we have independence is
>
> $$E = \frac{(\text{row total})(\text{column total})}{\text{grand total}}$$
>
> (8-3)

Table 8-4. Table of cell entries for fixed margins

	At goal	Not at goal	Total
Your patients	y^a	$r - y$	r
Dr. M.W.'s patients	$t - y$	$n - r - t + y$	$n - r$
Total	t	$n - t$	n

^{a} The expected count in the upper left corner is y.

Notice that the expected number of patients at goal in your practice is the number of patients in your practice, r, times the overall proportion of patients at goal, t/n.

Once we get the expected count for one cell, we can get the rest by subtraction, or the expected count for any cell can be calculated by formula (8-3). Indeed, the same formula works also for independence in larger tables with more rows and columns.

Now that we know how to get the expected counts, let us apply the method to computing the chi-square. We are given the table of observed counts. We compute the expected cell values. Both are shown in Table 8-5. For the upper left corner, the expected number of your patients at goal is

$$E = \text{number expected at goal}$$

$$= \frac{(\text{row total})(\text{column total})}{\text{grand total}} = \frac{(40)(38)}{70} = 21.7$$

We can also use formula (8-3) to compute the expected number of your patients not at goal.

$$E = \text{number expected to be not at goal} = \frac{(40)(32)}{70} = 18.3$$

Alternatively, we could get the 18.3 by subtracting 21.7 from the row total of 40, and so on for the other cells.

Now that we have the O's and the E's, we can compute chi-square. This measures the departure from independence. We get

$$\chi^2 = \sum \frac{(O - E)^2}{E}$$

$$= \frac{(18 - 21.7)^2}{21.7} + \frac{(22 - 18.3)^2}{18.3} + \frac{(20 - 16.3)^2}{16.3}$$

$$+ \frac{(10 - 13.7)^2}{13.7}$$

The numerators on the right-hand side are all alike, so we have

$$\chi^2 = (3.7)^2 \left(\frac{1}{21.7} + \frac{1}{18.3} + \frac{1}{16.3} + \frac{1}{13.7} \right)$$

$$= 3.22$$

Table 8-5. Observed and expected values in comparing two groups of patients

	Observed counts, O's			Expected counts, E's		
	AT GOAL	NOT AT GOAL	Total	AT GOAL	NOT AT GOAL	Total
Your patients	18	22	40	21.7	18.3	40
Dr. M.W.'s patients	20	10	30	16.3	13.7	30
Total	38	32	70	38	32	70

We can look up this value of χ^2 in Table 8-2 or in Table III-8. (As soon as we fill in one cell of the table, the rest can be gotten by subtraction from the totals, so we have only 1 degree of freedom.) The table gives a P value between .1 and .05. The probability of values of chi-square at least this large when we have independence is modest (about .07). Thus the different success rates of 45% for your patients and 67% for Dr. M.W.'s may represent sampling variability. Obtaining a larger sample may answer this question more definitely. Still, it would be well to see whether you and Dr. M.W. are doing something differently, or whether the populations differ.

Remembering that a chi-square distribution with 1 degree of freedom is the distribution of the square of a normally distributed variable, we could take the square root of the chi-square value and look it up in a normal table. The two-tailed normal value would be the same as the one-tailed chi-squared value. In our problem $\sqrt{3.22} = 1.79$ and from a normal table we get .074 as the two-tailed probability. Unfortunately, this device does not work for more than 1 df.

In Section 6-4 we introduced a formula for comparing the difference of two proportions,

$$z = \frac{\bar{p}_1 - \bar{p}_2}{\sqrt{\bar{p}(1 - \bar{p})(1/n_1 + 1/n_2)}}$$

where $\bar{p}_1$ and $\bar{p}_2$ are the observed proportions of successes in the two groups,

$$\bar{p} = \frac{n_1 \bar{p}_1 + n_2 \bar{p}_2}{n_1 + n_2}$$

is the overall proportion of successes, and n_1 and n_2 are the sample sizes.

In our notation of Table 8-4, and with d of your patients at goal,

$$\bar{p}_1 = \frac{d}{r} \qquad \bar{p}_2 = \frac{t - d}{n - r} \qquad \bar{p} = \frac{t}{n}$$

$$n_1 = r \qquad n_2 = n - r \qquad n_1 + n_2 = n$$

Although we do not prove it here, z^2 is exactly equal to our χ^2 with 1 degree of freedom for the 2 × 2 table.

8-4.   Larger Contingency Tables

If we had more than two groups of patients, perhaps data from three practices, we might ask whether all three have about the same percentage of patients at goal. Table 8-6 shows the two-by-three (2 × 3) table of counts summarizing the findings in samples from your practice, Dr. M.W.'s, and Dr. J.K.'s. The expected values would be calculated as before from

$$E = \frac{(\text{row total})(\text{column total})}{\text{grand total}}$$

In computing χ^2, we sum over all 6 cells. We have only 2 degrees of freedom because as we fill in the cells, once two are freely filled in, the values of the

Table 8-6. Chi-square calculation to test whether patients from three practices have the same proportion at goal

	Observed counts			Expected counts		
	AT GOAL	NOT AT GOAL	Total	AT GOAL	NOT AT GOAL	Total
Your patients	18	22	40	20	20	40
Dr. M.W.'s patients	20	10	30	15	15	30
Dr. J.K.'s patients	12	18	30	15	15	30
Total	50	50	100	50	50	100

$$\chi^2 = \frac{(18 - 20)^2}{20} + \frac{(22 - 20)^2}{20} + \frac{(20 - 15)^2}{15} + \frac{(10 - 15)^2}{15}$$
$$+ \frac{(12 - 15)^2}{15} + \frac{(18 - 15)^2}{15}$$

= 4.93 with 2 degrees of freedom
$P(\chi^2 \geq 4.93) = .09$.

rest are forced by the marginal totals. In our example, χ^2 with 2 degrees of freedom equals 4.93 and the P value is between .10 and .05, about .09. The evidence is not strong that the practices differ in performance.

We could also have more than two categories for columns as well as for rows in some problems.

> If there are R rows and C columns, the degrees of freedom are $(R - 1)(C - 1)$.

This formula does not work for the 1×2 table because we must get the expected values from a source outside the table. If we had to use only marginal totals, we would have no degrees of freedom.

8-5. The ½ Correction

Just as we had a correction for continuity when we approximated the binomial distribution with the normal, we also have one for chi-square.

1. *The 1 × 2 table.* When O differs from E by more than ½, we reduce the difference by ½ and get

    $$\chi^2 = \sum \frac{(|O - E| - ½)^2}{E}$$

    where $|O - E|$ is the absolute value of $O - E$. If the difference is less than ½, we regard the numerator as zero.

2. *The 2 × 2 table.* When the upper left corner cell, say y, differs from its expected value by more than ½, we diminish the difference by ½. If we let E_1, E_2, E_3, E_4 be the expected values for the four cells, the

corrected chi-square would be

$$\chi^2 = \left(\left|y - \frac{rt}{n}\right| - \frac{1}{2}\right)^2 \left(\frac{1}{E_1} + \frac{1}{E_2} + \frac{1}{E_3} + \frac{1}{E_4}\right)$$

where r and t are the totals for the row and column with y in them, and n is the grand total. If the difference $|y - E_1|$ is less than $\frac{1}{2}$, we set the difference to zero. (Note that $|y - E|$, the absolute value of y minus E, is the same for each of the 4 observed cells in a 2×2 table.)

3. *Larger tables.* Although there is a corresponding correction for larger contingency tables, it is complicated to explain and does not have such a large effect as we get in the 1×2 and 2×2 tables, so we do not give it.

4. *Summing chi-squares.* If we compute chi-squares for several tables, we can pool the chi-squares and the degrees of freedom if each table is independently measuring the same association. If we do this, we do not apply the correction term because it overcorrects and has a bad cumulative effect.

8-6.   Summary

The chi-square statistic can be applied to tables of counts. It is frequently used to test whether two or more characteristics are independent. In this chapter we tested whether being at goal diastolic blood pressure was independent of physician. We extended the example to the practice of three physicians.

In applying the statistic to contingency tables, one first computes the expected cell count from the row and column total by assuming independence. Chi-square is computed according to the formula

$$\chi^2 = \sum \frac{(O - E)^2}{E}$$

where O is the observed count and E is the expected count in the same cell. The sum is computed over all cells. The number of degrees of freedom of the chi-square statistic is determined by fixing the marginal totals and counting how many cells can be freely filled, given the marginal values.

Just as in the use of the normal distribution to approximate the binomial, the chi-square distribution is a continuous approximation to a discrete distribution in contingency table problems. In problems with 1 degree of freedom, we can use a $\frac{1}{2}$ correction for improved accuracy in computing the P value.

The chi-square statistic with k degrees of freedom can be viewed as the sum of squares of k independent values chosen from a normal distribution of mean 0 and variance 1. It has mean k and variance $2k$.

The chi-square statistic is usually applied to situations where the counts have substantial size. It is unusual to be able to apply it to the individual patient. However, the statistic is quite useful in comparing two or more groups of patients and may be of value in assessing the results of your practice.

Problems

8-1. What does the chi-square statistic measure?

8-2. How many degrees of freedom are there in testing for independence in a 2×2 contingency table?

8-3. What is a degree of freedom in a contingency table?

8-4. For a 2×2 contingency table, an important relation holds between the value of χ^2 and the standardized difference z comparing the two sample frequencies. What is it?

8-5. Continuation. If the value of z corresponds to the one-sided 10% level, to what percent level would χ^2 correspond?

8-6. For three successive 10-week periods, the number of weeks with migraine attacks has been:

	Period		
	FIRST 10 WEEKS	SECOND 10 WEEKS	THIRD 10 WEEKS
Attacks	6	2	1
No attacks	4	8	9

Use the χ^2 test to see whether the rates may reasonably be the same during the three periods.

8-7. At City Hospital, patients often do not show up for their clinic appointments. To determine whether a telephone reminder improves show-rate, you telephoned 25 patients the day prior to their appointment. Of the 25 patients who were telephoned, 20 kept their appointments. Only 8 of 20 who were not telephoned attended the clinic.
   a. Construct a 2×2 table of observed and expected counts assuming independence between telephone reminder and show-rate.
   b. Compute the chi-square statistic and P value to test for association.

References

Hypertension Detection and Follow-up Program Cooperative Group (1979). Five-year findings of the hypertension detection and follow-up program: reduction in mortality of persons with high blood pressure, including mild hypertension. *JAMA*, **242**:2562–71.

Additional Reading

Moses, L. E.; Emerson, J.; and Hosseini, H. Analyzing data from ordered categories. In Bailar, J. and Mosteller, F., eds. (1986) *Medical Uses of Statistics* (Chapter 11). Waltham, MA: New England Journal of Medicine.

CHAPTER 9

Regression: An Overview

OBJECTIVES

Uses of regression in clinical medicine

Basic concepts of regression:
1. *slope*
2. *intercept*
3. *residual*
4. *the assumption of linearity*
5. *the relation of regression to causation*

Deviation of individual patients from the usual regression relationship

Regression toward the mean

In this chapter we examine how the distribution of one variable may depend on the value of another. Examples with clinical implications could be given under many headings:

Disease prevention. What is the relation between dietary cholesterol and serum cholesterol, and is the relation causal?

Description of normal physiology and biochemistry. What is the relation between oxygen tension and rate of respiration? What is the relation between body size and cardiac output?

Disease detection and diagnosis. Why is an abnormally high systolic blood pressure (or serum glucose) for some patients likely to be followed by a value closer to "normal"?

Treatment. How should drug dosage be adjusted for patient weight, or for some measure of kidney function? What is the relationship between percent of skin surface burned and requirements for fluid replacement?

Prognosis. How (if at all) does the probability of survival of a transplanted kidney depend on the degree of match (0 to 4 antigens) in tissue type?

Medical practice. How does the length of the average hospital stay for total cystectomy depend on the age of a patient, and what are the problems in using this average to estimate length of stay for a specific patient?

Not all of these examples are discussed again in this chapter. Satisfactory solutions to some of them have been worked out in empirical ways, but the common threads in these and countless other clinical questions suggest that a unified approach to studying the relationship between two variables may be helpful.

A Preliminary Example

Mr. W. has a serious infection and is responding to gentamicin therapy. On clinical grounds, you believe his "trough" serum gentamicin concentration is near the therapeutically desirable level of 1 μg/ml. The laboratory reports 2 μg/ml—a toxic level. To evaluate your laboratory's results, you prepare 15 samples and submit them to the laboratory; 5 containing 1 μg/ml, 5 containing 2.5 μg/ml, and 5 containing 5 μg/ml. Figure 9-1 shows the values reported by the laboratory on the y axis plotted against the actual concentration on the x axis. It is evident that a value of 1 μg/ml may be reported as 2 μg/ml (and a value of 5 μg/ml reported as 7.5 μg/ml). Your clinical estimate of Mr. W's concentration may be correct.

Figures 9-1 and 9-2 illustrate several of the ideas of regression.

1. For each x value, more than one y value can occur. We say that y has a distribution for each x value. The y values plotted in Figure 9-1 rep-

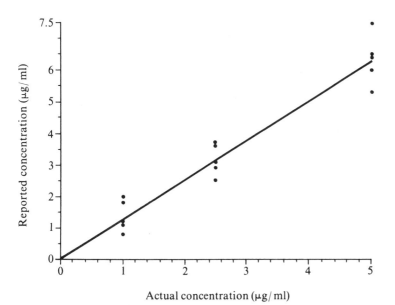

Figure 9-1. Plot of reported concentration of serum gentamicin against actual concentration for 15 prepared samples.

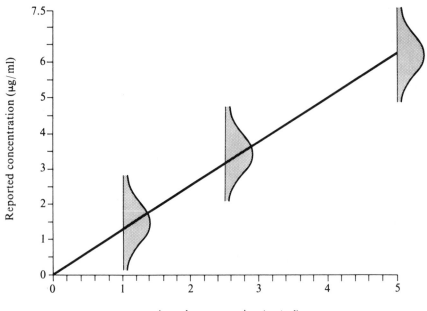

Figure 9-2. Plot of the probability distribution of reported concentration of serum gentamicin at actual concentrations of 1, 2.5, and 5 μg/ml.

resent samples of size 5 from the distributions at *x* equal to 1, 2.5, and 5. The three (hypothetical) probability distributions are shown in Figure 9-2 (laid down to avoid a third dimension).

2. The distribution of *y* at each *x* will have a mean and a variance. One or both of these may change with the value of *x*.

3. The plot of the mean value of *y* at each *x* is called the regression of *y* on *x*. When these mean values fall on a straight line, we say that *y* has a linear regression on *x*.

The line in Figure 9-1 was fitted by eye to the average value of *y* at each *x*. We will discuss the fitting of regression lines later in the chapter.

9-1.   Use of a Regression Line to Describe Relationships

Clinical Problem 9-1.   *The Effect of Chewable Isosorbide Dinitrate on Cardiac Index*

Mr. E.S. has recently begun to suffer from symptoms of congestive heart failure caused by ischemic cardiomyopathy. His cardiac index (in liters/min per square meter) is 1.68. What is likely to be the effect on cardiac index of 10-mg doses of chewable isosorbide dinitrate?

Discussion: Examining the Relationship Between Two Variables

Table 9-1 gives values of pretreatment and posttreatment cardiac index reported by Franciosa *et al.* (1978) for 11 male patients with congestive heart failure treated by 10-mg doses of chewable isosorbide dinitrate (CHIS). These data can also be plotted as shown in Figure 9-3. Two diagonal lines have been added. One (dashed) line shows points at which pretreatment and posttreatment values are equal; it rises at an angle of 45°. The other (solid) line was fitted by eye to approximate the relation between the pretreatment and posttreatment values. This figure shows two facts:

1. The relation between the two values for a given patient is approximated by the solid line, but the line does not rise very steeply and the points are scattered rather widely around it.
2. Cardiac index tends to be higher after treatment than before treatment (most of the points are above the dashed diagonal line). However, this improvement may be limited to patients whose pretreatment values were low (the three points below the 45° line are for patients whose initial values were highest).

From these data and the solid line, it looks as if treatment with 10-mg doses of CHIS could be expected to raise Mr. E.S.'s cardiac index from 1.68 by about .2 or .3, to between 1.88 and 1.98. The change might be much larger or smaller than this, and the cardiac index could even show a substantial fall (as in cases 2 and 14 of Table 9-1).

Clinical Problem 9-2. *Interpreting the Standard Cobalamin Assay*

Mr. W.S. is a 49-year-old man with a physical examination and blood tests strongly suggestive of pernicious anemia, a deficiency of cobalamin (vitamin B_{12}). However, measurements of his serum

Table 9-1. Pretreatment and posttreatment cardiac index for 11 male patients with congestive heart failure

Case number	Cardiac index (liters/min/m²)	
	PRETREATMENT	POSTTREATMENT
1	1.02	1.94
2	2.10	1.63
3	1.88	2.73
4	2.20	2.18
5	1.44	1.82
11	1.55	1.94
13	1.61	2.25
14	2.61	1.70
15	1.56	1.78
16	.99	1.52
22	1.53	1.97

Source: Reprinted by permission from J. A. Franciosa, R. C. Blank, and J. N. Cohn, *Am. J. Med.* **64**:207–13, 1978.

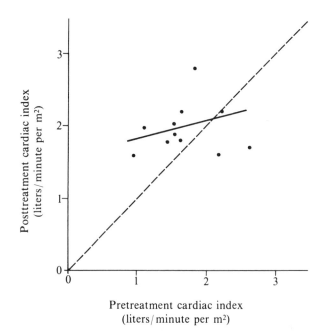

Figure 9-3.   Plot of posttreatment versus pretreatment cardiac index for 11 patients.

cobalamin using a standard cobalamin assay kit have produced re-sults at the lower end of the normal range. Is there reason to abandon the investigation of possible cobalamin deficiency?

Background: Comparing Two Different Cobalamin Assays —A Linear Relationship

Commercially available cobalamin kits are generally based on a method that measures both cobalamin and chemical analogs. Kolhouse *et al.* (1978) have reported a comparison of cobalamin levels determined by a method using an intrinsic factor (IF) which is highly specific for true cobalamin, and by a method using R protein which measures cobalamin plus analogs, as used in commercial kits. Data for 21 cobalamin-deficient patients and 74 normal subjects were reported.

Figure 9-4 shows a comparison of the two assays for the 74 normal sub-jects. The scales on the two axes are the same. All points are above the 45° (dashed) line, indicating that values from the R assay were always higher than those from the IF assay. Although the points show some degree of scatter, they seem to cluster fairly well around a straight line. Such a line could be drawn in by hand or (as here) calculated by a special method known as least squares from the numeric values of the original data points. In this set of data, the lines are not likely to differ by much. We discuss the least-squares method later in this chapter and in Chapter 9A.

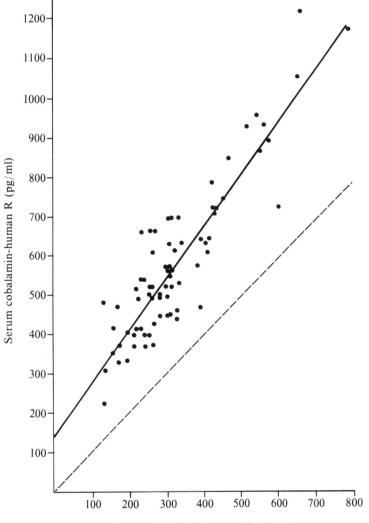

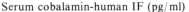

Figure 9-4. Plot of serum cobalamin measured by R protein (human R) against cobalamin measured by intrinsic factor (human IF) for 74 normal subjects. The dashed line connects points with equal values. The solid line is the least-squares regression line. [Reprinted by permission from Kolhouse, J. F.; Kondo, H.; Allen, N. C.; Podell, E.; and Allen, R. H. (1978). *N. Engl. J. Med.*, **299**:785–90.]

An Equation for a Straight Line

Recall that an equation for a straight line is

$$y = a + bx$$

where y is plotted on the vertical axis
   x is plotted on the horizontal axis
   b is the slope of the plotted line
   a is the point where the plotted line intercepts the y
      axis (the value of y when $x = 0$)

The Estimated Regression Line

It will now be convenient to distinguish between values of y actually observed and y values, designated $\hat{y}$, that are estimated by a regression line. The least-squares regression line shown in Figure 9-4 was calculated by the investigators as

$$\hat{y} = 147 + 1.34x$$

where $\hat{y}$ is the estimated result of the R protein assay and x the value of the IF assay (see the Appendix at the end of this chapter for computational details). We can use the estimated value of 147 to estimate the average concentration of analogs when cobalamin is completely absent ($x = 0$). The fact that the estimated slope of the line, $\hat{b}$, equals 1.34 (greater than 0) shows that the total (cobalamin plus analogs) tends to rise with increasing cobalamin levels. When this is true, we say that the two values are **positively correlated.** We can go a step further. The fact that $\hat{b}$ exceeds 1.00 means that the total concentration not only rises with increasing levels of true cobalamin, it actually rises **more** than true cobalamin. This implies that the analogs are also rising.

Figure 9-5 plots similar measurements for the 21 cobalamin-deficient subjects. The authors presented their data in a table, and Figure 9-5 has been plotted from these data. The least-squares linear regression of y on x (symbols defined as above for normal subjects) can be calculated as

$$\hat{y} = 154 + 1.41x$$

(In this calculation, six values reported as less than 10 have been treated as being equal to 10. They could just as well have been assumed to equal 0 or any value between 0 and 10, with only small effects on the computed line.)

This estimated regression line for cobalamin-deficient patients is very close to that for normal persons, suggesting that the relation between the two test values does not depend on whether the individual was deficient. (The test values themselves, of course, *do* depend on whether the individual was deficient.) Thus we can say that, on the average, a regression relation

$$\hat{y} = 150 + 1.4x$$

will be approximately correct for both normal and cobalamin-deficient persons.

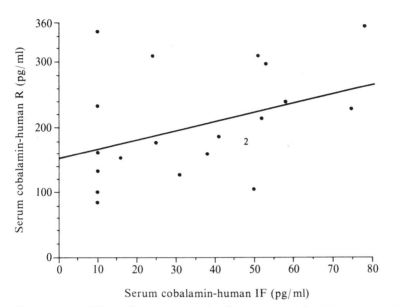

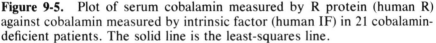

Serum cobalamin-human IF (pg/ml)

Figure 9-5. Plot of serum cobalamin measured by R protein (human R) against cobalamin measured by intrinsic factor (human IF) in 21 cobalamin-deficient patients. The solid line is the least-squares line.

Implications for Clinical Problem 9-2

To return to the clinical question of whether Mr. W.S. has pernicious anemia, the data clearly show a relation between the results of the R and IF assays, whether one looks at a figure (Figures 9-4 and 9-5) or a regression equation. However, the relation is not as close as one might like. Eight of the R values for cobalamin-deficient patients exceed the smallest R value for normal subjects, 220. Thus consistently low-normal results from cobalamin assays that use R protein do not rule out cobalamin deficiency. In contrast, none of the IF values for cobalamin-deficient patients exceed the smallest value for normal subjects, about 120. An IF assay seems indicated for Mr. W.S.

9-2. Regression to the Mean

Clinical Problem 9-3. *Should Your Patient with Large Bowel Diverticula Receive Wheat Bran?*

Mr. G.G. has multiple large bowel diverticula confirmed by radiographs and fiberoptic colonoscopy. He has troublesome constipation, with prolonged oral-anal transit time. Medical colleagues have told you that wheat bran is good for patients with diverticula because it reduces transit time; other colleagues have told you that it has an opposite effect, increasing transit time in patients with below-average time. Should your patient receive wheat bran?

Discussion: Experience with 39 Patients

An article by Brodribb and Humphreys (1976) reported the effects of daily doses of wheat bran on 39 patients with diverticular disease. After 6 months of treatment, oral-anal transit times were decreased in patients whose initial times were greater than 60 hours and increased in patients whose initial times were less than 36 hours. The data are shown graphically in Figure 9-6.

We find that the lines in the first figure slope upward, whereas those in the last figure tend to slope downward. The mean transit time in each group has moved toward the center of the overall distribution. At the same time, however, values originally near the center have tended to fan out into the extremes. Overall, pretreatment transit times (the collection of values shown on the left side of each graph) seem to have about the same distribution as posttreatment values (those on the right sides of the graphs). Patients as a group have derived little or no benefit from treatment, and treatment has not prevented the occurrence of wide deviations in the group originally classified as normal.

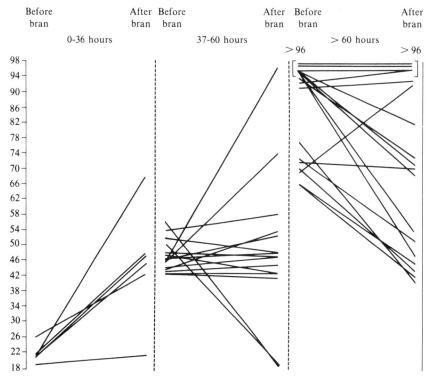

Figure 9-6. Oral-anal transit time in 39 patients before and after bran. Patients were divided into three groups according to initial transit times: 0–36 hours, 37–60 hours, and >60 hours. [Reprinted by permission from Brodribb, A. J. M., and Humphreys, D. M. (1976). *Br. Med. J.*, **1:**424–30.]

When Variables Have Substantial Day-to-Day Variation, the "Worst" Patients Will Improve

The data suggest that there may be considerable day-to-day variation in transit time (center panel) that is unrelated to treatment, so that patients classified according to low, medium, or high transit time do not remain in these categories. When there is considerable day-to-day variation, many of the patients classified in the high group are there because an unusually large value occurred by chance on that day. Subsequent values are more representative of the usual situation in these patients. Similar remarks apply to those with low values. However, this shift (or "regression") toward the mean was incomplete; patients whose initial values were markedly high tended to have high final values, although the latter were closer to the center than the former. A similar remark applies to patients whose initial transit time was markedly low.

This effect is discussed further in Section 9-3. It is already clear, however, that there may be problems with the "obvious" conclusion that bran treatment really had a substantial effect on bowel transit time. It is possible that treatment could be helpful, but more data will be necessary to see whether the apparent improvement reported by Brodribb and Humphreys represented a regression effect, an artifact of the grouping of patients.

9-3.   Uses of Regression

The regressions in Clinical Problems 9-1 and 9-3 were used to estimate the effects of a potential treatment, while that in Clinical Problem 9-2 was used to examine the relation between two diagnostic tests. Regression is, in itself, a way to describe just one aspect of the relation between two variables, although the regression may be used in any of a variety of ways. Some of the most common uses of regression in clinical medicine are:

1. *Description.* To describe, in compact form, the general level of one variable that is associated with each level of the other variable. Expressing cardiac output as cardiac index (liters/min per square meter) rests on a linear relation between output and body surface area.
2. *Adjustment.* To provide a means for "adjusting" two sets of y variables when the associated x variables tend to have different values. For example, in comparing the blood pressures of two groups of study participants, it may be important to adjust for body weight.
3. *Forecasting or prediction.* To aid in forecasting or predicting a level of some y variable for a new value of x. This does not require that the y value actually be obtained. A common example is the estimation of drug doses on the basis of the body weight. For many drugs it has been found that drug concentration is proportional to dose per unit weight, so that we can with some confidence predict drug concentration from dose adjusted for weight.

4. *Interpretation.* To aid in understanding and interpreting the mechanisms by which one variable is related to another. Although a regression relationship is never, in itself, sufficient evidence of causation, careful study of regressions may suggest or confirm hypotheses. One example arose from Figures 9-4 and 9-5, where the points cluster around a line with estimated slope greater than 1.0, suggesting that the cobalamin fraction and the analog fraction are positively correlated.

5. *Outlier detection.* To detect abnormal values, or outliers, that may merit detailed individual study. Thus whereas an R assay value of 350 and an IF value of 550 would not be unusual taken alone, their occurrence in the same patient would suggest several interesting possibilities, including serious laboratory error, presence of a previously unsuspected cobalamin antagonist, or a greater degree of variability in measurement than is suggested by Figures 9-4 and 9-5.

Perhaps point 4 above needs emphasis. Regression is, in itself, no more than a way to describe the relation between two variables. No matter how good the description, inferences of causation are not justified unless there is adequate supporting evidence. For example, there is a moderately close regression of height (*y* variable) on weight (*x* variable) for adults, but it would be nonsense to say that increasing weight "causes" increasing height. High systolic blood pressure is correlated with mortality from several cardiovascular diseases, but this alone is not sufficient evidence of causation. (Some diseases may indeed be causally related to hypertension, whereas for others, hypertension may serve only as a marker of atherosclerosis, the real causative element.) A third example, somewhat more subtle, is the relation between early detection of cancer or other diseases and the length of subsequent survival. A positive correlation between these variables, no matter how strong, does not, in itself, show that early detection causes or leads to prolongation of life. The length of survival from detection to death is fixed by its endpoints, detection and death. A longer interval is of little value to patients if it is attained only by moving detection forward in time; rather, we need to determine from other kinds of data whether the second endpoint, death, has been delayed. This effect of early detection has been called the lead-time bias.

Regression Effects

The data in Clinical Problem 9-3 illustrate a problem that is so common and so important that it is sometimes called "the" regression effect; it also gives regression its name. When the relation between two variables is affected by random variation, extreme values of one variable will rarely correspond to extreme values of the other, even when they are highly correlated. Thus very tall men tend to have sons shorter than themselves at adulthood, although taller than the average, whereas very short men have sons that are, on the average, taller than themselves, although shorter than the population mean. This phenomenon, and its analogs in other kinds of data, was called "regression toward the mean" because the original investigator, Francis Galton, was studying heredity.

For another example, consider the common observation that people with minor illnesses tend to get better, whereas perfectly healthy people tend to develop problems. Thus over a period of time we would expect some general measure of health, a "health index," to improve in people who have self-limiting diseases at the start, and to deteriorate in people who started healthy.

As a third example, consider the course of a chronic but highly variable disease such as asthma, which is characterized by remissions and exacerbations. A patient who has been getting worse in the recent past is likely to be getting better in the near future. If he was on treatment A, the patient and his physician may become discouraged by the downhill course and switch to treatment B. It would be a mistake to attribute later improvement to the change in treatment unless one had data from an experimental study designed to circumvent the pernicious and subtle effects of regression to the mean.

A final example comes from the Hypertension Detection and Follow-up Program (HDFP) (1977). One group of participants had diastolic blood pressure readings above 115 mm Hg at a screening in the HDFP clinic. The mean diastolic blood pressure in this group was 122.1 mm Hg. At the next clinic visit, the mean diastolic blood pressure was 114.7 mm Hg, even though intervention had not begun. Although this change could have resulted in part from increased familiarity of patients with the clinic environment, or increased use of previously prescribed medication, it is likely that most of the change is due to regression to the mean. When patients are selected because of extreme values of characteristics that change over time, subsequent values of those same variables tend to be closer to the population mean.

Warning: Regression to the mean is the source of many errors of interpretation in medical research. When individuals are selected according to the value of some time-varying characteristic, subsequent trends in the group of selected individuals may be the effects of selection, even though these trends are sometimes erroneously attributed to intervention.

Linearity

We often assume that the regression of y on x is a straight line; that is, the average value of the variable y increases or decreases by the same fixed amount for each change of one unit in x. In some instances, the assumption of linearity is justified by our prior understanding of the nature of the two variables, confirmed by preliminary study of the data. In other instances, we may not have specific reasons for believing that a linear relation holds (we may even have reasons to the contrary) but still find that a linear regression provides an adequate summary of the data within the range of x values of greatest interest. Thus we distinguish between **assuming** a linear relation and **using** a straight line. Either or both may be appropriate in the analysis of a specific set of data, although they should be carefully distinguished and separately justified.

9-4.   Using Transformations to Obtain Linear Relationships

Sometimes the relationship between two variables is not linear. Although methods for estimating nonlinear relationships have been developed, we often prefer, for the sake of simplicity, to use linear regression methods. Understanding of the underlying relationship between two variables, or evidence directly from the data, may suggest a way of reexpressing one or both of the two variables so that the relationship involving the new variable(s) is linear. Such reexpressions are called transformations and are a useful adjunct to linear regression methodology.

Clinical Problem 9-4.   *What Is the Patient's Glomerular Filtration Rate?*

Mr. C.L. has had diabetes for several years. He now has proteinuria. His plasma creatinine concentration is 2 mg%. What can we say about his glomerular filtration rate?

Background: The Relationship Between Plasma Creatinine and Glomerular Filtration Rate

Brochner-Mortensen and his colleagues (1977) studied the relationship between plasma creatinine (Cr) and glomerular filtration rate (GFR) in 200 females and 180 males. Data for 31 randomly selected men are given in Table 9-2 and plotted in Figure 9-7. Values of Cr, given by the authors in micromoles per liter, are expressed here in milligrams percent. We chose a subsample of men to allow easier and more detailed discussion of this example. The size of the subsample, 31, was randomly determined by the subsampling procedure.

It is apparent (from Table 9-2 and Figure 9-7) that this relationship is not linear. In fact, physiological considerations suggest that if GFR falls by 50%, Cr should double. Thus GFR should be linearly related to the reciprocal of Cr, that is, 1 divided by Cr. Figure 9-8 shows GFR plotted against the reciprocal of Cr for these 31 subjects. This relationship does appear to be linear.

Using these transformed values, the least-squares regression line can be estimated by standard methods. The least-squares line for the points in Figure 9-8 is

$$GFR = -2.44 + 87.9 \left(\frac{1}{Cr} \right)$$

Table 9-2. Glomerular filtration rate (GFR), plasma creatinine (Cr), and the reciprocal of plasma creatinine (1/Cr) for 31 male subjects selected at random from the study of Brochner-Mortensen *et al.* (1977)

GFR	Cr	1/Cr	GFR	Cr	1/Cr
90	.85	1.18	35	1.75	.57
45	.99	1.01	38	1.83	.55
103	1.13	.88	47	1.98	.51
100	1.13	.88	45	2.03	.49
93	1.13	.88	40	2.09	.48
90	1.13	.88	27	2.77	.36
70	1.13	.88	37	2.96	.34
77	1.27	.79	25	3.11	.32
47	1.41	.71	15	3.96	.25
45	1.47	.68	15	4.69	.21
60	1.47	.68	20	4.80	.21
53	1.56	.64	10	5.93	.17
35	1.56	.59	5	5.93	.17
63	1.70	.57	5	5.93	.17
55	1.75	.57	10	7.97	.12
			12	11.02	.09

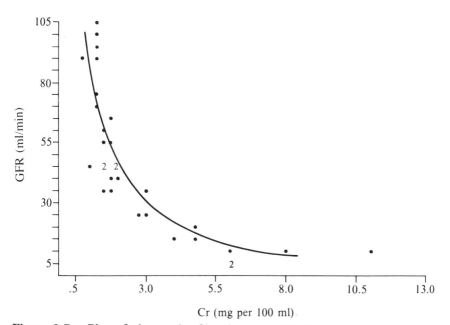

Figure 9-7. Plot of glomerular filtration rate (GFR) against plasma creatinine (Cr) for 31 men selected at random from the study of Brochner-Mortensen *et al.* (1977). The curved line is obtained from the least-squares regression line shown in Figure 9-8. The symbol 2 denotes two observations plotted at the same location.

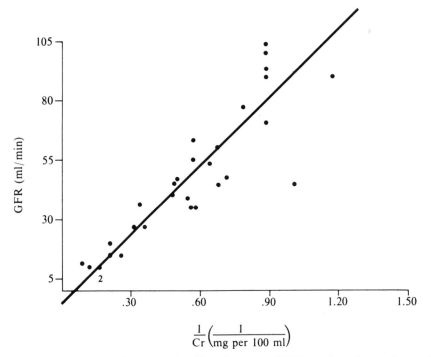

Figure 9-8. Plot of glomerular filtration rate (GFR) against the reciprocal of plasma creatinine (1/Cr) for 31 men selected at random from the study of Brochner-Mortensen *et al.* (1977). The solid line is the least-squares regression line.

and the least-squares line is shown in Figure 9-8. This nonlinear relationship between GFR and Cr can be plotted as a curve in the original units as shown in Figure 9-7.

Solution: Predicting the Glomerular Filtration Rate from Plasma Creatinine

Recall that Mr. C.L. had a Cr of 2 mg%. The reciprocal of this value is .50, so the estimated GFR for Mr. C.L. is

$$GFR = -2.44 + 87.9(.50)$$

$$= 41.51$$

For a second subject with Cr equal to 5.5, the reciprocal is

$$\frac{1}{Cr} = \frac{1}{5.5} = .182$$

and the estimated GFR is

$$GFR = -2.44 + 87.9(.182)$$

$$= 13.56$$

This point,

$$\frac{1}{Cr} = .182, \qquad GFR = 13.56$$

lies on the straight line in Figure 9-7, while the point

$$Cr = 5.5, \qquad GFR = 13.56$$

lies on the curved line in Figure 9-6. The regression equation

$$GFR = -2.44 + 87.9\left(\frac{1}{Cr}\right)$$

tells us that GFR is about 90 divided by plasma creatinine. (the exact estimate is 87.9 and the correction factor of -2.44 is close to 0 compared with the contribution from $87.9x$.)

9-5. Summary

Linear regression is widely used in medicine for describing the relationship between two variables. The linear relationship between two variables x and y is represented by the equation

$$y = a + bx$$

The intercept, a, is the value of y when x equals 0, and the slope, b, is the change in y for a 1-unit change in x. The usual method for estimating a and b is called least squares, and is described in the Appendix at the end of this chapter.

Regression lines can be used to summarize data, to adjust values of y for differences in x, to predict y at a new x value, to examine and explain relationships, and to detect abnormal values.

An important related phenomenon is called regression to the mean or "the regression effect." It is the tendency of individuals with extreme values of a variable to have values nearer the center of the distribution at subsequent measurement. Regression to the mean occurs when the variable displays significant variability between occasions. For example, regression to the mean is of trivial importance for height of adults but great importance for blood pressure measurement. Failure to recognize regression to the mean can easily lead to misinterpretation of changes in patient status.

When the relationship between two variables is not linear, it may be useful to consider transformations of one or both variables to achieve linearity. Regression analyses can be performed on the transformed scale and relationships reexpressed in terms of the original variables.

Problems

Problems 9-1 through 9-5 are based on the following clinical case: Mr. R.K. complains of migraine headache. He has about 10 attacks a month. In order to determine whether Mr. R.K. will be helped by propranolol and to find a good dosage for him, you conduct the following trial (after obtaining his approval).

Placebo and several different doses of propranolol are each prescribed for a 6-week period. Mr. R.K. records the number of headaches he has during each interval. You administer the drugs in a single blind manner so that you, the physician, know what is given, but R.K., the patient, does not. You start with placebo and slowly increase the dose in subsequent 6-week periods, then at the end of the trial give placebo and one of the lower doses again. Following are the actual results of such a trial:

Period	Propranolol dose (given TID)	Migraine rate (per 30 days)
1	Placebo	9.5
2	10 mg	7.5
3	40 mg	5.0
4	80 mg	4.5
5	120 mg	4.5
6	160 mg	3.5
7	Placebo	13.1
8	40 mg	4.7

9-1. Plot migraine rate (on the y axis) versus period. Fit a line by eye to the graph. Is there a relation between rate and period?

9-2. Plot migraine rate versus dose. Fit a line by eye to the graph. Is there a relation between migraine rate and dose? Is the relation linear?

9-3. Pharmacological principles suggest that response should be linearly related to the logarithm of the dose. Plot the migraine rate against the log of the dose. (Note that the log of 0 is not defined. In this instance it is convenient to assume that placebo contained some small dose, say 1 mg.) Fit a line by eye. Does the relation appear linear?

9-4. Propranolol comes in a 100-mg tablet. What migraine rate would you expect if you prescribed this to R.K.?

9-5. What dose would you prescribe to get maximum effect? What dose would you prescribe to obtain a substantial effect at a moderate dose?

9-6. The Hypertension Detection and Follow-up Program (1977) screened persons aged 30 to 69 for hypertension. About 10% of whites and 25% of blacks had diastolic pressures of 95 mm Hg or greater. Their average pressure was 103 mm Hg. These subjects were screened a second time. At the second screening, about one-third of these hypertensive subjects had diastolic pressures less than 90 mm Hg. The overall average diastolic pressure for those rescreened was 96 mm Hg. What might account for the lower pressures observed on the second screen?

9-7. Paradise et al. (1978) investigated the history of recurrent sore throat as an indication for tonsillectomy. They followed 65 children who gave an impressive history of recurrent sore throat. "Only 11 children had throat infections with . . . frequency conforming to their history." They concluded that an undocumented history of frequent sore throat was not a good predictor of future frequent sore throats. Do you think that a documented history would be a good predictor? Why?

References

Brochner-Mortensen, J.; Jensen, S.; and Rodbro, P. (1977). Assessment of renal function from plasma creatinine in adult patients. *Scand. J. Urol. Nephrol.*, **11**:263–70.

Brodribb, A. J. M., and Humphreys, D. M. (1976). Diverticular disease: three studies. *Bri. Med. J.*, **1**:424–30.

Franciosa, J. A.; Blank, R. C.; and Cohn, J. N. (1978). Nitrate effects on cardiac output and left ventricular outflow resistance in chronic congestive heart failure. *Am. J. Med.*, **64**:207–13.

Hypertension Detection and Follow-up Program Cooperative Group (1977). The Hypertension Detection and Follow-up Program: a progress report. *Circ. Res.*, **40**(Suppl. 1):106–9.

Kolhouse, J. F.; Kondo, H.; Allen, N. C.; Podell, E.; and Allen, R. H. (1978). Cobalamin analogues are present in human plasma and can mask cobalamin deficiency because current radioisotope dilution assays are not specific for true cobalamin. *N. Engl. J. Med.*, **299**:785–90.

Paradise, J. L.; Bluestone, C. D.; Bachman, R. Z.; et al. (1978). History of recurrent sore throat as an indication for tonsillectomy. *N. Engl. J. Med.*, **298**:409–13.

Additional Reading

Godfrey, K. Simple linear regression in medical research. In Bailar, J., and Mosteller, F., eds. (1986), *Medical Uses of Statistics* (Chapter 9). Waltham, MA: New England Journal of Medicine.

Appendix: Calculating Regression Coefficients

The usual method for estimating the intercept and slope of a linear regression is called least squares. In Figure 9-9 we display a plot of the following 3 data points:

x value	y value
3	5
4	8
5	7

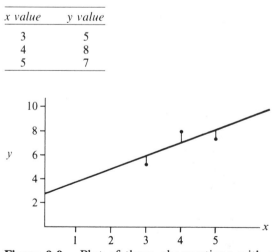

Figure 9-9. Plot of three observations with a straight line fitted by least squares. The value of each residual is the length of the line from the point to the least-squares line.

and a straight line fitted to these points by least squares.

The vertical distances from the 3 points to the line are called residuals. The least-squares line is defined by a single property: It minimizes the sum of the squares of these residuals. We denote the least-squares estimates of the intercept, a, and the slope, b, by $\hat{a}$ and $\hat{b}$.

Although we do not show the derivation here, the expressions for the least-squares estimates of a and b are easy to describe. If xy represents the product of one set of x and y values, and x^2 represents the square of the x value, the formulas for $\hat{a}$ and $\hat{b}$ are

$$\hat{b} = \frac{\text{sum of } xy - [(\text{sum of } x)(\text{sum of } y)]/n}{\text{sum of } x^2 - (\text{sum of } x)^2/n}$$

$$\hat{a} = \frac{\text{sum of } y}{n} - \hat{b}\frac{\text{sum of } x}{n}$$

where n is the number of observations. We illustrate with the same example. Note how we set up a table that facilitates calculation of xy and x^2 and the necessary sums.

EXAMPLE.   If $n = 3$ and the data are as follows:

	x	y	xy	x^2
First pair	3	5	15	9
Second pair	4	8	32	16
Third pair	5	7	35	25
Sums	12	20	82	50

then

$$\hat{b} = \frac{82 - [(12)(20)]/3}{50 - (12)^2/3} = \frac{2}{2} = 1$$

Similarly,

$$\hat{a} = \frac{20}{3} - 1 \times \frac{12}{3} = 6.67 - 4 = 2.67$$

The fitted line is

$$\hat{y} = 2.67 + x$$

in this example.

LINES THROUGH THE ORIGIN.   Often in biomedical problems it is reasonable to have a line through the origin—when x equals 0, y also equals 0. The fitted line has the form

$$y = bx$$

The formula for $\hat{b}$ is then

$$\hat{b} = \frac{\text{sum of } xy}{\text{sum of } x^2}$$

In our example we find that

$$\hat{b} = \frac{82}{50} = 1.64$$

and the estimated regression line through the origin is

$$\hat{y} = 1.64x$$

We have used an example with 3 observations to simplify computing. When the data set contains more points, a table such as the one in the preceding example is still constructed, having four columns but as many rows as the number of observations. Calculations follow the same steps outlined in the example.

More About Regression: A Closer Look at the Regression of Glomerular Filtration Rate on Plasma Creatinine

> **OBJECTIVES**
>
> *Computer programs for regression*
> *Interpreting computer output*
> *Regression coefficients*
>   *slope*
>   *intercept*
> *Testing for a significant regression*
> *Analysis of variance*
>
> *Predicted and residual values*
> *The correlation coefficient*
> *Examining residuals*
>   *curvature*
>   *nonnormality*
>   *nonconstant variance*
>   *outliers*

Because preprogrammed hand calculators and computer packages are now widely available, regression analysis no longer requires a detailed understanding of the underlying calculations. This chapter describes how to read and interpret the output from a computerized regression analysis, using as our example the data on plasma creatinine levels and glomerular filtration rates described in Chapter 9. We emphasize graphical and exploratory techniques, including plots of the regression lines and residuals, and use some of the methods for summarizing data described in Chapter 5A.

The output presented here was generated by the MINITAB system (Ryan *et al.*, 1976) and modified slightly for clarity. Output from other popular computer packages is similar in style and content. In a textbook emphasizing regression analysis, Kleinbaum and Kupper (1978) use output from the SPSS system in some examples.

9A-1. Reviewing the Printout from a Standard Regression Program

Figure 9-7 plots plasma creatinine levels (Cr) and glomerular filtration rates (GFR) for 31 men chosen at random from the study of Brochner-Mortenson *et al.* The Cr and GFR values are listed in Table 9-2. Although the relationship is distinctly nonlinear, one could fit a straight line to these observations

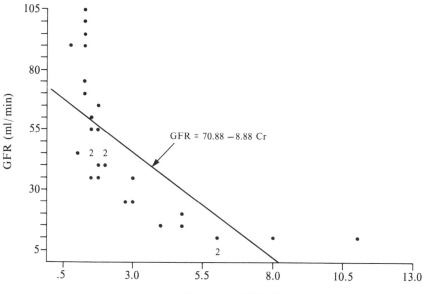

Figure 9A-1. Plot of glomerular filtration rate (GFR) against plasma creatinine (Cr) for 31 men, with the least-squares regression line. The symbol 2 denotes two observations plotted at the same location.

using the least-squares method. We performed this analysis and present it here to illustrate some of the output from a standard regression program.

Figure 9A-1 shows these 31 points along with the least-squares line for the regression of GFR on Cr. Table 9A-1 shows the corresponding tabular output from the MINITAB system.

The Least-Squares Line

The first entry in Table 9A-1 specifies the equation for the least-squares line of y (GFR) on x (Cr):

$$\hat{y} = 70.88 - 8.88x$$

Often, packaged programs report the computed values with more precision than we find useful for describing the relationship. In reading the output, it can be useful to round some of these values to highlight the magnitude of regression coefficients and other quantities. For example, we might express the regression equation as

$$\hat{y} = 71 - 8.9x$$

or even

$$\hat{y} = 70 - 9x$$

without sacrificing the important information. In this instance we conclude

Table 9A-1. Output of least-squares regression of y (GFR) on x (Cr) using the MINITAB system

the regression equation is
$$\hat{y} = 70.88 - 8.88x$$

	coefficient	standard deviation of coefficient (SD)	t ratio = coefficient/SD
intercept	70.88	5.72	12.39
x	−8.88	1.55	−5.73

analysis of variance due to	df	ss	ms = ss/df
regression	1	13,460.2	13,460.2
residual	29	11,869.5	409.3
total	30	25,329.7	

F test for regression = 32.89 with 1 and 29 df
r-squared = 53.1%

that the least-squares line has an intercept approximately equal to 70 and a reduction of 9 ml/min in GFR for every increase of 1 mg% of Cr.

Below the regression equation in Table 9A-1, we find a panel giving additional information about the coefficients in the regression equation. Notice that the values of 70.88 and −8.88 are repeated in the column labeled "coefficient."

The first coefficient is the intercept and the second, labeled the coefficient of x, is the slope. The next column gives the estimated standard deviations of these coefficients. To test whether the population value of either coefficient is equal to 0, we divide the estimate by its standard deviation to obtain the t ratio, given in the last column. When the true coefficient is 0, and other assumptions described later also hold, this ratio has a t distribution with $n - 2$ degrees of freedom, in this case 29. Large positive or negative values of the t ratio provide evidence against the hypothesis that the true value is 0.

From the t tables given in Chapter 5 (or Table III-2) we can establish that both −5.73 and 12.39 are very extreme values, occurring with probability less than .001. Thus, using the hypothesis testing procedure described in Chapter 7, we reject both the hypotheses of 0 intercept and of 0 slope ($P < .001$). That is, the linear regression of GFR on Cr has negative slope and does not go through the origin. We emphasize the dependence of this analysis on assumptions to be described in Section 9A-2.

The Analysis of Variance Table

The next entry in Table 9A-1 is the analysis of variance table. The last line of the analysis of variance table gives the total sum of squares (denoted ss) of the GFR values, equal to 25,329.7 with $n - 1$ degrees of freedom, here 30. If y_i is the GFR for the ith man, where i is a value between 1 and 31, and $\bar{y}$ is the average GFR for all 31 men, the total sum of squares is computed from the expression

$$\sum_{i=1}^{31} (y_i - \bar{y})^2$$

Calculations like this one are discussed in Chapter 5. Once again, 25,329.7 provides too much precision for most purposes. It may be sufficient to say that the total sum of squares is about 25,000.

We use the next two examples to explain the decomposition of the total sum of squares into the regression sum of squares and the residual sum of squares. To do so, we first describe the calculation of predicted GFR for each man, based on his Cr value and the least-squares line.

EXAMPLE 1.   One man has a Cr of 1.27. Find his predicted GFR, $\hat{y}$.

SOLUTION.   We compute the predicted value by substituting 1.27 into the regression equation. Thus

$$\hat{y} = 70.88 - 8.88 \times 1.27$$
$$= 59.6$$

The **regression sum of squares** appearing in the analysis of variance table is the sum of squares of deviations of these predicted values from $\bar{y}$, and can be expressed as

$$\sum_{i=1}^{31} (\hat{y}_i - \bar{y})^2$$

In this example the regression sum of squares is 13,460.2, or about 13,000.

The difference between the observed GFR, y_i, and the predicted GFR, $\hat{y}_i$, is called the **residual**, that which remains after subtracting the predicted value from the observed value.

EXAMPLE 2.   The man whose Cr is 1.27 has a GFR of 77. What is his residual?

SOLUTION.   Since $\hat{y} = 59.6$, the residual is

$$y - \hat{y} = 77 - 59.6$$
$$= 17.4$$

Each observation has an associated residual. As the name suggests, the **residual sum of squares** appearing in Table 9A-1 is computed by squaring these 31 residuals and summing them. In this example,

$$\sum_{i=1}^{31} (y_i - \hat{y}_i)^2 = 11,869.5$$

about 12,000. Although we do not prove it here:

> The regression and residual sums of squares must sum exactly to the total sum of squares.

The analysis of variance table also contains mean squares (ms), obtained by dividing each sum of squares by its degrees of freedom. The regression mean square, 13,460.2, measures the contribution of the regression of GFR on Cr to the total sum of squares, while the residual mean square, 403.08, estimates the variability of GFR values around the regression line. The ratio of these two mean squares is called an F statistic. It appears in Table 9A-1 as 32.89 with 1 and 29 degrees of freedom. This value of 32.89 is equal to $(5.73)^2$, where 5.73 is the t statistic for testing that the true slope equals 0. Thus we can also use the F statistic to test the hypothesis of 0 slope. Critical values of the F distribution for significance levels of .05 and .01 are given in Table III-6. For a significance level of .01, the critical value for 1 and 29 degrees of freedom is 7.60. Since 32.89 is greater than 7.60, we reject the hypothesis of 0 slope ($P < .01$). Although the F statistic is used interchangeably with the t ratio for testing that the slope equals zero in regression analysis with one independent or x variable, it plays a more important role in multiple regression analyses that use several independent variables in the regression model.

The last entry in Table 9A-1 is labeled r-squared (r^2). This number expresses the percentage of the total variance explained by regression. In this instance,

$$r\text{-squared} = \frac{13,460.2}{25,329.7} \times 100\%$$
$$= 53.1\%$$

Using our rounded values, we find that

$$r\text{-squared} \simeq \frac{13,000}{25,000} \times 100\%$$
$$\simeq 52\%$$

When we compute the square root of .53 and give the result the same sign as the slope, we obtain $-.73$. This value, denoted r, is called the correlation coefficient and is often reported as a measure of the strength of association between y and x.

We have no universal standard for evaluating the adequacy of a regression analysis by the value of r^2. Typical values vary from a few percent to almost 100% depending on the setting. However, we can examine the appropriateness of the assumption of linearity. The next section describes some useful techniques for doing so.

9A-2. Examining Residuals

Residual analysis can help us detect four features that affect the regression analysis.

1. *Curvature*. When the regression of y on x is not linear, we would like to model that nonlinearity in the analysis, either by fitting a nonlinear regression equation or by making a transformation. Curvature is sometimes easier to detect by plotting the residuals against x than by plotting the original data.

2. *Nonnormality*. Some elements of a standard regression analysis rest on the assumption that y has a normal distribution at each value of x. The distribution of the t ratio described in Section 9A-1 depends on this assumption, as do the confidence intervals for the regression line and for new y values which are often reported with regression analyses. Analysis of the residuals helps determine the reasonableness of the assumption of normality.

3. *Constant variance*. The t ratios and confidence intervals depend also on the assumption that the distribution of y has the same variance for every value of x. Residual analysis helps us to evaluate this assumption.

4. *Outliers*. Because least-squares regression analysis minimizes the sum of squares of residuals, a few wild values can dramatically affect the regression analysis. Large residuals flag these unusual points and give us the opportunity to validate the data and examine the records for unusual cases. As described in Chapter 5B, it is often useful to rerun a regression analysis omitting valid but atypical observations.

In this section we illustrate all these uses of residuals as we continue to examine the data on GFR and Cr. Figure 9A-2 shows the plot of 31 residuals from the regression of GFR on Cr. Each residual is plotted against the Cr value for that individual.

The two largest Cr values yield large positive residuals, the residuals for moderate Cr values are uniformly negative, and those for small Cr values scatter around the line in both directions. The large number of small Cr values is very noticeable. These points tie down one end of the regression line. We can see that a line with some curvature would improve the fit for the larger Cr values.

The following is a stem-and-leaf diagram (as defined in Chapter 5A) for the 31 residuals from the regression of GFR on Cr:

```
 4 | 2
 3 | 299
 2 | 79
 1 | 07
 0 | 279
-0 | 0468888
-1 | 1233347789
-2 | 012
```

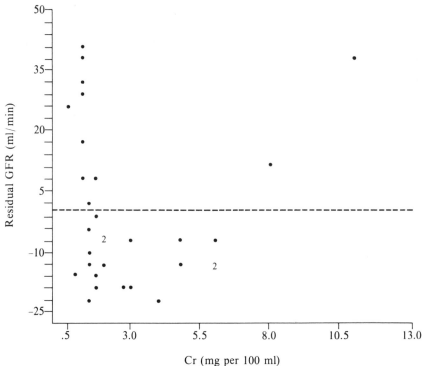

Figure 9A-2. Plot of residual GFR against Cr for 31 men.

For instance, the residual 17.4, computed in Example 2, is expressed as 17 and appears as a 7 in the row beginning with 1. In Chapter 5A we defined the fourths and the interquartile range, *I*, of a set of observations. We also defined outliers as values more than 2*I* above the upper fourth or more than 2*I* below the lower fourth. For this set of 31 points, the fourths, obtained by counting in a depth of 8 from above and below, are 10 and −14, and the interquartile range is

$$10 - (-14) = 24$$

Consequently, the interval extending from the lower fourth −2*I* to the upper fourth +2*I* is

$$-14 - 48 \quad \text{to} \quad 10 + 48$$

or

$$-62 \quad \text{to} \quad 58$$

All the residuals fall inside the interval.

A more dramatic feature of this stem-and-leaf diagram is its asymmetry, with negative values concentrated just below the median and positive residuals straggling above the median. The normality assumption would appear to be suspect, particularly because of the long-tailed distribution of positive residuals. On the whole, the residual plots suggest shortcomings in the linear regression analysis. Fortunately, we saw in Chapter 9 that we can do better.

9A-3.   Transforming Plasma Creatinine to Improve Linearity

The Reciprocal of Creatinine

We suggested in Chapter 9 that physiologic considerations imply a linear relationship between GFR and the reciprocal of Cr, defined as 1 divided by Cr. That relationship was shown in Figure 9-8, and Table 9A-2 gives the regression analysis for the same data.

The least-squares line is

$$\hat{y} = -2.44 + 87.9z$$

where z represents the reciprocal of Cr. From the coefficient list in Table 9A-2, we note that the t ratio for the estimated intercept is $-.47$, a value not at all extreme for the t distribution with 29 degrees of freedom. This suggests that the true intercept is close to 0. We also note that r-squared has improved to 79%, compared to 53% in the regression of GFR on Cr. The unexplained variation has been reduced from 12,000 to 5000, indicating that the regression of GFR on the reciprocal of Cr gives a substantially better fit to the observations.

Comparing Residual Plots

Figure 9A-3 shows the residual plot for the same 31 men, plotted against the reciprocal of Cr. The residuals now scatter vertically around 0 over the entire range of Cr values. The residuals in Figure 9A-3 are standardized, as will be explained shortly. On the whole, we see significant improvement in the linear relationship. However, two large positive residuals at $z = .88$ and a very large negative residual at $z = 1.01$ stand out.

Table 9A-3 gives results individually for each of the 31 patients. The second and third columns give the reciprocal of Cr (z) and the GFR (y) values. The fourth column gives the predicted GFR and the fifth the residual.

Table 9A-2.   Output of least-squares regression of GFR (y) on the reciprocal of Cr (z) using the MINITAB system

the regression equation is
$$\hat{y} = -2.44 + 87.9z$$

	coefficient	standard deviation of coefficient (SD)	t ratio = coefficient/SD
intercept	−2.44	5.20	−.47
z	87.90	8.41	10.45

analysis of variance due to	df	ss	ms = ss/df
regression	1	20,014.1	20,014.1
residual	29	5,315.6	183.3
total	30	25,329.7	

F test for regression = 109.2 with 1 and 29 df
r-squared = 79.0%

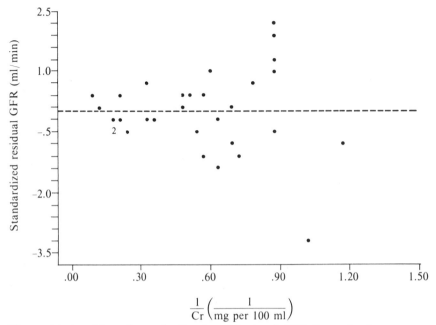

Figure 9A-3. Plot of standardized residuals of GFR from the least-squares regression of GFR on the reciprocal of Cr.

The predicted and residual values sum to the observed value. The last column gives the standardized residual, obtained by dividing each residual by its estimated standard deviation. These standardized residuals are plotted in Figure 9A-3. Although we do not explain how to calculate standardized residuals here, most packaged programs provide them. They are useful for detecting outliers, as we now explain.

The standardized residuals are approximately distributed according to the standard normal distribution (with 0 mean and standard deviation 1) under the usual assumptions of linear regression analysis. In that instance, standardized residuals greater than 2 in absolute value should occur about 5% of the time, and residuals exceeding 3 are rare. The standardized residual of -3.21 corresponds to the low point seen in Figure 9A-3. The large residual indicates that this observation is far from the main body of data and influences the regression analysis more than other single observations.

The following is the stem-and-leaf diagram for the 31 unstandardized residuals given in Table 9A-3:

```
 2 | 58
 1 | 00458
 0 | 01344567
-0 | 111225778
-1 | 12339
-2 |
-3 |
-4 | 1
```

Table 9A-3. Predicted values, residuals, and standardized residuals from the regression of GFR (y) on 1/Cr (z) based on 31 observations

Patient	z	y	Predicted y value, $\hat{y}$	Residual $y - \hat{y}$	Standardized residual
1	1.18	90.00	101.28	−11.28	−.92
2	.88	103.00	75.35	27.65	2.13
3	.88	100.00	75.35	24.65	1.89
4	.88	93.00	75.35	17.65	1.36
5	.88	90.00	75.35	14.65	1.13
6	.88	70.00	75.35	−5.35	−.41
7	1.01	45.00	85.96	−40.96	−3.21
8	.79	77.00	67.02	9.98	.76
9	.71	47.00	59.79	−12.79	−.97
10	.68	45.00	57.40	−12.40	−.93
11	.68	60.00	57.40	2.60	.20
12	.64	53.00	53.93	−.93	−.07
13	.64	35.00	53.93	−18.93	−1.42
14	.59	63.00	49.42	13.58	1.02
15	.57	55.00	47.75	7.25	.54
16	.57	35.00	47.75	−12.75	−.96
17	.55	38.00	45.58	−7.58	−.57
18	.51	47.00	42.01	4.99	.37
19	.49	45.00	40.78	4.22	.32
20	.48	40.00	39.61	.39	.03
21	.36	27.00	29.31	−2.31	−.17
22	.34	37.00	27.25	9.75	.74
23	.32	25.00	25.85	−.85	−.06
24	.25	15.00	19.79	−4.79	−.37
25	.21	15.00	16.31	−1.31	−.10
26	.21	20.00	15.86	4.14	.32
27	.17	10.00	12.38	−2.38	−.18
28	.17	5.00	12.38	−7.38	−.57
29	.17	5.00	12.38	−7.38	−.57
30	.13	10.00	8.59	1.41	.11
31	.09	12.00	5.54	6.46	.51

$2I$ above the upper fourth and $2I$ below the lower fourth is

-35 to 35

The distribution is symmetric, centered at zero and approximately bell-shaped, suggesting that the normality assumption is justified. The residual of -41 is detected as an outlier.

One possible contribution to this large negative residual as well as the two large positive residuals is error of Cr measurement. Errors of constant size represent a large percentage of small Cr values and lead to large errors when the reciprocal is computed.

To see this, notice that a laboratory error of .5 mg% associated with a Cr value of 1.13 leads to values of the reciprocal of either 1.59 ($= 1/.63$) or .61 ($= 1/1.63$), depending on the direction of the error, instead of the correct value, .88. Thus the error is either $-.27$ or .71 (mg%)$^{-1}$. If the correct value

Table 9A-4.   Output of least-squares regression of GFR
(y) on the reciprocal of Cr (z) after deleting observation 7

the regression equation is
$\hat{y} = -5.43 + 96.11z$

	coefficient	standard deviation of coefficient (SD)	t ratio = coefficient/SD
intercept	−5.43	4.31	−1.26
z	96.11	7.18	13.39

analysis of variance due to	df	ss	ms = ss/df
regression	1	21,907.0	21,907.0
residual	28	3,422.4	122.2
total	29	25,329.4	

F test for regression = 179.27 with 1 and 28 df
r-squared = 86.5%

is 6 mg%, an error of the same size yields a reciprocal of .154 or .182, when
the correct value is .167. The error is only − .013 or .015 (mg%)$^{-1}$. If larger
errors were associated with the reciprocals of small Cr values through a
mechanism such as this, observations corresponding to large reciprocals
would tend to fall farther from the regression line.

9A-4.   Further Analysis After Omitting a Suspicious Value

The residual analysis for the regression of GFR on the reciprocal of Cr, especially the stem-and-leaf diagram shown on page 219, suggests that the observation associated with a residual of − 41 is atypical. In this section we explore the effect on the regression analysis when this point is omitted.

   Table 9A-4 gives the MINITAB regression output. The slope of the regression line is somewhat greater than before.

   The intercept is still not significantly different from 0 in this regression
analysis, and r-squared is increased from 79% to 87% after omitting this
point. The stem-and-leaf diagram for the 30 residuals from the regression
line given in Table 9A-4, shown below, suggests that the residuals from this
fit are approximately normally distributed about 0 with no outliers.

```
   2 | 03
   1 | 0023
   0 | 03345669
  −0 | 01112346699
  −1 | 4568
  −2 | 1
```

9A-5.   Summary

The printout from a standard regression program contains the regression
coefficients (the slope and intercept) and the standard deviation of the coefficients. The t ratio can be used to test whether the coefficients are different

from 0. Printouts also contain an analysis of variance. Total variance is divided between variance due to regression and residual variance. The F test is also used to test whether the slope is significantly different from 0. The percentage of the total variance explained by the regression is called r^2, or r-squared. Its square root with the same sign as the slope is the correlation coefficient.

The difference between an observed y value and the predicted y value from a regression equation is called a residual. Residual analysis is an important part of regression analysis. It can help detect curvature, nonlinearity, nonconstant variance, and outliers. Analysis of residuals showed that the regression of glomerular filtration rate on serum creatinine was not linear but curved and assisted in detecting an outlier in the data.

Problems

9A-1. When the data are curvilinear, what can be done to straighten them out?

9A-2. In a sample of 50 (x, y) pairs of measurements, how many degrees of freedom are there for assessing the significance of the slope as compared with zero?

9A-3. What is an F test used for in a regression problem?

9A-4. If the number of points is 52 and the sum of squares for regression is 100 and the sum of squares of residuals is 300, find the mean squares and carry out an F test for regression.

9A-5. In Problem 9A-4, find the value of r-squared.

9A-6. Examine Figure 9A-3. Does the variability of the residuals increase as $1/Cr$ increases?

References

Kleinbaum, D. G., and Kupper, L. L. (1978). *Applied Regression Analysis and Other Multivariable Methods*. North Scituate, Mass.: Duxbury Press.

Ryan, T. A.; Joiner, B. L.; and Ryan, B. F. (1976). *MINITAB Student Handbook*. North Scituate, Mass.: Duxbury Press.

Additional references on transforming variables are:

Emerson, J. D., and Stoto, M. A. (1983). Transforming Data, Chapter 4 and Section 83. In D. C. Hoaglin, F. Mosteller, and J. W. Tukey, eds., *Understanding Robust and Exploratory Data Analysis*, John Wiley, New York, pp. 97–127, 264–67.

Mosteller, F., and Tukey, J. W., (1977). *Data Analysis and Regression*. Addison-Wesley, Reading, Massachusetts, Chapter 4, pp. 79–88.

Introduction to Multiple Regression

OBJECTIVES

Multiple correlation coefficient

Standard deviation of residuals

Testing regression coefficients

Standardized residuals

Fitting attributes

Proxying

Curved surfaces

Modeling

Stepwise regression

Uses of multiple regression

In Section 9-4, we used a linear regression based on the reciprocal of the amount of creatinine (Cr) to help estimate the patient's glomerular filtration rate (GFR). In Chapter 9A, we examined the problem in more detail. The goal was to find a mathematical form for the relationship between GFR and Cr and then use that relation to predict the GFR.

If one variable can aid a forecast, why not use more than one? **Multiple regression** is one method for using several variables to improve the estimation procedure.

Given variables $x_1, x_2, \ldots, x_k$ to be used for prediction of y, the observed value, we can generalize the method of least squares described in Chapters 9 and 9A to obtain an estimate, $\hat{y}$, of the form

$$\hat{y} = b_0 + b_1 x_1 + b_2 x_2 + \cdots + b_k x_k, \qquad (9B\text{-}1)$$

where b_0 is the intercept and $b_1, b_2, \ldots, b_k$ are the weights or regression coefficients associated with the predictor variables. For any data set, the **least squares estimates** are the values of the b's that minimize the sum $\Sigma(y - \hat{y})^2$, where the sum runs over all data points. This approach also has the property that the average of the residuals, $y - \hat{y}$, is zero. We say that $\hat{y}$ is an unbiased estimate of y.

Standard computer packages provide the least squares estimates, their standard errors, and other information helpful in analyzing the data and in reporting the results of the analysis. This chapter discusses the fitting and interpretation of multiple regression models.

9B-1. Fitting a Multiple Regression

Earlier we analyzed a sample of 31 cases from the Brochner-Mortensen *et al.* (1977) study of renal function and Cr. The study actually included 180 men and 200 women, and the investigators recorded the height, weight, and age, as well as the GFR and serum Cr of every participant. Those authors have kindly made these data available to assist us in illustrating the ideas of multiple regression.

We shall assume that the data have been entered into the computer and that standard programs give appropriate printouts, which we will now explain.

First, we get the regression of GFR on the inverse of Cr (1/Cr) for the 180 men. We find

MEN: GFR = $-2.17 + 89.8(1/Cr)$

which is close to the results we found earlier for a sample of 31 subjects (p. 194). The value of r^2 is 76.7%. The hope is that by using additional information about the patient we can improve the estimate by reducing the variability of the errors in the forecasts. In general, older patients have lower GFR's than younger patients with the same serum Cr; also muscular men have higher GFR's than less muscular men with similar serum Cr values. Introducing age and weight, therefore, may improve our predictions.

Adding Variables Improves Prediction

For the same 180 men, we use 1/Cr, age in years, and weight in kilograms to estimate GFR. The new estimate is

MEN: GFR = $-31.5 + 81.5(1/Cr) - .318$ AGE $+ .679$ WEIGHT.

Some features of this equation agree with our earlier remark: **GFR decreases with age and increases with weight**. The new r^2 (or R^2, as we will explain shortly), adjusted to take account of fitting more variables, is 82.1%, and so the additional variables have reduced the variance of the deviation scores, $y - \hat{y}$, by 5.4%, from 23.3% to 17.9%. On one hand, this does not seem like a great improvement, but on the other, it is about a quarter of the conceivable improvement, which is 23.3% (i.e., $100 - 76.7$).

The corresponding regression for the 200 women is

WOMEN: GFR = $-9.1 + 62.1(1/Cr) - .408$ AGE $+ .451$ WEIGHT.

Multiple Correlation Coefficient

When we had one predictor x predicting y, the correlation, r, between y and x, or equivalently, except for sign, between y and $\hat{y}$, measured the degree of linear relationship between them. In multiple regression, the corresponding correlation between y and $\hat{y}$ is called R instead of r. Just as r^2 measures the amount of variance of y accounted for by x (or $\hat{y}$), R^2 measures the amount accounted for by the multivariate forecast $\hat{y}$. When more variables are used in the forecast, more variance is inevitably accounted for, and some adjustment is required to pay for the number of variables used. We call the adjustment an **adjustment for degrees of freedom**.

The residuals, $y - \hat{y}$, have reduced variability when R^2 increases. Indeed their variability $\Sigma(y - \hat{y})^2/(n - k - 1)$, where k is the number of predictors, equals the variance of y, s_y^2, multiplied by $1 - R^2$, where R^2 has been adjusted for degrees of freedom and is measured as a proportion instead of a percentage. Thus high r^2 or R^2 implies better precision in estimation. For instance, if r^2 or R^2 is .75, the mean squared error of prediction using the regression model is one-fourth as large as the prediction using only the mean.

The Residual Standard Deviation

Returning to our example for men, using only 1/Cr, we got

MEN: GFR $= -2.17 + 89.8(1/\text{Cr})$

with a residual standard deviation of 18.81 (printed by the package), associated with an $R^2 = .767$ (adjusted), expressed as a proportion instead of as a percentage. Our result when we introduced age and weight was

MEN: GFR $= -31.5 + 81.5(1/\text{Cr}) - .318\ \text{AGE} + .679\ \text{WEIGHT}$

with adjusted $R^2 = .821$ and residual standard deviation 16.49. The residual standard deviation went down as the R^2 went up. We find the ratio of the residual standard errors to be approximately

$$\sqrt{\frac{s_y^2(1 - .767)}{s_y^2(1 - .821)}} = \sqrt{\frac{.233}{.179}} = 1.141 \left(= \frac{18.81}{16.49} \right)$$

The ratio of the residual s's is 1.141.

Thus if we had to estimate GFR only on the basis of the distribution of GFR for men, we would use the grand mean as the estimate, and our residual errors would have a standard deviation of 38.96 or about 39. By introducing 1/Cr, age, and weight, we have cut the standard deviation of the residuals to 16.49 or about 16.5. Our accuracy, therefore, has more than doubled. **The accuracy of estimation being described here is for individuals near the center of the distribution**. As the person departs from the centroid, we are more nearly engaged in extrapolation; there are additional losses in accuracy stemming from this source. Computer programs can report on such reliability problems for individuals, but we shall not go into it here.

Reliability of Regression Coefficients

Parallel to Tables 9A-1 and 9A-2, the multiple regression program also gives information about the reliability of the multiple regression coefficients and the significance of the regression.

Table 9B-1. Analysis of fit of multiple regression equation for estimating GFR from 1/Cr, age, and weight for 180 men

Variable	Coefficient	SD of coefficient	t-ratio = coef/SD
intercept	− 31.5	8.62	− 3.66
1/Cr	81.5	3.43	23.76
age	− .318	.085	− 3.74
weight	.679	.102	6.64
analysis of variance			
due to	df	ss	ms = ss/df	F-ratio
regression	3	223,967	74,656	274
residual	176	47,791	272	
total	179	271,758		

1/Cr = reciprocal of creatinine value; gfr = glomerular filtration rate; sd = standard deviation; df = degrees of freedom; ss = sum of squares; ms = mean square

Table 9B-1 shows the coefficient values for men in the regression equation that includes 1/Cr, age, and weight, together with the standard deviation of those coefficients; that is, their unreliability as estimated by the regression program. The ratio of the coefficient and its standard error is a *t*-statistic, which is approximately normally distributed. We are asking whether each coefficient is far from zero. Each is more than 3 standard deviations away, and we would reject the idea of a zero coefficient. We note that 1/Cr and weight are the most significant, with age and the intercept far behind in statistical significance.

Analysis of Variance for Significance of Regression

The **analysis of variance** given in the bottom panel of Table 9B-1 separates the sum of squares into two parts—the sum of squares associated with the regression equation and the residual sum of squares left after the regression equation is fitted. Their sum must equal the total sum of squares. The ratio of the mean square for regression to that for residuals gives the *F*-ratio, which can be used to test whether the regression is making a difference. Our *t*-statistics have already convinced us. Our .01 level *F*-table (Appendix Table III-6B) gives 3.95 as the critical value for 3 and 120 d.f. and with 3 and ∞ d.f., 3.78, Thus our ratio of 274 is enormous. We are convinced again that the regression equation is predicting something. This is to be expected here because we already know that 1/Cr was a significantly good predictor by itself. We have added to the strength of the predictor equation by using age and weight as well.

Some programs print out all the residuals, $y - \hat{y}$, or just the extreme residuals standardized according to the uncertainty associated with them. From data not given here, one observation seems particularly wild—the largest residual for an individual was 5.8 standard errors. In all, three individuals have residuals larger than 3 standard errors; the other two are 3.2 and 3.6. In 180 observations, it would not be unusual to have 5 to 10 wild observations, and here we have only 3.

9B-2.   Fitting Using an Attribute Rather Than a Measured Variable

It is sometimes convenient to include an attribute in a regression equation. For example, the intuitive way of handling sex would be to produce separate equations—one for men and one for women. At the same time, if we could describe its effect by merely adding one term for sex, our description of the process would be more parsimonious. Let us see how that would turn out in our problem. For the 180 men, we have again

MEN: GFR $= -2.17 + 89.8(1/Cr)$

explaining 76.7% of the GFR variance for men. For 200 women, we have

WOMEN: GFR $= -4.6 + 66.0(1/Cr)$

explaining 76.6% of the GFR variance for women.

Extra Variables to Replace Stratification

Now we add sex as an extra variable, arbitrarily assigning 1 to male and 2 to female. (Any distinct pair of numbers would do, and so they might be chosen for convenience. Often 0 and 1 are the most convenient pair, sometimes -1 and 1.) The multiple regression equation turns out to be

BOTH SEXES: GFR $= 28.2 + 76.0(1/Cr) - 20.6$ SEX,

$$\text{SEX} = \begin{cases} 1 \text{ male} \\ 2 \text{ female} \end{cases}$$

with $R^2 = 75.4\%$.

The prediction for the pool of men and women is about as good as before. This equivalence is achieved in a surprising way. The coefficient for the reciprocal of Cr falls between those for men and women (weighted a little toward women because there are more women in the data), while the intercept term is no longer practically negligible as it was for men and women taken separately. To compare with the old intercepts, we need to take account of sex. By substituting 1 for men we get 7.6 ($= 28.2 - 20.6$), and from substituting 2 for women we get -12.6 ($= 28.6 - 41.2$). Thus, the intercept partly compensates for the intermediate character of the slope for 1/Cr. On the average, for women, the right-hand side term 76.0(1/Cr) is contributing

$(76.0 - 66.0) = 10$ times the average of $1/Cr$, compared with the contribution to the equation for women only.

Sometimes several dichotomous variables need to be taken into account, and it may then be inconvenient to produce a separate equation for each possible pattern of attributes. For example, with sex (male–female), age (old–young), and weight (high–low), we have eight patterns

male, old, high	female, old, high
male, old, low	female, old, low
male, young, high	female, young, high
male, young, low	female, young, low

If m is the number of dichotomous variables, we have 2^m patterns. So $m = 5$ yields 32 equations, and the data often run out when we make so many distinct fits. If we use the regression approach for $m = 5$, we might just fit 6 regression coefficients—one for each dichotomous variable and one constant term.

Three or More Categories

The device of assigning distinct numbers to the two categories works well for situations where attributes are dichotomous. When three or more categories have been used, a variety of possibilities present themselves.

> **Dummy variables**: For some reason, when we assign numbers to two categories, such as male-female, and these create a predictor variable where we did not have a numerical variable before, the new variable is called a dummy variable.

A. Pool into a dichotomy. If three categories are young, middle-aged, and old, pool the middle category with either the young or the old group. This step loses some information. It is sometimes not clear, however, that any of the categories is a "middle" category—single, widowed, and divorced, for example.

B. Assign ordered numbers. When the order is clear, as in young, middle-aged, and old, it can be convenient and useful to assign equal-spaced numbers, such as 0, 1, and 2, to the categories. The choice of two of the numbers is free, as in the case of the dichotomies, and the third forces a spacing relation among the categories. The order of the categories is often clear from the context. If there are many categories, there are spacings that are better than equal spacings for some purposes, but this problem does not often arise.

C. Fit a number. With three categories, we could assign two of the categories scores, such as 0 and 1, and let the score for the other category be determined by the data by methods we shall not pursue here.

9B-3.   Proxy Variables

An annoying, confusing, yet useful feature of multiple regression is the proxying or substituting that variables do for one another. When two variables are highly correlated, as are height and weight in adult humans, then either may be profitably used in an appropriate prediction equation. If both are used, then their coefficients may be rather uncertain because the variables provide nearly equivalent information, and they can trade off. The point is that the equations are trying to estimate y well, and they have no way of attending to the proper coefficients from the point of view of causation. This is what is annoying and confusing. On the other side, we can often use a variable as a substitute for another we wish we had. So if we knew height but not weight, we might use it as a proxy, even when weight is the causal variable.

We can use our GFR example to illustrate proxying. We might wish to use weight to help estimate GFR. Height does not seem causally relevant. Attending to weight, we find for men

MEN: GFR $= -47.6 + .661$ WEIGHT $+ 84.4(1/Cr)$

with an R^2 adjusted for degrees of freedom of 80.8%.

If we had used **height** instead of **weight**, we would have

MEN: GFR $= -122 + .707$ HEIGHT $+ 86.1(1/Cr)$

with adjusted $R^2 = 77.9\%$. The use of weight in addition to $1/Cr$, therefore gives slightly better predictive power than height. The difference in R^2 would make an encouraging but not compelling case for the causal role of weight. Moreover, we cannot know from the statistics given that the fundamental cause is not some other variable related to height and weight.

When we use both height and weight in addition to $1/Cr$, we get

MEN: GFR $= -90.1 + .609$ WEIGHT $+ .273$ HEIGHT $+ 83.4(1/Cr)$

with adjusted $R^2 = 80.9\%$. Essentially we have gained nothing in linear predictive power by using both weight and height because 80.9 and 80.8 are so close. Weight apparently contains the useful information in height and weight combined.

9B-4.   Curved Regression Surfaces

Our original regression formula

$$\hat{y} = b_0 + b_1 x_1 + b_2 x_2 + \cdots + b_k x_k$$

did not put constraints on the x_i. They could be correlated or algebraically related variables. We could, for example, have only the variables x_1 and x_2 with

$$x_1 = x$$

$$x_2 = x^2$$

$$\hat{y} = b_0 + b_1 x + b_2 x^2$$

Although we have only one variable, namely x, we have a multiple regression fitting problem. Least squares programs make no distinctions between a situation where the x_i's are functionally related and one where they are only statistically related, and so they solve for the b's in the same manner. When the x's are all of one sign, inevitably x and x^2 are highly correlated. We find ourselves in the proxy situation where x and x^2 may tend to trade off, so that the coefficients are poorly determined even if the estimates of y are very close to the observed values.

Not only can we introduce quadratic or higher order polynomial terms, we could introduce other mathematical functions such as exponentials (e^x) or trigonometric ones (such as $cos\ x$) when called for. Still another sort of structure may be useful. Sometimes the idea of the interaction of two variables may be expressed as their product. For example, we might have the form

$$\hat{y} = b_0 + b_1 x_1 + b_2 x_2 + b_3 x_3$$

where

$$x_3 = x_1 x_2$$

From the point of view of the fitting, nothing special has occurred; but this approach allows multiple regression to be used in fitting more complicated functions than may at first appear possible.

9B-5.   Modeling

When we deal with mathematical formulas to forecast outcomes, each term or component often has a physical interpretation. For example, from examining airline schedules for large passenger airlines we might develop the scheduled time of flight in hours as

SCHEDULE: $H = .5 + D/500 + 1.0S,$

where D is the distance in miles (and the planes cruise at 500 mph) and S is the number of stops other than the final destination. The constant .5 hour comes from takeoff and landing time at place of origin and destination, the $D/500$ the hours in flight, and the $1.0S$ an hour extra for each intermediate stop. This equation resembles a multiple regression equation of form

$$H = b_0 + b_1 D + b_2 S. \tag{9B-2}$$

If we obtained values for actual travel times, rather than scheduled times, and fitted the b's by least squares, the veteran physician traveler would not be surprised to find a result something like

OBSERVED: $\hat{H} = 1 + D/250 + 1.5S,$ $\tag{9B-3}$

which suggests that trips may take substantially longer than scheduled. Although the terms retain their potential interpretation, the proxying may trade some b_1 for b_2 or some of b_2 for b_0. In other words, although the average fitted H will equal the average of the right-hand side of (9B-3), we do not promise that the takeoff and landing times average 1 hour as the constant term suggests, nor that flight time will average $\bar{D}/250$, where $\bar{D}$ is the average distance.

Exploring for the Model

For many regression efforts, we do not automatically have a chosen form such as (9B-2); and we must explore a variety of possible forms before deciding which to choose. When many choices for x_i's are available, exploring every possible set of variables can be very expensive. If we have k x_i's that we could use, then we have 2^k linear equations each with a constant term. To see why we have so many, note that in the form

$$\hat{y} = b_0 + \boxed{}\, x_1 + \boxed{}\, x_2 + \cdots + \boxed{}\, x_k,$$

the right hand side has k boxes. We can fill each box with either a zero or b_i, where i is the subscript of the x associated with the box. Consequently, there are 2 ways to fill each box; and therefore, $2 \times 2 \ldots \times 2$, k times, ways in all. In a complicated medical investigation, between basic patient properties like age, height, weight, sex, and blood pressure, and the many biochemical measures that are available, $k = 10$ is not even unusual, leading to $2^{10} = 1024$ linear regression models. None of this includes the possibility that we might wish to consider alternate forms of a variable such as $1/Cr$ instead of Cr itself. In this direction, we may also want to consider using powers and crossproducts of variable in the regression. If we have four variables, x_1, x_2, x_3, x_4, and wish to consider all their squares and all their crossproducts as well, then we have four terms of form x_i, four of form x_i^2, and six of form $x_i x_j$, making 14 in all, and offering a choice of 4096 models. It would be good to have a system that explores these models and picks one or a few out more or less automatically to avoid eye strain.

Automatic Science: Stepwise Regression

Stepwise regression: In an automated stepwise regression program, the first variable to be chosen is that x_i with the highest correlation with y. Among all one-variable predictors, that one is best. The second variable (x) to be chosen need not be that with the second highest correlation with y because the second x may be too highly related to the first variable chosen, and so be redundant. Thus, the second variable has to have the effects of the first variable removed from it, as does y, before their correlation can properly be assessed. Once the second x is chosen, we have no guarantee that the first two variables (x's) chosen are the best pair. That would require investigating all $k(k-1)/2$ pairs instead of k for the first and $k - 1$ for the second for a total of $2k - 1$ investigations. Stepwise regression is a way of getting a good answer, not the best.

Such methods go by the name **stepwise regression**. One type primarily considers adding terms one at a time until it gives up the search. First, it chooses the term that maximizes the correlation coefficient between y and $\hat{y}$. Then it tests whether adding one more term would help. If so, it tries still another, continuing until the added precision is not worth the additional degree of freedom lost—or until some other stopping rule applies. This method steps up by adding terms. A parallel step-down method starts with all the terms, and considers discarding them with similar rules one by one. In either approach, we try to get along with relatively few terms without having to examine all 2^k possible regressions.

These stepwise methods need not be blind to science. If, in estimating GFR, we want $1/Cr$ and weight to be in the final equation, we can insist on them and the automatic exploration will range over the other variables we choose to consider for entry into the equation. We have already seen that once $1/Cr$ and weight were in the equation, it was not worth adding height. This is the sort of choice the stepwise systems make.

Proxying with Many Variables

Proxying often shows up very clearly in studies that use stepwise regression. The Coronary Drug Project Research Group (1974) illustrated this beautifully in their study of factors influencing long-term prognosis after recovery from myocardial infarction. They explored the effects of 40 variables measured at baseline on the three-year mortality from all causes. To do this, they used a stepwise regression program to choose the 10 variables best at predicting mortality two ways.

Method 1. They chose variables and derived regression coefficients on the basis of 760 placebo patients in 15 clinical centers in four eastern seaboard cities.

Method 2. They chose variables and derived regression coefficients on the basis of 830 placebo patients in 15 clinics associated with university medical centers, none of which are in the eastern seaboard cities.

Their first finding was that of the 10 variables chosen, only 2 were common to both sets. They applied the two very different resulting predictive equations to predict mortality for the 1199 placebo patients not used to develop either equation. Their second finding was that both equations gave good predictions. Thus, even though different patients were used to choose the variables and even though largely different variables were used, the predictions still worked. The results remind us that proxying occurs a great deal and that in regression we primarily study prediction not causation.

Attempting to Control for Interfering Variables

Let us consider a couple of additional examples where causation plays a role in our modeling. If we want to estimate the effect of particulates in the air on the death rate, we may do it by setting up a regression equation fitted to the death rates associated with many geographic regions, using the amounts of particulates as the forecasting variables. If these are the only predication variables used, then critics will complain that potent causes of death—old age, poverty, and lack of education—have not been included in the equation, and that without them all the blame is being put on the particulates. If you now enter these in the equations, other critics will want ethnic group and sex entered. Indeed, the amount of information the critics will wish to have added has no end. And we have no natural end to the problem because we do not have a closed system as we have in some chemical reactions. That is, we do not have a few variables that we think account for all or practically all of the variation. The most we can hope is that the regression will give us some notion of the sizes of effects that may be occurring. One difficulty is that the coefficients that we get in the equations, with occasional exceptions in some special physical science or contrived experimental situations, depend on the other coefficients that are allowed to enter the equation. Consequently, we cannot get uniqueness.

9B-6. Review of Uses of Multiple Regression

It may be useful to close by considering again the uses of regression as discussed on pages 191–192 for simple regression with one predictor.

Description

We try to get a form of relationship between the variable and its predictor. Much of this was accomplished in the Cr study when we were able to decide that $1/Cr$ was a useful way to model the relationship. This choice flowed in part from biochemical considerations, but in other circumstances it might have been discovered by direct analysis of the data [For methods, see Mosteller and Tukey (1977)]. Beyond this, we entered age and weight, and ultimately discarded height. We achieved an equation that accounted for a large portion (82.1%) of the variance of GFR. Thus, at the same time that this description gave us a way of adding up terms, it also gave us a way of making adjustments.

Adjustments

In the Cr example, we adjusted for age and weight, as well as for $1/Cr$. In the air pollution example, we would adjust for variables other than particulates that we thought influential in causing death. Note that the concept of adjustment is somewhat circular unless we include all the predictor variables. Otherwise, we talk of predicting GFR from age adjusted for $1/Cr$ and weight, or predicting GFR from $1/Cr$ adjusted for age and weight, while the same equation is doing both things. The equation does not distinguish between predicting y from x_1 adjusted for x_2 and x_3 and predicting y from x_2 adjusted for x_1 and x_3. All it does is predict y from x_1, x_2, and x_3. It is worth repeating that the effect of a variable depends on what other variables are present in the equation.

Forecasting or Prediction

Once an equation is established, it can be used on the other individuals from the same population for forecasting or predicting. Sometimes, the purpose is to decide what dose of a medication should be given in the light of several physical variables. Sometimes, the estimate tells us about expected length of life or time in hospital. These estimates may then be compared with actual performance. When individual observations are far from the forecast, this may cause us to investigate the reasons.

Interpretation

Sometimes one or a few predictors account for all or nearly all of the variation in the system. Then they may suggest that other variables are not important in the system, perhaps because they do not vary enough to matter. In our Cr problem, height seemed not important after weight was taken into account. We believed that muscle mass, not stretched-outness, was the appropriate correlate; the regression results encouraged us to cling to that position, while recognizing that other causal variables might be relevant.

Outlier Detection

In the National Halothane Study (Bunker *et al.,* 1969), death rates following surgery associated with the several institutions participating in the study had very large differences, a factor of 27 from lowest to highest. After adjusting for the patient mix (that is, the severity of the operation, the physical status of the patient, age, and sex) the ratio was reduced to a more modest 3. This still left some outliers to be explained, but the adjustment for patient mix and allowance for sampling variation beyond this was most instructive. The printout for observations in standardized scores helps the investigator explore for outliers.

Discriminating Between Groups

To sort individuals into two groups, the multiple regression equation can be used to establish a systematic procedure. If the physician has a set of variables that is regarded as useful for making the prediction, then we can form a multiple regression. First, we need a set of individuals for whom the outcome is already determined (appendicitis vs. not appendicitis). Let us assign members of the appendicitis group the score 1 (the value of the *y* variable) and of the nonappendicitis group the score 0 (the value of the *y* variable). Then we take measurements on these patients for the variables that would be assessed in advance such as temperature, tenderness, and white-cell count. This leads to a regression equation

$$\hat{y} = b_0 + b_1 x_1 + \cdots + b_k x_k$$

which we fit to the data. Each individual then has a forecast value of $\hat{y}$. the individuals with high forecasts are associated with appendicitis, while those with low forecasts are associated with no appendicitis. There will be a distribution of scores (see Figures 9B-1 and 9B-2). When the distributions overlap a great deal (Figure 9B-1), the forecast will not be worth much (R^2 will be low). When the distributions do not overlap or scarcely overlap (Figure 9B-2), then they may be very helpful (R^2 will be high). Once the b's are determined, then the equation can be used to classify new individuals from the original population.

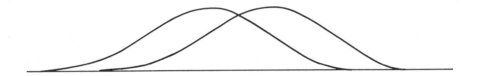

Figure 9B-1.   Distributions with large overlap

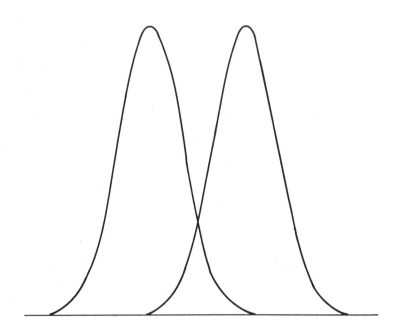

Figure 9B-2.   Distributions with small overlap

Logistic Regression

In regression analysis of an outcome scored as 0 (failure) or 1 (success), the fitted value is interpreted as the predicted probability of success. The trouble with ordinary regression analysis of such outcomes is that the predicted probabilities may fall below 0 or exceed 1.

> **Logistic regression analysis** is a special method that keeps the predictions between 0 and 1 by transforming the probability of success using the logit transformation
>
> $$\log \left(p/(1-p) \right),$$
>
> so that transformed values range between minus infinity and plus infinity.

Multiple logistic regression offers procedures and interpretations equivalent to those of multiple linear regression, but in the transformed scale of the logit of success (or disease) probability. Although a complete discussion of multiple logistic regression is outside the scope of this book, an excellent introduction to this subject is given by Kleinbaum *et al.* (1982, Chapters 20–23).

In the study of factors influencing survival after recovery from myocardial infarction mentioned earlier, the Coronary Drug Project Research Group (1974) analyzed their mortality data using both ordinary multiple regression and also a multiple logistic regression with the same variables. When the same number of variables was used, the multiple correlation coefficients were almost identical using the two methods. So the methods had equal predictive power. The logistic prediction equation did have the stated advantage of keeping the death rate between 0 and 100, whereas the ordinary regression slipped into a negative prediction for the bottom members in the lowest risk group, going down to -7% when 10 variables were used.

9B-7.   Summary

> Multiple regression generalizes simple regression with the role of $a + bx$ for predicting y replaced by $b_0 + b_1 x_1 + \cdots + b_k x_k$. This generalization allows the use of several variables for forecasting and we can introduce polynomial and other terms to improve the forecast, $\hat{y}$. Just as r^2 measures the proportional reduction of variance of deviations, $y - \hat{y}$, from using simple regression, R^2 measures the reduction from multiple regression.

Standard computation packages carry out the estimation of the b's and provide informative data about the closeness of the estimates to the observed values. An F-ratio offers a test of the improvement in forecasting over chance after allowing for the number of variables fitted.

In addition to continuous variables like height and weight, attributes (male–female) can be introduced into the prediction equation by giving each category an arbitrary score. With more than two categories, we need the order of the categories to help assign scores to them or a special statistical package to compute useful scores. We may introduce the attributes to avoid having to fit a model to each category separately based on little data. With many attributes, the number of possible models increases exponentially.

Highly correlated variables proxy for one another. Random fluctuations in the data may make them trade off in the sizes of their coefficients. Thus even though the forecast of y may be excellent, the b's for proxying variables may be individually ill-determined. For example, $3.31 - 3.67x + 2.38x^2$ does not look much like $1 + x^2$, but gives very similar fitted values for x between .9 and 1.5. Proxying has advantages, too, because we may be able to use a variable as a substitute for one we cannot measure.

When choosing one from among many possible regression models, special statistical packages called stepwise regression programs can run through the many regressions automatically to get a forecast parsimonious in number of variables retained. The investigator can retain preferred variables throughout the search.

Good prediction does not assure causation. In most uses of regression, we do not have a closed system of variables as occurs in mathematics or some physical science situations: the area of a cylindrical can of radius, r, and height, h, is $A = 2\pi r^2 + 2\pi rh$. Thus r, h, and A are all the variables in the problem (unless we introduce measurement error). In medical and social problems, other variables usually lurk; the issue is whether they can be neglected.

We use multiple regression for description, for adjustment, for forecasting, for outlier detection, and for discriminating between similar groups.

Problems

9B-1. How are r and R similar?

9B-2. If $R = 0.8$, how much is the variance of the residuals $y - \hat{y}$ reduced compared with the variance of y?

9B-3. How many models of the form $\hat{y} = b_0 + b_1x_1 + \cdots + b_kx_k$ do we have if we can choose which b's to make zero and which to fit, with $k = 11$?

9B-4. With three predictor variables x_1, x_2, and x_3, if all linear, squared, and cross-product terms are used, how many terms does the regression have?

9B-5. Why might you fit a dichotomous attribute instead of fitting an equation for each category separately?

9B-6. Suppose that, for 12-year-old boys, lung volume, y, is predicted by

$$\hat{y} = b_0 + b_1x_1 + b_2x_2$$

where x_1 and x_2 are measures of chest circumference taken a few days apart.

What does our discussion of proxying suggest about the fitted values of b_1 and b_2? How would you expect them to relate to c_1 where

$$\hat{y} = c_0 + c_1 \bar{x},$$

and $\bar{x} = (x_1 + x_2)/2$?

9B-7. Table 9B-2 gives data for a random sample of 30 men from the Brochner-Mortensen *et al.* (1977) study. Enter the data into a computer with a regression package and use age and weight to estimate GFR. What R^2 do you get? Compare with the use of $1/Cr$ in the chapter.

Table 9B-2. Data from a random sample of 30 men from the 180 in the Brochner-Mortensen *et al.* study of renal function

Patient	Age (yr)	Weight (kg)	Height (cm)	Cr	GFR
1	25	68.2	179.0	.71	167.0
2	25	76.5	182.0	.45	103.0
3	28	92.0	171.0	1.20	95.0
4	21	65.5	178.0	1.10	80.0
5	30	65.5	168.0	.99	133.0
6	32	75.0	169.0	.80	89.0
7	38	77.0	180.0	4.35	19.0
8	44	68.5	167.5	.98	92.0
9	49	64.0	170.0	.85	102.0
10	42	81.6	170.0	1.31	88.0
11	41	84.7	182.0	1.11	68.0
12	47	52.5	175.0	1.77	29.0
13	44	74.5	170.0	8.21	6.2
14	43	63.9	158.0	4.50	7.7
15	43	80.4	173.0	2.47	27.0
16	57	77.0	178.0	.64	111.0
17	56	62.5	174.0	.75	116.0
18	51	94.8	182.0	1.02	113.0
19	51	75.2	162.0	1.00	68.0
20	52	50.7	171.0	5.43	10.0
21	56	64.0	165.0	9.75	10.0
22	54	71.0	168.0	9.15	7.5
23	66	75.8	168.0	1.48	68.0
24	64	75.4	175.5	.81	88.0
25	62	76.0	181.5	1.50	63.0
26	64	80.8	174.0	1.65	56.0
27	65	66.8	184.0	2.00	36.0
28	61	82.2	166.0	10.75	6.3
29	69	60.7	163.5	6.84	9.6
30	67	50.0	168.0	3.80	12.0

Cr = creatinine; GFR = glomerular filtration rate

Source: These data were graciously provided by J. Brochner-Mortensen, S. Jensen, and P. Rodbro.

9B-8. Use the data in Table 9B-3 to estimate GFR based on 1/Cr, age, and weight. Compare your results with those given in the chapter for the whole 200 women.

Table 9B-3. Data from a random sample of 30 women from the 200 in the Brochner-Mortensen *et al.* study of renal function

Patient	Age (yr)	Weight (kg)	Height (cm)	Cr	GFR
1	28	63.0	170.0	.70	137.0
2	21	58.5	167.5	1.04	122.0
3	27	57.9	169.0	1.08	68.0
4	29	55.0	156.0	1.20	57.0
5	31	91.0	173.0	1.21	104.0
6	30	55.0	170.0	.93	125.0
7	34	54.8	168.0	.82	68.0
8	32	44.5	155.0	1.20	37.0
9	39	54.0	157.0	7.20	3.5
10	41	53.3	167.0	.83	82.0
11	47	64.2	163.0	.90	74.0
12	45	64.4	157.0	1.27	53.0
13	45	42.5	151.0	.90	51.0
14	47	47.0	162.5	1.99	28.0
15	48	59.3	164.0	9.79	3.6
16	48	60.8	157.0	2.77	22.0
17	50	59.7	160.0	.66	80.0
18	52	58.3	162.0	.91	62.0
19	53	65.7	165.0	4.06	9.6
20	59	70.0	160.0	2.64	16.0
21	55	47.5	161.0	2.57	5.0
22	64	56.3	160.0	.75	66.0
23	61	92.0	170.5	.87	68.0
24	68	71.3	144.0	1.55	29.0
25	69	49.3	154.0	2.13	15.0
26	64	46.7	156.0	2.94	11.0
27	70	61.9	166.0	1.40	40.0
28	74	56.2	159.0	2.00	19.0
29	72	47.1	154.0	11.17	1.0
30	70	51.0	164.0	5.20	9.0

Cr = creatinine; GFR = glomerular filtration rate

Source: These data were graciously provided by J. Brochner-Mortensen, S. Jensen, and P. Rodbro.

9B-9. Using the prediction equation

$$\hat{y} = -47.6 + .661 \text{ WEIGHT} + 84.4(1/\text{Cr})$$

calculate the estimated GFR for the first two men listed in Table 9B-2. Then calculate the estimates using the equation

$$\hat{y} = -122 + .707 \text{ HEIGHT} + 86.1(1/\text{Cr}).$$

How well do the fitted values agree? Explain the result.

9B-10. Compare the schedule (9B-2) and observed (9B-3) forecasts of air travel time when $D = 3000$ and $S = 3$.

References

Brochner-Mortensen, J.; Jensen, S.; and Rodbro, P. (1977). Assessment of renal function from plasma creatinine in adult patients. *Scand. J. Urol. Nephrol.* **11**:263–70.

Bunker, J. P.; Forrest, W. H., Jr.; Mosteller, F.; Vandam, L. D., eds. (1969). *The National Halothane Study: A study of the possible association between halothane anesthesia and post-operative hepatic necrosis.* Report of The Subcommittee on Anesthesia, Division of Medical Sciences, National Academy of Sciences-National Research Council. National Institutes of Health, National Institute of General Medical Sciences. Washington, D.C.: U.S. Government Printing Office.

Coronary Drug Project Research Group (1974). Factors influencing long-term prognosis after recovery from myocardial infarction—three-year findings of the Coronary Drug Project. *J. Chron. Dis.* **27**:267–85.

Kleinbaum, D. G.; Kupper, L. L.; and Morgenstern, H. (1982). *Epidemiologic Research: Principles and Quantitative Methods.* Belmont, CA: Lifetime Learning.

Mosteller, F., and Tukey, J. W. (1977). *Data Analysis and Regression.* Reading, MA: Addison-Wesley Publishing Company, pp. 79–118.

CHAPTER 10

The Management of Stable Angina:
Interpreting Life Tables

OBJECTIVES

Recognizing survival data in clinical information
Interpreting life tables and survival distributions
Estimating survival distributions
Comparing survival distributions
Comparing results from different clinical trials

Since 1960, medical investigators have initiated an increasing number of comparative clinical trials to evaluate new treatments for cardiovascular diseases, cancer, and other life-threatening diseases. These trials often require several years of patient follow-up. Investigators use the long-term survival experience of patients participating in these trials to assess the performance of therapies. The analysis of long-term survival requires special methods, one of which is called life-table analysis. This chapter discusses life-table analysis, required for critical interpretation of important clinical trials, in the context of a clinical problem involving the treatment of a patient with stable angina.

10-1. Using a Life Table to Assess Risk

Clinical Problem 10-1. *Should the Patient With Stable Angina Undergo Coronary Bypass Surgery?*

Mr. J. R. is a 55-year-old self-employed business man. He has had high blood pressure for several years and, 3 years ago, had a myocardial infarction. Eight months ago he developed exertional angina, which has been fairly well controlled with beta-blockers and long acting nitrates. Would coronary artery surgery prolong Mr. J. R.'s life?

Discussion: Interpreting the Evidence from the Veteran's Administration Trial of Coronary Bypass Surgery for Stable Angina

To answer this question, we review the 11-year survival data from the Veteran's Administration's randomized trial comparing surgical and medical management of stable angina (Veteran's Administration Coronary Artery Bypass Surgery Cooperative Study Group, 1984), hereafter referred to as the VA trial. Between 1972 and 1974, the VA trial enrolled 686 patients with stable angina pectoris of more than 6 months duration and resting electrocardiographic evidence of myocardial ischemia. These patients were randomly assigned to surgical or medical therapy.

An early report from the VA study described a statistically significant survival difference in favor of surgery in patients with left main coronary-artery disease (Takaro *et al.*, 1976). For the report considered here, the 595 patients without left main disease were classified into three "subgroups with a low, middle, or high clinical risk on the basis of a multivariate risk function [used] to predict 5-year mortality using four established clinical-risk variables measured at base line: the New York Heart Association classification, a history of hypertension, a history of myocardial infarction, and an ST-segment depression on the resting electrocardiogram. . . . Patients in the low-risk subgroup included those with none or only one of the four risk factors except for ST depression. The high-risk subgroup consisted of patients with combinations of two or three of the strongest predictors (ST depression, a history of myocardial infarction, and a history of hypertension—i.e., those with multiple clinical risk factors."

The 11-year cumulative survival rates for patients without left main coronary-artery disease, grouped by clinical risk, are shown in Figure 10-1. We concentrate on the first panel, which shows the survival experience of surgically and medically treated patients in the high-risk group. The abscissa in this figure is the elapsed time in years since randomization. The ordinate is the proportion of patients in the treatment group who are still alive. Because all patients are alive at the start of the investigation, each curve begins at time 0 and proportion 1.00 (the top left of each figure). Because the proportion of patients who remain alive decreases over time, each curve slopes downward from left to right.

The figure displays 5-, 7-, and 11-year survival rates for each treatment group. For the 83 high-risk patients randomized to surgical treatment, the survival rate 5 years after randomization was 0.86. The rates at 7 and 11 years were 0.72 and 0.49, respectively. Survival rates for the medically treated patients were uniformly lower over the 11 years of follow-up. Rates for the 5, 7, and 11 year follow-up intervals were 0.63, 0.52, and 0.36, respectively. An analysis comparing the survival experience in the two treatment groups showed statistically significant differences in survival for all three follow-up intervals. The *P* values for the comparisons at 5, 7, and 11 years are given at the top of the leftmost panel.

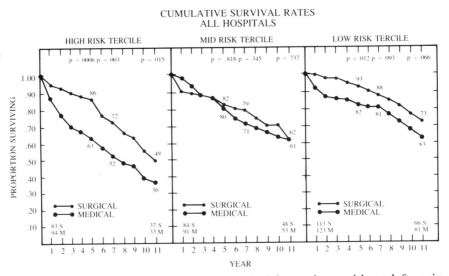

CUMULATIVE SURVIVAL RATES
ALL HOSPITALS

Figure 10-1. Eleven-year cumulative survival for patients without left main coronary-artery disease, according to clinical risk. Numbers of patients at risk are given at bottom of figure. M = medical; S = surgical. [Reprinted by permission from The Veterans Administration Coronary Artery Bypass Surgery Cooperative Study Group (1984) *N. Engl. J. Med.*, **311**:1333–39.]

Because Mr. J. R. has a history of hypertension and a previous myocardial infarction, he satisfies the VA criteria for the high-risk group. In the VA trial, high-risk patients randomized to surgery experienced 14% mortality in 5 years, about one death in seven patients, while patients randomized to medical care experienced 37% mortality, about one death in three patients. The relative risk, or ratio of mortality rates was 37/14, or 2.64 to 1. The experience from the VA trial suggests that Mr. J. R.'s risk could be reduced by surgery, provided that he has operable disease. To investigate this question, you recommend coronary angiography.

10-2.   Estimating and Comparing Survival Distributions

In anticipation of further use of life tables in choosing treatment for Mr. J. R., we digress from the clinical problem to give a more detailed explanation of life tables and the statistical methods used to study them. We begin by introducing the idea of a **survival distribution**.

The survival experience of any group of patients—the high-risk patients receiving coronary bypass surgery, for example—can be summarized by the survival probabilities for each follow-up time after randomization. For instance, at 5, 7, and 11 years of follow-up the survival probabilities in the surgically treated high-risk group are 0.86, 0.72, and 0.49. The survival probability is 1 at the beginning of follow-up and decreases over time. When these probabilities are plotted against time as in Figure 10-1, the resulting curve is called a survival distribution. To summarize, we have

> Definition: The **survival distribution** for a group of patients consists of a set of survival probabilities and a rule specifying one of these probabilities for each follow-up time.

If each patient had been followed from randomization until death or the 11-year follow-up examination, the calculation of the survival distributions in Figure 10-1 would be straightforward. In the VA trial, however, a few patients were lost to follow-up and some patients still on study had not completed 11 years of follow-up at the time of the 1984 report. We describe the "life-table" method for calculating a survival distribution when survival information is incomplete in the following paragraphs.

Consider the surgically treated high-risk patients. The basic data for each patient includes the date follow-up began, the date of death when applicable, and, for patients not known to have died, the date last observed alive. Table 10-1 gives these data for 3 patients in this group. (We are indebted to Dr. Katherine Detre and the Veterans Administration Study Group for providing the data on individual patients.)

This information can be summarized for each patient by the duration of follow-up and the survival status at the end of follow-up. Patient 11500, for example, died during the 10th year of follow-up. Another, more useful, way of summarizing the survival data focuses on the numbers of deaths and patients lost to follow-up in each year. These data are given in Table 10-2.

The first row of Table 10-2 indicates that 83 patients began the first year of follow-up and 4 of these patients died during that year. No patients were lost to follow-up, that is, all of the remaining 79 patients were known to have survived for at least 1 year after randomization. Since 79 of 83 patients survived the interval, the proportion surviving the interval is

$$P(\text{alive at 1 year}) = 79/83$$

$$= 0.952.$$

This number appears in the penultimate column of Table 10-2.

The second row of the table indicates that 79 patients began the second year of follow-up and that 2 of these patients died during the second year. Again, all other patients were followed for at least 2 years and were known to have been alive 2 years after randomization. Thus, the proportion of pa-

Table 10-1. Date follow-up began and date of death or date last known to be alive for 3 high-risk patients in the VA trial

ID	Date follow-up began	Date of death	Date last known alive
11500	3/17/72	12/21/81	
11501	12/03/73		6/30/84
12502	12/10/73	12/09/77	

Source: These data were graciously provided by Dr. K. Detre on behalf of the Veterans Administration Coronary Artery Bypass Surgery Cooperative Study Group.

Table 10-2. Life table for surgically treated high-risk patients in the VA trial

Interval (years)	No. at risk	No. of deaths	Lost to FU	Withdrawn alive	Prop surviving	Cum prop surviving
0–1	83	4	0	0	0.952	0.952
1–2	79	2	0	0	0.975	0.928
2–3	77	2	0	0	0.974	0.904
3–4	75	2	0	0	0.973	0.880
4–5	73	2	0	0	0.973	0.855
5–6	71	8	0	0	0.887	0.759
6–7	63	3	0	0	0.952	0.723
7–8	60	5	0	0	0.917	0.663
8–9	55	3	0	2	0.944	0.626
9–10	50	6	0	7	0.871	0.545
10–11	37	3	0	10	0.906	0.494

Source: These data were graciously provided by Dr. K. Detre on behalf of the Veterans Administration Coronary Artery Bypass Surgery Cooperative Study Group.

tients surviving the second year, **among patients surviving the first year of follow-up** was

$$P(\text{alive at 2 years}\,|\,\text{alive at 1 year}) = 77/79$$

$$= 0.975$$

The cumulative proportion surviving two years is the product of these two proportions. Thus,

$$P(\text{alive at 2 years}) = P(\text{alive at 1 year}) \times$$
$$P(\text{alive at 2 years}\,|\,\text{alive at 1 year})$$

$$= (0.952)(0.975)$$

$$= 0.928.$$

This number appears in the last column of the life-table.

In general, the cumulative proportion of patients surviving to any follow-up time is calculated from the proportion surviving each of the intervals up to that time. For example,

$$P(\text{alive at 5 years}) = (0.952)(0.975)(0.974)(0.973)(0.973)$$

$$= 0.855.$$

Since no patients were lost to follow-up or withdrawn from study during the first 5 years of follow-up, this proportion could be calculated directly as the number of patients alive at 5 years, 71, divided by the number at risk, 83, or

$$P(\text{alive at 5 years}) = 71/83$$

$$= 0.855$$

The power of the life-table method lies in its ability to handle incomplete data. Consider the ninth year of follow-up. In that year, 55 patients were alive at the beginning of the interval, and 3 of these patients died during the year. Of the remainder, 50 patients survived the year, and 2 had not completed the year of follow-up when the analysis was performed. Patients failing to complete the interval are classified as "withdrawn alive" or lost to follow-up and require special treatment. One popular method, used in the VA Study, treats each such patient as contributing one half-year of follow-up to that interval (Anderson *et al.*, 1980). This approach is justified by the argument that such patients are followed on the average for about half of the year before being lost to follow-up. Thus, the number of patient-years of follow-up for the ninth year is calculated as

$$53 + (2/2) = 54$$

and the probability of surviving the interval is calculated as

$$(54 - 3)/54 = 0.944$$

This quantity enters the calculation of the survival distribution at year 9 and all subsequent years. In general, the number of patients followed over the interval decreases as the follow-up time increases because patients can either die or withdraw. When patients are entered over a period of enrollment, the longest follow-up will be provided by the patients first entered into the study.

To summarize,

To calculate the **survival distribution** for a group of patients by the life-table method, one first divides the follow-up time into a set of adjacent intervals. The **probability of surviving each interval** is calculated as the proportion of patients beginning the interval alive that remain alive at the end of the interval. If a patient fails to complete the interval through loss to follow-up or the end of the study, that patient contributes 1/2 to both the number of patients at risk and the number surviving the interval. For any time T, the probability of surviving to T is set equal to the product of the probabilities of surviving the intervals ending before T. This is a conditional probability calculation of the following form:

$$P(\text{alive at } t_k) = P(\text{alive at } t_1) \times P(\text{alive at } t_2 | \text{alive at } t_1)$$
$$\ldots \times P(\text{alive at } t_k | \text{alive at } t_{k-1})$$

The foregoing discussion has focused on the calculation of the "empirical" survival distribution, the observed survival distribution for a group of patients under study. Typically, these patients are regarded as a sample from a specified patient population. In our discussion, the surgically treated high-risk patients in the VA study are regarded as a sample from the population of patients satisfying the criteria specified by the VA study group and receiving surgical treatment. As a result, the empirical survival distribution can be regarded as an estimate of the *expected* survival distribution of patients in the specified population. A full discussion of the variability of the estimated survival probability and the comparison of two survival distributions is beyond the scope of this book, but has been treated at an introductory level elsewhere (Anderson *et al.*, 1980; Peto *et al.*, 1977). It is useful to remember, however, that the variability of the empirical survival probability increases with follow-up time because the number of patients entering the interval decreases as the follow-up time increases. Thus, relatively large differences in the "tails" of survival distributions may not be statistically significant.

10-3. Implications of Angiographic Results for Treatment

Because your review of the VA trial suggested that Mr. J. R. may benefit from surgery, you recommended coronary angiography. The angiography shows operable three-vessel disease, an end-diastolic pressure of 16 mm Hg, and an ejection fraction of 0.40. Returning to the results from the VA trial, you note that the investigators defined an angiographic high-risk subgroup consisting of patients with three-vessel disease and impaired left ventricular function. Survival experience for medically and surgically treated patients in the high-risk group, and for patients in the angiographic low-risk group, are shown in Figure 10-2. The 5-, 7-, and 11-year survival rates for surgically treated high-risk patients are 0.83, 0.76, and 0.50, respectively. These values are almost identical to the survival rates for the clinical high-risk group, presumably in part because the two groups have many patients in common. The same is true of the medically treated high-risk groups. Table 10-3 compares the survival rates of surgically and medically treated patients, cross-classified by medical and angiographic risk, at 7 and 11 years. For patients like Mr. J. R., who are in the high-risk group by both criteria, the 7-year survival rates are 0.36 for the medically treated group and 0.76 for the surgically treated group. Table 10-3 suggests that patients with a combination of clinical and angiographic risk may derive an even greater advantage from surgical intervention than those in other pairs of risk categories.

10-4. Considering the Experience from Other Randomized Trials

Although the original protocol for the VA trial outlined the analytic steps for development of high-risk groups (The Veterans Administration Coronary Artery Bypass Surgery Cooperative Study Group, 1984, p. 1334), some reviewers have criticized such "post hoc" subgroup analysis (Braunwald,

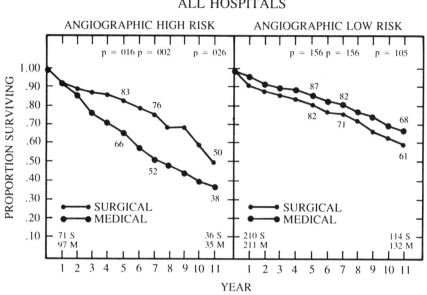

CUMULATIVE SURVIVAL RATES
ALL HOSPITALS

Figure 10-2. Eleven-year cumulative survival for patients in the VA trial without left main coronary-disease, according to angiographic risk. High risk was defined as three-vessel disease plus impaired left ventricular function. [Reprinted by permission from The Veterans Administration Coronary Artery Bypass Surgery Cooperative Study Group (1984) *N. Engl. J. Med.*, **311**:1333–39.]

Table 10-3. Cumulative survival rates at 7 and 11 years in patients without left main coronary-artery disease, according to angiographic and surgical risk[a]

	No.	7-year rate ±s.e. (%)		11-year rate ±s.e. (%)	
		MED	SURG	MED	SURG
High angio risk					
High clinical risk	67	36 ± 7	76 ± 9	24 ± 7	54 ± 10
Middle clinical risk	49	50 ± 9	79 ± 9	40 ± 9	43 ± 13
Low clinical risk	50	79 ± 8	77 ± 8	62 ± 10	52 ± 10
Low angio risk					
High clinical risk	108	67 ± 7	70 ± 6	46 ± 7	46 ± 7
Middle clinical risk	124	82 ± 5	78 ± 5	73 ± 6	68 ± 6
Low clinical risk	184	90 ± 3	81 ± 4	76 ± 5	66 ± 5

[a]A total of 13 patients could not be classified—4 had missing data for the number of diseased vessels, 2 for left ventricular function, and 7 for clinical-risk subgroup.

Source: Reprinted by permission from The Veterans Administration Coronary Artery Bypass Surgery Cooperative Study Group, *N. Engl. J. Med.* **311**:1333–39, 1984.

1983). The use of study outcomes to define patient subgroups can result in biased comparisons. (The issue of subgroup analysis of clinical trial results is discussed extensively in Chapter 12). Further, the results from the VA trial refer to surgical (and medical) treatment strategies employed between 1972 and 1974, and treatment has changed substantially in the interim. Although subgroup analysis is an important part of a thorough analysis of any clinical trial, and changing medical practice arises in the interpretation of any report of long-term follow-up, some doubt remains about the applicability of the VA experience to your patient. You decide to review the results from two other major trials of coronary artery surgery, the European Coronary Surgery Study (European Coronary Surgery Study Group, 1980) and the Coronary Artery Surgery Study (CASS), sponsored by the National Heart, Lung, and Blood Institute (CASS Principal Investigators and their Associates, 1983), before recommending surgery for Mr. J. R.

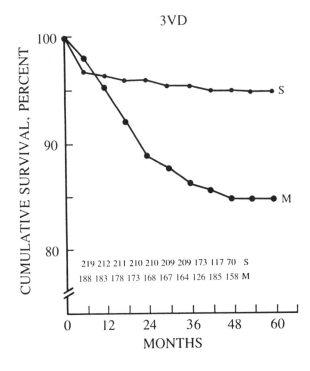

Figure 10-3. Five-year cumulative survival for patients in the European Coronary Surgery Study with three-vessel disease treated surgically (S) and medically (M). [Reprinted by permission from The European Coronary Surgery Study Group (1980) *Lancet*, **2**:491–95.]

The European study did not enroll patients with ejection fraction under 50%, so the results of that trial do not apply directly to Mr. J. R. The relative survival experience of medically and surgically treated patients with three-vessel disease and good left ventricular function may, however, be of some value in making a treatment decision for Mr. J. R. In the European trial, the survival rates at 5 years for 219 surgically treated and 188 medically treated patients with three vessel disease were 0.95 and 0.84, respectively, and the difference in survival rates was statistically significant ($P < .001$, European Coronary Surgery Study Group, 1982). The survival curves, shown in Figure 10-3, indicate an early advantage in the medical group due to two surgical deaths among the 219 patients randomized to surgery. By 12 months, however, the surgical group had a better survival rate, and the relative survival advantage of surgically treated patients increased steadily thereafter. Thus, the European trial would support a recommendation of surgery for Mr. J. R. on the basis of improved survival beyond a year.

The Coronary Artery Surgery Study (CASS) provides especially useful information. From August 1975 to May 1979, CASS randomized 780 patients with stable ischemic heart disease to surgical ($n = 390$) and medical ($n = 390$) therapy at 11 sites. Surgical mortality was low, only 1.4% (CASS Principal Investigators and Their Associates, 1983), and patients with ejection fraction between 30% and 50% were eligible for randomization. Six-year survival rates for all randomized patients were 92% for patients randomized to surgery and 90% for the group randomized to medical care. No significant differences in survival rates were reported either overall or in any of the clinical and angiographic subgroups reported. Subgroup analysis of the 6-year survival experience, however, revealed a trend in favor of surgery in both patients with three-vessel disease and patients with ejection fraction below 50%. Moreover, among 78 patients with three-vessel disease *and* ejection fraction below 50%, the 5-year survival rates were 90% for the 42 patients randomized to surgery and 80% for the 36 patients randomized to medical care. The difference in 5-year survival rates approached statistical significance ($P = .06$, Figure 10-4). A subsequent report (Passamani *et al.*, 1985) confirmed this trend. Seven-year survival rates of patients with low ejection fractions and three-vessel disease were 88% in the surgery group and 65% in the medical care group ($P = .009$). Two aspects of these results are noteworthy. First, surgical mortality was very low, even in this high-risk group of patients. Although the exact number of surgical deaths cannot be ascertained from Figure 10-4, it appears that no more than one surgical death occurred in this group. Second, CASS obtained an excellent survival rate in medically treated patients. Even in this high risk group of patients, the mortality rate in the medical group was only 4% per year. This excellent experience with medical treatment is the major difference between the results of the European trial and CASS. Moreover, the results from CASS may be more reflective of recent experience in the major surgical centers in the United States than the VA trial that began prior to some important changes in the medical and surgical management of coronary artery disease.

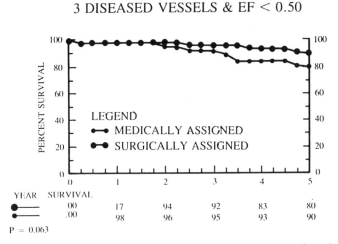

Figure 10-4. Five-year cumulative survival rates for CASS patients with ejection fractions of less than 50% and three-vessel coronary artery disease. [Reprinted by permission from CASS Principal Investigators and Their Associates (1983) *Circulation*, **68**(5):939–50.]

10-5. A Treatment Recommendation for Mr. J. R.

Although two American trials reported no significant differences between survival in surgical and medical groups as a whole, and the results of subgroup analysis in any single trial should be interpreted cautiously, the three major studies of coronary bypass surgery have reported remarkably consistent results in patients with three-vessel coronary artery disease. The European trial reported 5-year mortality rates of 6% in the surgical care group and 16% in the medical care group, a relative risk of 2.7. The VA trial reported 7-year mortality rates of 24% (surgery) and 64% (medicine) among patients with three-vessel disease and high clinical risk, again a relative risk of 2.7. Finally, CASS reported 6-year mortality rates of 10% (surgery) and 20% (medicine) among patients with three-vessel disease and poor left ventricular function, a relative risk of 2.0. Taken together, these studies suggest that Mr. J. R. can reduce his risk of death in the next 5 years by a factor of 2 or more by undergoing coronary bypass surgery. Although there is some risk of perioperative mortality, the CASS trial estimates that risk at less than 3%. Furthermore, the efficacy of bypass surgery in the alleviation of the pain of angina is well established. Based on these considerations, many physicians would feel that surgery is the best treatment strategy for Mr. J. R.

10-6. Summary

Life tables have become a standard part of the presentation of data from studies of the natural history and treatment of chronic diseases. They permit the calculation and comparison of survival distributions among groups of patients, even when some patients are lost to follow-up. The proportion surviving each interval is calculated as the number surviving divided by the number of persons beginning the interval minus one-half the number lost to follow-up. The cumulative proportion surviving is calculated as the product of the proportions surviving in each interval up to the time of interest. Review of the survival distributions reported from three major clinical trials of coronary bypass surgery indicates that patients with three-vessel coronary artery disease and high clinical risk experience substantially better survival in 5 to 11 years of follow-up when treated surgically. Furthermore, the CASS study, which is most directly relevant to current surgical practice, suggests that surgical mortality is low even in this high-risk group of patients. We therefore conclude that, in the absence of other significant medical considerations, surgical therapy is indicated in this patient subgroup.

Problems

10-1. Estimate by eye the value shown in Figure 10-3 for the 5-year survival rate among medically treated patients.

10-2. If no patients were lost to follow-up, the data given in Figure 10-4 are sufficient to determine the number of deaths occurring in each year of follow-up. Given that assumption, how many deaths occurred in the medical group during the third year of follow up?

10-3. Assuming no patients were lost to follow-up, how many deaths occurred prior to the 5-year examination in each angiographic high-risk treatment group shown in Figure 10-2? Use this information to construct a 2×2 table comparing the proportions of deaths in the two treatment groups. Calculate the chi-square test for association. How does the P-value compare to the value given in Figure 10-2? Discuss the differences between the two tests and how these differences might result in differences of significance level.

10-4. Table 10-4 gives survival data for 10 patients enrolled to a follow-up study lasting 1 year. Construct the life table by month for these patients. To simplify calculations, times are recorded only to the nearest month. Assume that follow-up began on the first day of the month listed and that death or last contact occurred during the month listed.

Table 10-4. Date follow-up began and date of death or date last known to be alive for 10 patients enrolled in a follow-up study

ID	Date follow-up began	Date of death	Date last known alive
1	3/81	6/81	
2	5/81	10/81	
3	6/81		3/82
4	7/81	1/82	
5	8/81	10/81	
6	8/81	2/82	
7	10/81		2/82
8	1/82	2/82	
9	2/82	2/82	
10	2/82		3/82

References

Anderson, S.; Auquier, A.; Hauch, W. W.; Oakes, D.; Vandaele, W.; and Weisberg, H. I. (1980). *Statistical Methods for Comparative Studies*. John Wiley and Sons, New York.

Braunwald, E. (1983). Effects of coronary-artery bypass grafting on survival: implications of the randomized Coronary Artery Surgery Study. *N. Engl. J. Med.*, **309**:1181–84.

CASS Principal Investigators and their Associates (1983). Coronary Artery Surgery Study (CASS): a randomized trial of coronary artery bypass surgery. *Circulation*, **68**(5):939–50.

European Coronary Surgery Study Group (1980). Long-term results of a prospective randomised study of coronary artery bypass surgery in stable angina pectoris. *Lancet*, **2**:491–95.

Passamani, E.; Davis, K. B.; Gillespie, M. J.; Killip, T; and the CASS Principal Investigators and their Associates (1985). A randomized trial of coronary artery bypass surgery: survival of patients with a low ejection fraction. *N. Engl. J. Med.*, **312**:1665–71.

Peto, R.; Pike, M. C.; Armitage, P.; Breslow, N. E.; Cox, D. R.; Howard, S. V.; Mantel, N.; McPherson, K.; Peto, J.; and Smith, P. G. (1977). Design and analysis of randomized clinical trials requiring prolonged observation of each patient. Part 2. Analysis and examples. *Br. J. Cancer*, **35**:1–39.

Takaro, T.; Hultgren, H. N.; Lipton, M. J.; Detre, K. M., *et al.* (1976). The VA Cooperative Randomized Study of Surgery for Coronary Arterial Occlusive Disease. II. Subgroup with significant left main lesions. *Circulation*, **54** Suppl(3): III-107–17.

The Veterans Administration Coronary Artery Bypass Surgery Cooperative Study Group (1984). Eleven-year survival in the Veterans Administration randomized trial of coronary bypass surgery for stable angina. *N. Engl. J. Med.*, **311**: 1333–39.

CHAPTER 11

Reading a Report
of a Clinical Trial

Why read a report of a clinical trial? The practicing physician wants to treat patients. Clinical trials can be guides to therapy. This chapter and the next discuss how clinical trials bear on individual patient care. Applying the results of a clinical trial to the individual patient requires two steps. First, the physician must decide whether the clinical investigation provided strong evidence that the experimental treatment benefited the patients under investigation. Second, the physician needs to determine whether the results generated from the patient population and treatment used in the clinical investigation can be appropriately applied in his or her patient and practice setting. This chapter discusses aspects of clinical trials which bear on whether the results observed actually reflect treatment effect. The next chapter considers applying the results of the trial to the individual patient.

11-1.   Treating Duodenal Ulcer

Clinical Problem 11-1.   *Should You Prescribe Antacid Therapy as Recommended in a Clinical Trial Report?*

Your patient, Mr. E.I., has a duodenal ulcer. On your desk is a copy of the *New England Journal of Medicine* containing a report by Peterson *et al.* of the effect of antacid on duodenal ulcer healing. The abstract reads:*

To determine whether a large-dose antacid regimen is effective in promoting healing of duodenal ulcer, 74 patients with endoscopically proved duodenal ulcer completed a 28-day double-blind clinical trial comparing such a regimen with an inert placebo. The ulcer healed completely in 28 of the 36 antacid-treated patients as compared to 17 of the 38 placebo-treated patients ($P < 0.005$). The antacid regimen was not more effective than placebo in relieving ulcer symptoms. Presence or absence of symptoms during the fourth treatment week was a poor predictor of presence or absence of an ulcer crater. Ulcers of placebo-treated patients who smoked cigarettes were less likely to heal than those of nonsmokers ($P = 0.03$). Except for mild diarrhea, no side effects of the antacid regimen were observed. We conclude that a large-dose antacid regimen hastens the healing of duodenal ulcer.

Should you start your patient on antacid treatment?

11-2.   Bias, Chance, and Treatment Effect

In the Peterson study, patients treated with the antacid regimen had a higher rate of healing than did placebo-treated patients. Can there be any doubt that the improved healing was due to antacid therapy? Unfortunately, there are other explanations for the different healing rates observed in the antacid and placebo-treated patients.

The greater healing in the antacid group might reflect sampling variability. That is, both placebo and antacid treatment produce about the same healing rate. The observed difference in healing under this hypothesis reflects chance variation in two samples from a population with a common healing rate.

A second possible explanation for the observed difference might be that the two groups of patients came from populations with different healing rates caused by some factor other than antacid treatment. For example, the antacid-treated patients might have been more healthy to start with, they might have been given special treatment during the trial in addition to antacids, or they might have been followed by the investigators in a particular way which systematically excluded observing some of the bad events following therapy.

> We will use the term **bias** for something other than the experimental therapy that causes a difference between the treatment groups.

Thus, in this ulcer trial, as in any clinical investigation, we have three broad categories of explanation of the observed results: treatment effect,

* Reprinted by permission from W. L. Peterson, R. A. L. Sturdevant, H. D. Frankl, *et al.*, *N. Engl. J. Med.*, **297**:341–45, 1977.

chance, and bias. Well-conducted clinical investigations attempt to minimize bias and assess sampling variability. When clinical investigations offer such control, the trial provides good evidence about the treatment effect.

11-3. Basic Design of Clinical Trials

In spite of the many possible designs of clinical trials, if we concentrate on the simplest problem of comparing two therapies using a single measure of outcome, two designs cover most investigations: two independent groups and crossover designs.

Independent Groups Design

The independent group or parallel design compares an experimental group of patients treated with a new therapy to a control group treated with standard therapy or placebo. (Or there may be two experimental groups, each acting as a control for the other.) The Peterson study employed two independent groups, one treated with antacid, the other with placebo. Ideally, as we discuss below, patients are randomly assigned to the different treatments.

Crossover Design

Sometimes both treatments can be used on the same patient, and when they are, the design is called a crossover experiment. When appropriate, this design has several statistical advantages including better control and more reliability. Here, randomization would decide the order of treatments within a patient.

Chapter 4 discussed two crossover trials involving patients with angina pectoris. One trial treated every patient with nitroglycerine, isosorbide dinitrate, and two different placebos and measured exercise tolerance. The other measured nitroglycerine consumption in patients treated with propranolol and placebo. In both, the order of treatment was randomized.

11-4. Allocation of Patients to Treatment Groups

Perhaps the greatest potential source of bias in a clinical trial arises from the possibility that the patients assigned to the experimental treatment may differ from the controls before therapy is instituted. To minimize the occurrence of pretreatment differences, patients should be randomly assigned to the experimental and standard therapies. Consider some nonrandomized designs:

1. *No controls*. Occasionally, it is argued that the course of the disease treated with standard therapy is well known. If so, the argument goes, con-

trols need not be allocated to standard therapy because the success of the new treatment can be judged against the well-known standard prognosis. As a dramatic example, Morton established the efficacy of anesthesia by demonstrating that one anesthetized patient felt none of the excruciating pain which had previously invariably accompanied surgery. Indeed, such an approach has merit if we know what to expect from the illness. All too often, problems arise as the anticipated uniformity of the disease is questioned in hindsight. Critics will argue that diagnostic criteria may have been changed, crucial evaluations were performed differently, or that the investigator may have chosen only patients with a good chance of surviving an arduous treatment.

2. *Historical controls*. When controls are chosen from patients treated at another time, perhaps by another physician at another institution, we have the danger that the controls differed from the experimental group prior to treatment. Although it is possible and even probable that in some instances historical controls were sampled from a population of patients with the same initial prognosis as the experimental group, we have no way to be certain of this despite a detailed description of the patients in each group. Admittedly, historical controls are appealing and efficient, but skeptics point to the possibility of subtle differences in diagnosis, disease severity, outcome measurement, and ancillary therapy as explanations for any observed differences in outcome. The problems with using historical controls can be appreciated when one considers that different series of patients with the same illness often have markedly different outcomes. For example, Moertel and Reitmeir (1969) reported results of 20 studies of fluorouracil for advanced large bowel carcinoma. In these studies of the same treatment for the same disease, response rates ranged from 8% to 85%.

3. *Concurrent nonrandomized controls*. Even if the experimental and control treatments are offered by the same physician at the same institution during the same time period, any assignment to therapy which is not random may introduce dissimilarities in the two treatment groups. Consider a treatment assignment based on patient preference for the new or standard therapy, or assignment by the patient's personal physician—haphazard as it may seem. These procedures for assigning therapy may not purposely construct patient groups that are dissimilar, but it is always possible, if not probable, that patient preference or physician judgment is systematically related to the patient's state of well-being. Then the treatment assigned to patients who are in better health is likely to appear to be preferable whether it is or not.

Sometimes clinical investigations assign patients to treatment based on hospital number, birth date, or day of the week. Although it may seem that such assignment is arbitrary, bias can still arise. In these treatment schemes, it is possible to ascertain the treatment before a particular patient is enrolled. It is conceivable that those physicians in the know will actively recruit some patients and hold back others based on the anticipated therapy. For this reason, a patient should be entered in the trial before the therapy for that patient is known. In addition, those assigning hospital numbers may have special practices far from random since a number can provide helpful information beyond identifying the patient.

Analysis of Prognostic Factors in Nonrandomized Trials

Investigators who employ nonrandomized controls often argue that their controls are satisfactory because the distributions of prognostic factors in the control and experimental groups are similar prior to therapy. Although such similarities are reassuring, many important prognostic factors which are unmeasured, unmeasurable, or even unknown at present may lead to bias. Of course, it is impossible to determine the equivalence of the distribution of these unmeasured or unknown factors in the two groups.

One possible example of the influence of unrecognized prognostic factors may be found in the relation of treatment adherence to mortality in patients treated with placebo (The Coronary Drug Project Research Group, 1980). As part of the Coronary Drug Project, over 8300 men with a history of myocardial infarction were assigned in a random manner to various therapies. Over 2700 were assigned to placebo. Those patients who actually took the placebo had about one-half of the 5-year mortality of those patients assigned to placebo who did not take at least 80% of the prescribed medication. This striking difference persisted even when mortality was adjusted on the basis of 40 prognostic factors in the two groups of patients. It seems probable that many factors associated with pill-taking behavior influence mortality from coronary disease, factors we do not measure at present.

Despite the fact that nonrandomized controls may be biased, they offer distinct advantages in terms of sample size, ease of investigation, and possibly fewer ethical dilemmas. These virtues should be exploited if historical controls can be shown to provide a satisfactory estimate of the efficacy of standard treatment. Rather than debate the issue, some randomized controlled trials might be run in parallel with historical or other nonrandomized controls to demonstrate those circumstances in which such controls are adequate.

Randomized Controls

In the absence of empirical validation for the satisfactory use of nonrandomized controls, the randomized trial remains the only method of treatment assignment which provides strong assurances about the comparability of the treatment groups. However, it does not guarantee their equivalence, as we discuss later in this chapter.

First, patients should agree through informed consent to participate and be entered into the trial; then they should be randomized. The actual randomization process should rely on a table of random numbers rather than such things as coins or dice. In a double-blind trial, pill bottles are consecutively numbered so that active drug and placebo are randomly interspersed—the next patient gets the next pill bottle.

In the Peterson study, the following description of the randomization process is given:

> At each of the (3) institutions involved in the study, medications were randomized in sets of 10 (five antacids and five placebo). This procedure ensured

that patients were equally likely to be assigned to antacid or placebo, regardless of when or where they entered the trial.

> We feel that random allocation of treatment is almost always a prerequisite for a convincing study. Without randomization, results are at best "suggestive." Randomization is so important that the method of randomization should be briefly described, as in the Peterson study, to lend credence to statements that treatment is random.

Blinding or Masking

A second source of bias in clinical trials arises from the possibility that the treatment of the groups differs not only with respect to the particular experimental drug, procedure, or therapeutic strategy under investigation, but also in other important and perhaps subtle or unrecognized ways. For example, patients receiving a new investigational treatment may be followed closely, receive exemplary care for other medical problems, or generally benefit from their special status as patients receiving experimental treatment. Or, in evaluating response to therapy, the investigator may subconsciously apply one set of criteria to patients receiving experimental therapy and a different set to patients on standard treatment. An investigator's desire to show the new treatment to be beneficial may unintentionally make him or her more likely to ignore adverse occurrences in the experimental group, or to avoid this prejudice, the investigator may tilt clinical judgments in the opposite direction. Finally, a patient who is aware of the treatment may anticipate certain treatment effects or adverse reactions. This anticipation may influence the patient's perception and reporting of illness, even if it does not influence the disease process itself.

These differences between treatment groups can be eliminated by **blinding** the investigator and patient, that is, concealing the treatment assignment. When the assignment is withheld only from the patient, the treatment assignment is single-blind, when withheld from both the patient and the physician, the method is called double-blind.

> Blinding eliminates bias in ancillary care, patient response, and physician evaluation. It should be used in comparative studies whenever it is both possible and compatible with optimal patient care.

Even though clinical investigations go to great lengths to make the experimental and standard therapies appear indistinguishable, unblinding may occur. The extent of unblinding is difficult to assess. Unblinding is especially likely to occur if one therapy has particular side effects, or even beneficial effects that alter the disease process.

The Peterson study attempted to blind both investigators and patients. Control patients received placebos. One physician managed the patient during the month-long treatment and recorded information about ulcer symptoms and side effects. A different physician, who was "unaware of the patient's progress," performed the final endoscopy and judged healing.

Sometimes, blinding is extremely difficult or impossible—consider comparing surgical and medical treatment. In this instance it is difficult to eliminate bias due to differences in ancillary therapy and patient expectation. Nevertheless, some measures can be taken to ensure uniform evaluation of outcome. It may be possible to blind the investigators who perform tests critical to judging outcome. For example, in determining whether bile acid treatment can dissolve gallstones, the radiologist reading the cholecystogram can be blinded to the treatment even if the physician and patient are not. Finally, investigations can focus on relatively unambiguous clinical events, such as death or hospitalization, as measures of efficacy.

Judgments as to why events occurred appear more ambiguous. Investigator bias could influence a decision to attribute death or hospitalization to a particular disease, say heart disease. The total death count in an unblinded study is probably unbiased; the subgroup of deaths attributed to a specific cause may or may not be.

Complete Follow-up

Even when a trial is blinded or focuses on unambiguous events, bias may arise if events that occur in some randomized patients are not counted. For example, investigators may not report events that occur after patients discontinue treatment. They may be unable to report events in patients who stop treatment and move away. Of course, the clinical events occurring in such patients may be unrelated and irrelevant, especially if ceasing treatment had nothing to do with the disease or the therapy. Including these unrelated events only decreases the statistical power of the investigation, and we have seen that the power of most investigations is already too low. This is the motivation to exclude apparently unrelated events which occur before treatment is started or after it has stopped.

On the other hand, if patients stopped taking treatment because of side effects or a feeling that the treatment was not helping, subsequent events become relevant and crucial to the investigation. Systematic exclusion of such patients would give an erroneous picture of the treatments under investigation. Difficulties arise because we cannot determine whether the events in dropouts relate to the therapy or disease.

A clinical investigation that follows all patients assigned to experimental and standard therapy for the full duration of the planned follow-up period, and reports all events in adherents and nonadherents, provides some assurance that bias from selective dropouts did not influence the results.

In the Peterson ulcer study, 44 patients were entered into each treatment. The abstract mentions 38 placebo and 36 antacid patients. The report says that 8 patients withdrew (2 placebo and 6 antacid) and could not be re-evaluated for ulcer healing. They were not counted in the results. Some bias may have been introduced by the exclusion, especially if side effects or poor healing led to withdrawal. An additional 6 patients (4 placebo and 2 antacid) were hospitalized, reevaluated, and terminated from the study before 28 days had elapsed. These patients were excluded from the observations reported in the abstract, but the body of the paper reports the healing observed in these patients, and including them does not alter the study results. Although the lack of complete follow-up in a few patients introduces the possibility for bias in the Peterson study, it appears only a remote possibility.

Some investigators have suggested that dropouts should be assigned outcomes so as to reduce the observed differences between treatments. In the Peterson study, they would count the 2 placebo withdrawals as healed, and the 6 antacid withdrawals as not healed. We feel that this "worst case" approach is excessively conservative. Reporting the number of patients who could not be evaluated in each treatment group is often sufficient.

To summarize this discussion, bias offers a potential explanation for the results of any clinical trial. Three design features can minimize this possibil-

Table 11-1. Comparison of treatment groups[a] in the Peterson study

Characteristic	Placebo[b]	Antacid[c]
Age (yr)	$47 \pm 2.6 \ (26 - 74)^d$	$45 \pm 2.3 \ (22 - 69)$
Number of male patients	33	28
Duration of symptoms (yr)	8.4 ± 1.4	8.9 ± 1.6
Number of cigarette smokers	25	28
Number of aspirin users	2	5
Number of alcohol users	12	11
Basal acid output (mEq/hr)	$5.8 \pm .9$	$4.8 \pm .7$
Peak acid output (mEq/hr)	35.7 ± 2.6	38.5 ± 2.9
Fasting serum gastrin (pg/ml)	49.4 ± 5.2	49.9 ± 4.2
Fasting serum pepsinogen (ng/ml)	106.7 ± 8.2	105.5 ± 7.8
Initial ulcer size (mm)[e]	$8.1 \pm .7$	$7.3 \pm .5$
Number of patients with 2 ulcers	7	4

[a] None of the differences between groups was statistically significant at the 5% level.
[b] 38 patients.
[c] 36 patients.
[d] Mean ± SE, with range in parentheses.
[e] Ulcer size in patients with 2 ulcers is the larger of the 2.

Source: Reprinted by permission from W. L. Peterson, R. A. L. Sturdevant, H. D. Frankl, et al., N. Engl. J. Med., **297:**341–45, 1977.

ity. Random allocation of treatment ensures that patients in each treatment group are initially sampled from the same population. Blinding of patients and observers to the actual treatment ensures that ancillary therapy and outcome evaluation are equally applied. Complete follow-up of both adherents and nonadherents avoids bias in reporting of relevant clinical events. Clinical investigations with these features are well designed by current standards.

11-5. Analysis of Clinical Trials and Chance

Once bias is well controlled, any difference between treatment groups can arise only through chance or treatment effect. Since the P value of the observation is the probability of observing a difference as great as or greater than that actually observed, given the chance model (see Chapter 7), accurate and appropriate calculation of the P value is very important.

> Minimally, a clinical trial should report the P values for the observations and cite the statistical tests used. A clinical trial that reports results with a very low P value provides strong evidence that chance does not explain the difference between the treatment groups.

The Peterson study used the chi-square test (see Chapter 8) to compute the P value for the difference in healing between the groups. The value was $P < .005$.

Imbalance in Randomly Generated Treatment Groups

Even if the P value is low, chance may have played a role. As mentioned above, random allocation of subjects to treatment and control therapy does not guarantee that the treatment groups are equivalent. Random allocation guarantees only that the treatment groups were sampled from the same population initially.

Since it is recognized that random allocation may, by chance, generate nonequivalent groups, extensive and perhaps excessive attention is frequently devoted to reviewing and analyzing the pretreatment characteristics of the patients. Table 11-1 compares the antacid- and placebo-treated patients with respect to 12 characteristics. Although the table shows no statistically significant difference between the two groups, some differences might have influenced outcome. Consider patients with two ulcers: 7 in the placebo and 4 in the antacid group. If multiple ulcers heal slowly or not at all, some of the difference in healing may be due to this imbalance. How can we tell?

To explore this possibility further, it is necessary to know whether multiple ulcers do in fact heal slowly. If not, this imbalance could not explain the results, even if there were 10 patients with multiple ulcers in the placebo and only 1 in the treatment group.

> An imbalance in baseline characteristics can explain results only if the characteristics are correlated with the outcome of the study.

The present study did not report whether multiple ulcers healed slowly, so one cannot tell whether this imbalance was important.

Looking again at the table of baseline characteristics, we see that the placebo group had, on the average, larger ulcers. This probably did not contribute to the results, as the report notes that ulcer size did not correlate with healing.

What if there were a statistically significant difference in baseline characteristics? Would this negate the study results? First, we recall from Chapter 7 that it is not unusual to find statistically significant differences when several characteristics are analyzed. If the 12 characteristics reported in Table 11-1 are independent, the chance of finding none statistically significant at the 5% level will be $(.95)^{12} = .49$. Thus, if many characteristics are reviewed, some will be statistically significantly discordant even when the populations are concordant.

Since many characteristics, both known and unknown, probably correlate with outcome, some must be imbalanced in a way that tends to explain any observed difference.

How, then, can we ever be confident that the results are not due to an imbalance of baseline prognostic factors, even if treatment groups have been generated by a random allocation process? One answer is that:

> The P value provides an upper bound for the probability of an imbalance in baseline characteristics large enough to produce the observed results.

Suppose that, several years from now, a serum factor is discovered which precisely predicts ulcer healing in conventionally treated patients (the placebo group in the Peterson study). Could the results of the Peterson study be explained solely by an imbalance in the assignment of patients whose ulcers were predicted to heal? If antacid treatment had no effect, the serum factor would predict healing in $28 + 17 = 45$ patients, and the results could be explained by assignment of 28 of these patients to the antacid group when the expected number is only

$$\frac{45 \times 36}{74} = 21.9$$

The probability of an imbalance this extreme or more extreme resulting from randomization is given by the P value (see Chapter 7). In the ulcer study, the

P value is less than .005. This is true whether or not the relevant characteristics are known or even measurable. In practice, baseline characteristics will not predict patient outcomes perfectly. In that situation, the imbalance in baseline characteristics must be even greater to produce the apparent treatment effect. Thus the P value provides an upper bound to the probability that the apparent treatment effect would arise from an imbalance in baseline characteristics.

One approach to exploring the importance of baseline characteristics looks first at the P value for the outcome difference between treatment groups. This provides the probability that imbalance arising from randomization explains the result. Then one can look at the baseline characteristics and try to determine whether an extreme imbalance occurred. If an extreme imbalance is found in a characteristic associated with the outcome, one can generally form only an intuitive estimate of whether the imbalance explains the results.

Adjustment

Sometimes, informative adjustments can be made, particularly when the trial is large and many endpoints have been observed. It is often possible to quantify the association between an imbalanced prognostic factor and outcome in order to adjust the morbidity or mortality rates in each group. If one group is older, one can compute an age-adjusted mortality rate after observing the relation between age and mortality in the trial. If adjusted rates are still statistically different, the imbalance of the prognostic factor did not account for the observations.

Adjustment has some limitations. One can adjust only for assessed prognostic factors. Excessive adjustment may wash out a real effect, particularly if an investigator searches through all baseline variables with an eye to reducing observed differences. Because there is usually some chance association between baseline patient characteristics, treatment group, and outcome, assiduous adjustors can minimize differences.

It seems prudent to specify a few potentially important prognostic factors in advance and adjust for these whether or not they minimize differences. If there are one or two striking dissimilarities in other baseline characteristics, additional adjustment may be informative. However, each adjustment "costs" in terms of sample size (degrees of freedom of the statistical test) and further reduces the statistical power of the investigation. In large investigations, this loss has little importance for modest numbers of adjustments.

P Values and Multiple Comparisons

Problems in computing and interpreting the P value arise when investigators perform multiple comparisons, raising the possibility of selection effects. When multiple comparisons are made, finding a statistically significant result is more likely than the P value may lead us to believe.

In Chapter 7 we considered an example with four opportunities to test active treatment against placebo and concluded that the probability of observing one or more comparisons to be significant at the .05 level was in fact about .20 under the chance model. Investigators sometimes perform many analyses but select for publication those with low P values.

Multiple comparisons are expected when more than two therapies are investigated. Surprisingly, there are many other opportunities for multiple comparisons when subgroups of patients, endpoints, or time periods are examined.

At first glance, the Peterson study seems to be free of multiple comparisons. However, the methods section describes attempts to measure the actual size of the ulcers. Instead of categorizing ulcers as healed or not, the investigators could have presented change in ulcer size. Patients were questioned about pain and other symptoms related to ulcer disease. Here is another opportunity for testing statistical significance. Several subgroups might be examined. For instance, healing in men, smokers, or patients that used alcohol are all of interest. Indeed, the healing rate of smokers was singled out for special interest, partially because the results were statistically significant.

Not only do multiple comparisons arise from the many possible subgroups or endpoints of a trial, but one comparison may be tested with several different statistical tests. Each time a different statistical test is used, even on the same data, it offers a chance for a significant result.

> Multiple testing of multiple endpoints and subgroups increases the probability of finding chance differences with low P values.

How can we guard against viewing such findings as treatment effect? Statistical tests do exist for comparing several treatment groups. Chi-square and analysis of variance can be applied to three or more independent groups. Unfortunately, tests have been developed to account for multiple analysis and comparisons within the same data for only a few situations.

In theory, several devices could limit the possibility of selection effects from multiple analyses of data. The most straightforward would require clinical investigators to specify **in advance** a small number of analyses that will be performed as the primary assessment. (Although this may have been done in the ulcer trial, it was not mentioned in the report.)

A second approach requires that a prespecified analysis demonstrate a statistically significant effect of therapy before subgroups are singled out for separate scrutiny. The Canadian Cooperative Study Group (1978) trial of aspirin and sulfinpyrazone in threatened stroke first demonstrated that aspirin reduced stroke and death when all subjects were considered. Only then was it noted that aspirin benefited men but not women. This analysis makes one confident of a positive effect in males in contrast to an analysis which reported no effect overall, and a statistically significant effect when the sexes were scrutinized separately.

A third approach looks for evidence from other sources which corroborates the effects observed in a particular subgroup. For example, an earlier investigation of aspirin prophylaxis of venous thrombosis after hip replacement noted that aspirin was effective in males but ineffective in females (Harris *et al.*, 1977). This "outside" evidence suggesting a sex difference in aspirin's effect on thrombotic disease supports the finding of the Canadian Cooperative Study Group and corroborates the suggestion that aspirin (in the given doses) had no beneficial effect on thrombotic disease in women. Unfortunately, corroborative evidence from outside the trial may be nonexistent and there may be little prospect that a further trial will be held that might corroborate the suggestion.

At present, most clinical trials do not prespecify a limited number of analyses in advance of looking at the data. After the data have been analyzed, it is difficult to estimate how many separate analyses were undertaken in the course of the data review. One might adopt the view that the P value of the overall result provides a good estimate of the potential contribution of chance. P values associated with subgroup analysis probably underestimate the frequency with which the chance model could account for the observations, and post hoc subgroup results should be viewed as hypothesis-generating rather than proof. Other evidence must be assembled from outside the clinical trial before such hypotheses can be firmly accepted.

The problems generated by multiple comparisons remain one of the unsolved areas of clinical trials. Many of the controversies arising from randomized trials reflect this difficulty. For example, the Anturane Reinfarction Trial (Anturane Reinfarction Trial Research Group, 1980), which focused on the subgroup of patients with "analyzable" sudden death in the first 7 months of therapy, the Canadian Cooperative Study Group trial (1978), which focused on stroke and death in males, and the University Group Diabetes Program (1970), which reported an increased cardiovascular mortality in patients treated with oral hypoglycemics, have all led to controversies partly because of multiple comparisons and subgroup analysis.

The problems of multiple comparisons should not be taken as a special shortcoming of randomized trials. Studies employing historical controls have perhaps an even greater potential for selection effects from multiple comparisons. Nonrandomized trials offer more flexibility for the investigator to select the cases and controls through "arbitrary" limits on the age and sex of patients, or date or place of treatment. That these limits are often constructed during a review of clinical material may not be appreciated from the language of the final report. Perhaps one reason for the inattention to multiple comparisons and selection effects in the nonrandomized trial arises because even if the results are statistically significant, bias is never adequately eliminated, and it is difficult to consider in detail the role of chance.

Solution to Clinical Problem 11-1: Evaluating the Clinical Trial

Does the Peterson study of antacids for duodenal ulcer provide convincing evidence that antacids were effective in the study patients? The study was randomized and double-blind. The only apparent source of bias that might have contributed to the results arose from the 8 patients who dropped out without a reassessment of healing. Chance appears to be an unlikely explanation, as the P value is low ($P < .005$). The apparent slow healing rate in smokers treated with placebo may represent a chance observation arising from multiple analyses. Other evidence is needed to corroborate this finding. We conclude that this study provides convincing evidence that the antacid regimen used contributed to healing in the study patients.

11-6.   Implications of a Nonsignificant P Value

We now turn to a brief consideration of other aspects of clinical trials that bear on the interpretation of the results.

A clinical trial that shows no statistically significant difference between treatments does not, of course, prove that the therapies are equivalent. As discussed in Chapter 7, if the sample size is small, failure to reject a null hypothesis of no difference can be attributed either to the low statistical power of the investigation or to the approximate correctness of an hypothesis of no treatment effect.

> As the statistical power of an investigation is an important aspect of the study which bears on the interpretation of results, we believe that power should be accurately calculated and reported together with the P value.

It is worth repeating from Chapter 7 that one common outcome of an examination of the power of a statistical investigation is that the scientific tools are too insensitive for a study of feasible size to detect an effect that we consider desirable. If patients assigned to the experimental therapy did substantially worse than those on standard therapy (even though this difference was not statistically significant), one can be reasonably sure that the experimental therapy is not much better than standard.

11-7.   Implications of Inaccurate Diagnosis

What effect would it have on the Peterson trial if a large percentage of the patients did not have peptic ulcer disease? Clearly, the effect of antacids on peptic ulcer would probably be greater than that reported. Although including patients without peptic ulcer disease would be a flaw in the study design, it seems reasonable to assume that antacids would have no effect on patients without ulcer disease (say patients with cancer). Thus any observed difference between the two groups would have to be due to a large treatment effect on a small subset of the population.

For instance, suppose that only $2/3$ of the randomized patients actually had an ulcer at the beginning of the Peterson study. Then the observed failure rate of 55% (21 of 38) in the placebo group would represent a failure rate of about

$$\frac{55\%}{2/3} = 83\%$$

among patients with ulcers. Similarly, the failure rate of 22% (8 of 36) in the antacid group would represent a failure rate of about 33% among ulcer patients. Thus the difference in cure rates between the antacid and placebo groups would be

$$83\% - 33\% = 50\%$$

rather than the difference of

$$55\% - 22\% = 33\%$$

estimated from all randomized patients. The important point is that, in randomized studies, inaccurate diagnosis probably results in a diminution of treatment effect. Inaccurate diagnosis followed by randomization can result in finding no benefit when in fact there are benefits in properly diagnosed patients. It cannot often explain away observations suggesting benefit.

11-8. Implications of Suboptimal Therapy

Trials are sometimes criticized for using suboptimal therapy. Patients may not have taken all their medicine, or less than ideal doses may have been prescribed. As with poor diagnosis, suboptimal therapy in a randomized trial can only explain a study which fails to find a treatment effect when optimal use of a drug therapy would have provided one. It cannot explain away results which suggest that therapy improves outcome.

11-9. Which Events Should Be Counted in a Clinical Trial?

Medical researchers hold divergent views as to which outcome events should be counted in a clinical trial. The first view holds that only events obviously related to the disease or therapy should be counted as the outcome—that events which are unrelated to the disease or treatment, or which occur before therapy is initiated or after it is terminated, should not be included in the final tally. Consider, as an example, the Anturane Reinfarction Trial, which investigated sulfinpyrazone in the prevention of death after myocardial infarction. This study randomly assigned approximately 1500 patients with recent myocardial infarction to sulfinpyrazone or placebo and followed them for several months. In the final analysis, the investigators counted only those cardiac deaths which occurred 7 days after starting therapy and within 7 days of terminating the protocol treatment.

The second view disagrees for the following reasons. Counting a restricted number of events focuses the analysis on a particular aspect of the disease and its treatment. In the present example, it focuses attention on sulfinpyrazone and its effect on cardiac mortality. Such an analytic focus may establish an important scientific fact, but it may not be appropriate when the trial is used as a guide for therapy. A clinically useful therapy not only reduces morbidity or mortality from a particular disease, but also does not counterbalance its good effects by increasing other bad events. Therefore, it seems appropriate to count all deaths and all other clinical events of importance to patients if one is using the trial as a guide to therapy. One could even say that deaths occurring after the therapy is discontinued also need to be recorded, because we cannot be sure when the effects of therapy have ceased.

The problem with counting the entire range of clinical events is that if some of them are unrelated to therapy, the power of the statistical investigation is reduced. Although ideally, one would count only relevant events, practically speaking, this is impossible, since one cannot be sure which events are relevant. Perhaps being killed by a meteorite could be an exception, but automobile accidents could be related to the patient's disease or its treatment.

Many studies report a statistically insignificant benefit when all events are considered, but demonstrate a significant effect in one particular class of events. For example, a multicenter trial of low-dose heparin for prophylaxis for surgical patients reported 100 deaths in the control group but only 80 in an equal-sized treatment group, $P > .10$. Pulmonary embolism caused 16 deaths in the control group and only 2 in the treated group, $P < .001$. As the overall death rate favored the treated group, it seems reasonable to conclude that there were no major bad effects of low-dose heparin and that the drug certainly reduces death due to embolism. This seems almost as persuasive as a trial that demonstrates a statistically significant reduction on overall mortality. Of course, if the overall death rate in the heparin group were greater, even if this were not statistically significant, there would be considerable doubt as to the benefits of therapy.

> When trials are used as guides to therapy, all clinically important events should be reported. This may reduce the statistical power of the investigation, but the investigation will provide an estimate of the total risks and benefits associated with therapy. We may have to be satisfied with results that show a trend toward total benefit and a statistically significant reduction in a specific type of morbidity or mortality.

11-10. Summary

The outcome of a clinical trial can reflect treatment, chance, and bias effects. Several design features of "good" clinical trials minimize the opportunity for bias and analytic devices provide an estimate of the potential contributions of chance.

Design Features to Minimize Bias

1. *Random allocation.* Patients should be assigned to treatment (or treatment order) randomly, after they have been recruited for the trial. The method of randomization should be briefly described.
2. *Blinding.* When possible, patients, investigators, and/or evaluators should be kept unaware of the treatment.
3. *Unambiguous endpoints if not blinded.* Blinding is least important when the trial focuses on clear-cut endpoints such as death. When the endpoint depends on a judgment, such as a determination of the cause of death, blinding is more important.
4. *Complete follow-up.* Complete follow-up for all patients who enter the trial for the planned duration of the study, whether or not they are still following the protocol therapy, will limit the opportunity for bias. It may also reduce the statistical power of the investigation.

Potential Contribution of Chance

1. *Calculate* P *values.* Studies should provide *P* values for (important) observations. The method of calculation and the quantities compared should be stated.
2. *Report the statistical power of the trial.* Trials with low power may not detect even treatment effects that are large.
3. *Specify major endpoints in advance.* Investigations should be planned in advance to focus on one or two major endpoints. The method of analysis should also be prespecified. Statistically significant results of other analyses should be regarded as hypothesis-generating, particularly if they appear different from the main finding.
4. *Attend to the main effect in the entire study population.* Consider the treatment effect to be that observed when the entire study population is analyzed as a whole. Results that suggest qualitatively different effects in a particular subpopulation should be viewed as hypothesis-generating.

Checklist for reading a clinical trial: Did the trial

1. Randomly allocate treatment?
2. Employ blinding?
3. Employ unambiguous endpoints?
4. Report complete follow-up?
5. Calculate P values (or other appropriate statistics)?
6. Discuss power?
7. Specify endpoints in advance?
8. Emphasize the main effect in the entire study population?

Many studies cannot or will not contain the design features necessary to eliminate bias and/or accurately estimate the P value. Even so, the results may still reflect treatment effect. However, the results are difficult to interpret because the reader remains unsure of the contribution of chance and/or bias.

The following ideas can help one's reading and interpreting of a clinical trial.

1. Even when treatments are randomly assigned, there are bound to be initial differences between the study groups. It is unlikely that such differences, even if statistically significant, caused results with a low P value.
2. Results with a nonsignificant P value still allow the possibility of a treatment effect.
3. In a randomized trial, inaccurate diagnosis and suboptimal treatment can only explain findings of no treatment effect, except in the presence of bias.
4. Trials that focus on limited aspects of morbidity or mortality may not be adequate guides to the therapy of the individual patient.

Problems

11-1. Describe parallel and crossover study designs. How is randomization accomplished? What is randomized in each design?

11-2. What are the advantages and disadvantages of nonrandomized controls?

11-3. How can bias be minimized in nonblinded randomized trials? Suggest three endpoints that would not be subject to biased assessment and three more that would.

11-4. How might counting events only in patients who adhere to therapy bias study results? What is the advantage of this kind of analysis?

11-5. Why should major endpoints be specified in advance?

11-6. Does finding statistically significant differences between baseline characteristics of the treatment and control groups invalidate the study? Why?

11-7. Read "Lidocaine in the prevention of primary ventricular fibrillation" (Lie *et al.*, 1974), which appears in the Appendix to this chapter. Discuss the features of this trial which minimize bias and make it possible to estimate the contribution of chance. Do you think this is a "good" trial? Why? Do you think that lidocaine prevents death? Why? (This trial will also be referred to in the problems for Chapter 11.)

11-8 Read a current clinical trial. Analyze the design features and discuss whether bias was minimized and whether the *P* value is a good estimate of the potential role of chance.

References

The Anturane Reinfarction Trial Research Group (1980). Sulfinpyrazone in the prevention of sudden death after myocardial infarction. *N. Engl. J. Med.,* **302**:250–56.

The Canadian Cooperative Study Group (1978). A randomized trial of aspirin and sulfinpyrazone in threatened stroke. *N. Engl. J. Med.,* **299**:53–59.

The Coronary Drug Project Research Group (1980). Influence of adherence to treatment and response of cholesterol on mortality in the Coronary Drug Project. *N. Engl. J. Med.,* **303**:1038–41.

Harris, W. H.; Salzman, E. W.; Athanasoulis, C. A.; Waltman, A. C.; and DeSanctis, R. W. (1977). Aspirin prophylaxis of venous thromboembolism after total hip replacement. *N. Engl. J. Med.,* **297**:1246–49.

International Multicenter Trial (1975). Prevention of fatal postoperative pulmonary embolism by low doses of heparin. *Lancet,* **2**:45–51.

Lie, K. I.; Wellens, H. J.; van Capelle, F. J.; and Durrer, D. (1974). Lidocaine in the prevention of primary ventricular fibrillation. *N. Engl. J. Med.,* **291**:1324–26.

Moertel, C. G., and Reitmeier, R. J. (1969). *Advanced Gastrointestinal Cancer; Clinical Management and Chemotherapy.* New York: Hoeber.

Peterson, W. L.; Sturdevant, R. A. L.; Frankl, H. D.; et al. (1977). Healing of duodenal ulcer with an antacid regimen. *N. Engl. J. Med.,* **297**:341–45.

University Group Diabetes Program (1970). A study of the effects of hypoglycemic agents on vascular complications in patients with adult-onset diabetes. *Diabetes,* **19**(Suppl. 2):747–830.

Additional Reading

Bailar, J., and Mosteller, F. (1986). *Medical Uses of Statistics.* Waltham, MA: New England Journal of Medicine.

Moses, L. E. The series of consecutive cases as a device for assessing outcomes of intervention. Chapter 6.

DerSimonian, R.; Charette, L. J.; McPeek, B.; and Mosteller, F. Reporting on methods in clinical trials. Chapter 13.

Mosteller, F. Writing about numbers. Chapter 15.

Bailar, J. Communicating about statistics with a scientific audience. Chapter 16.

Halvorsen, K. Combining results from independent investigations: meta-analysis in medical research. Chapter 20.

Department of Clinical Epidemiology and Biostatistics, McMaster University Health Sciences Center (1981). How to read clinical journals: V. To distinguish useful from useless or even harmful therapy. *Can. Med. Assoc. J.*, **124**:1156–62.

Shapiro, S., and Louis, T. (1982). *Clinical Trials: Issues and Approaches.* New York: Marcel Dekker.

Warren, K. S. (1981). *Coping with the Biomedical Literature.* New York: Praeger.

Appendix: Lidocaine in the Prevention of Primary Ventricular Fibrillation

A Double-Blind, Randomized Study of 212 Consecutive Patients[*]

K. I. Lie, M.D., Hein J. Wellens, M.D., Frans J. van Capelle, Ph.D., and Dirk Durrer, M.D.

ABSTRACT.   To assess the efficacy of lidocaine in preventing primary ventricular fibrillation in acute myocardial infarction a double-blind, randomized study was performed in 212 consecutive patients under the age of 70 years admitted to the hospital within six hours of infarction.

Group A (107 patients) received an intravenous bolus injection of 100 mg of lidocaine followed by an infusion of lidocaine (3 mg per minute) for 48 hours. Group B (105 patients) received 5 per cent glucose and water. The groups were comparable in age, sex, site and size of infarction, admission time and mortality rate.

Ventricular fibrillation did not occur in Group A but did in nine patients of Group B ($P < 0.002$).

Side effects, including drowsiness, numbness, speech disturbances and dizziness, developed in 16 patients (15%).

These findings indicate that lidocaine in the dosage given was highly effective in preventing primary ventricular fibrillation, but rigid observation of patients and control of infusion rates are required to decrease the likelihood of side effects. (N Engl J Med 291:1324–26, 1974)

Although lidocaine has been shown to be an effective and safe drug in suppressing ventricular ectopic activity and ventricular tachycardia[1–5] recent investigations have suggested that its prophylactic administration did not prevent primary ventricular fibrillation in acute myocardial infarction.[6,7] Since other studies have indicated that the antiarrhythmic effect of lidocaine might be dose related[2,8] we devised a double-blind, randomized study of intravenous lidocaine with a higher dosage than that used previously.[6,7] In view of the high incidence of primary ventricular fibrillation during the very first hours of acute myocardial infarction,[9] the present study was performed

[*] Reprinted by permission from *N. Engl. J. Med.*, **291**:1324–26, Dec. 19, 1974.

From the Department of Cardiology and Clinical Physiology and the Interuniversity Cardiological Institute, University Hospital of Amsterdam, Wilhelmina Gasthuis, Amsterdam, Netherlands (address reprint requests to Dr. Lie at the Department of Cardiology and Clinical Physiology, Wilhelmina Gasthuis, Eerste Helmersstraat 104, Amsterdam, Netherlands).

in patients admitted to the hospital within 6 hours of the onset of infarction. In the light of the present controversy concerning the predictability of primary ventricular fibrillation[10-12] patients with so-called warning arrhythmias were also included in the study. These were defined as ventricular extrasystoles fulfilling one of the following criteria: occurring with a frequency of more than 5 beats per minute; falling in the vulnerable phase of the cardiac cycle; being multifocal in origin; or being coupled or occurring in runs.

Material and Methods

The study was carried out in patients under the age of 70 years admitted to the hospital within six hours of the onset of symptoms. Excluded from the study were patients with either congestive heart failure, cardiogenic shock, complete atrioventricular block, persistent bradycardia (rate of less than 50 beats per minute), persistent ventricular tachycardia or ventricular fibrillation on admission. Patients were initially admitted to the study on the basis of a typical history of chest pain with suspicious electrocardiographic changes. Patients were retained in the study if the diagnosis of acute myocardial infarction was established by the appearance of diagnostic Q waves with evolutionary ST-segment or T-wave changes and serial rise of serum enzymes (creatine phosphokinase, glutamic oxalacetic transaminase and lactic dehydrogenase). Immediately after admission to the coronary-care unit, an intravenous infusion was started, and a bolus injection of 100 mg of lidocaine or 5% glucose and water was given. The choice of injection was based on complete randomization. Depending on the type of bolus injection given, patients then received an infusion of either lidocaine at a rate of 3 mg per minute or 5% glucose and water. The infusion was continued for 48 hours at a constant rate by use of an infusion pump. The patients were not informed whether they might or might not receive lidocaine on the basis of randomization. The nature of the infused solution was unknown to both the medical and the nursing staff. Subsequent development of any of the complications listed above led to discontinuation of the infusion but retention in the study. Other conditions for termination of the infusion consisted of severe respiratory depression, development of confusion or occurrence of epileptic seizures. Continuous electrocardiographic monitoring was performed in all patients during their stay in the unit, with continuous tape recording during the first 12 hours of infusion. Blood levels of lidocaine were determined by gas chromatography.[13]

Results

During the study period from June, 1973, to September, 1974, 716 patients were admitted consecutively to the coronary-care unit with the suspected diagnosis of acute myocardial infarction. According to the previously listed criteria, 225 patients entered the trial. Of these patients, 13 had to be with-

Table 1. Characteristics of the 107 treated and 105 control patients

Characteristic	Treated patients	Controls
Mean age (yr)	58.1	59.0
Standard deviation	7.1	7.7
Sex		
Male	84	83
Female	23	22
Localization		
Anterior	48	51
Inferior	56	49
Anterior and inferior	3	5
Mean peak SGOT (IU)	111	118
Admission time[a]		
0–2 hr	49	51
2–4 hr	38	36
4–6 hr	20	18

[a] Interval between onset of infarction & admission to the coronary-care unit.

drawn from the study because diagnosis of acute myocardial infarction could not be proved by subsequent serial enzyme studies. Of the remaining 212 patients, 107 received lidocaine, and 105 5% glucose and water. The two groups were comparable in age, sex, site and size of infarction and admission time (Table 1). As shown in Table 2, primary ventricular fibrillation occurred only in the patients not receiving lidocaine; these differences were significant ($P < .002$, Fisher's exact test). In four of nine patients warning arrhythmias were not registered before ventricular fibrillation. The mortality rate in both groups was equal (Table 3), cardiac rupture being the most frequent cause of death. Although eight patients with primary ventricular fibrillation could be successfully defibrillated, one patient died from recurrent attacks of ventricular fibrillation, the attacks being unresponsive to either antiarrhythmic therapy, cardiac pacing or repeated defibrillation. Thirteen patients did not complete 48 hours of infusion: in four congestive heart failure or cardiogenic shock developed, two died of cardiac rupture, four had complete atrioventricular block, and in three, persistent bradycardia (rate under 50 per minute) developed.

Table 2. Frequency of ventricular arrhythmias in 107 treated and 105 control patients

Type of arrhythmia	Treated patients	Controls
No ventricular extrasystole	27	14
Occasional ventricular extrasystole (<5/min)	46	30
Warning arrhythmia	34	61
Ventricular tachycardia	2	6
Ventricular fibrillation	0	9 + 2[a]

[a] Paroxysmal ventricular fibrillation.

Table 3. Causes of death

Cause	Treated patients	Controls
Cardiac rupture	3	5
Cardiogenic shock	3	2
Pulmonary edema	2	2
Ventricular fibrillation	0	1
Totals	8	10

Side effects occurred in 16 patients receiving lidocaine (15%) and included drowsiness (11 patients) numbness of tongue and lips (three patients), speech disturbances (two patients) and dizziness (two patients). In seven of the 16 it became necessary to halve the rate of infusion.

Blood levels of lidocaine (measured six hours after onset of infusion) ranged from 1.5 to 6.4 μg per milliliter (3.5 $\pm$ 0.9 μg per milliliter, mean $\pm$ S.D.). The blood lidocaine levels in the 16 patients with side effects ranged from 2.9 to 6.3 μg per milliliter (mean, 4.2 μg per milliliter).

Discussion

Lidocaine has been shown to increase the fibrillation threshold and to reduce premature ventricular contractions after acute ischemia in the dog heart.[14] After the demonstration by Lown et al.[1] that lidocaine was effective in suppressing ventricular ectopic activity in acute myocardial infarction, it became the drug most frequently used in coronary-care units all over the world. Recently, however, Darby and her co-workers[6] and Church and Biern[7] reported on the ineffectiveness of lidocaine in preventing ventricular fibrillation; in both studies the treated group received an intravenous infusion at a rate of 2 mg per minute. As suggested by Gianelly and his associates[2] and Harrison and Alderman,[8] the clinical effectiveness of the drug might be closely related to the dosage used.

Our results indicate that lidocaine in a dosage of 3 mg per minute preceded by an intravenous bolus of 100 mg is effective in preventing primary ventricular fibrillation in the coronary-care unit. At this dosage however, we noticed a rather high rate of side effects (15%), which forced us to diminish the infusion rate in half the patients showing such effects. These side effects were more common in the older age group, since they occurred in 12 of 52 patients 60 years of age and older, in comparison to four of 55 under the age of 60 years. This observation indicates the necessity of paying special attention to patients in the older age group when the dosage of lidocaine presently recommended (3 mg per minute) is used. The use of an infusion pump seems mandatory with this dosage to prevent life-threatening side effects such as convulsions and respiratory depression. It is conceivable that side effects would have been more frequent if we had not excluded from the study patients above 69 years of age. During the study period, primary ventricular fibrillation developed in only one out of 141 patients more than 69 years old. The lower rate of primary ventricular fibrillation after acute myocardial infarction with increasing age has previously been reported.[12,15] This finding suggests that the benefits of prophylactic treatment of lidocaine in the group over 69 years of age do not outweigh the risk of side effects.

The importance of the so-called "warning arrhythmias" as an indicator of impending ventricular fibrillation has recently been challenged.[10–12] In the present study four of nine patients with primary ventricular fibrillation did not exhibit warning arrhythmias before the onset of the fibrillation. Furthermore, these warning arrhythmias were registered in 59% of the patients in the control group who did not have ventricular fibrillation (Table 2). These data suggest that warning arrhythmias are not helpful in the decision whether or not antiarrhythmic therapy should be instituted.

Since the incidence of primary ventricular fibrillation falls exponentially with time after onset of infarction,[9] there is an urgent need for a safe and effective drug that can be given as a prophylactic measure outside the hospital.[5] The present study may indicate that lidocaine is such a drug. However, the characteristics and mechanisms of ventricular arrhythmias occurring during the earliest phase of infarction seem to differ from those seen in the coronary-care unit.[16] Recent studies[9,17] have suggested that lidocaine was only moderately effective in suppressing ventricular ectopic activity seen during the first hour of infarction. Furthermore, the high incidence of bradyarrhythmias immediately after onset of infarction[9] may prohibit the administration of the drug. To our knowledge, no double-blind study has been performed in patients seen during the earliest phase of infarction in which the efficacy of lidocaine in preventing primary ventricular fibrillation outside the hospital has been determined.

In the absence of such data we do not consider it advisable to recommend prophylactic administration of lidocaine outside the hospital to every patient suspected of having an acute myocardial infarction.

We are indebted to Dr. A. F. Willebrands and Mr. E. A. Ottevanger (Astra) for co-operation in the project and to Miss S. Tasseron for technical assistance.

References

1. Lown, B.; Fakhro, A. M.; Hood, W. B.; et al. (1967). The coronary care unit: new perspectives and directions. *JAMA*, **199**:188–98.
2. Gianelly, R.; von der Groeben, J.O.; Spivack, A. P.; et al. (1967). Effect of lidocaine on ventricular arrhythmias in patients with coronary heart disease. *N. Engl. J. Med.*, **277**:1215–19.
3. Chopra, M. P.; Portal, R. W.; and Aber, C. P. (1969). Lignocaine therapy after acute myocardial infarction. *Br. Med. J.*, **1**:213–16.
4. Mogensen, L. (1970). Ventricular tachyarrhythmias and lignocaine prophylaxis in acute myocardial infarction: a clinical and therapeutic study. *Acta. Med. Scand.*(Suppl.)**513**:1–80.

5. Koch-Weser, J. (1971). Antiarrhythmic prophylaxis in acute myocardial infarction. *N. Engl. J. Med.*, **285**:1024–25.

6. Darby, S.; Bennet, M. A.; Cruickshank, J. C.; et al. (1972). Trial of combined intramuscular and intravenous lignocaine in prophylaxis of ventricular tachyarrhythmias. *Lancet*, **1**:817–19.

7. Church, G., and Biern, R. (1972). Prophylactic lidocaine in acute myocardial infarction. *Circulation,* **46**:(Suppl. 2):139.

8. Harrison, D. C., and Alderman, E. I. (1971). Relation of blood levels to clinical effectiveness of lidocaine, Lidocaine in the Treatment of Ventricular Arrhythmias. Edited by D. B. Scott, D. G. Julian, Edinburg, E., and S. Livingstone. 178–89.

9. Adgey, A. A. J.; Allen, J. D.; Geddes, J. S.; et al. (1971). Acute phase of myocardial infarction. *Lancet*, **2**:501–4.

10. Lawrie, D. M.: Higgins, M. R.; Godman, M. J.; et al. (1968). Ventricular fibrillation complicating acute myocardial infarction. *Lancet*, **2**:523–28.

11. Dhurandhar, R. W.; MacMillan, R. L.; and Brown, K. W. G. (1971). Primary ventricular fibrillation complicating acute myocardial infarction. *Am. J. Cardiol.*, **27**:347–51.

12. Lie, K. I.; Wellens, H. J.; and Durrer, D. (1974). Characteristics and predictability of primary ventricular fibrillation. *Eur. J. Cardiol.*, **1**:379–84.

13. Keenaghan, J. B. (1968). The determination of lidocaine and prilocaine in whole blood gas chromatography. *Anesthesiology*, **29**:110–12.

14. Spear, J. F.; Moore, E. N.; and Gerstenblith, G. (1972). Effect of lidocaine on 'the ventricular fibrillation threshold in the dog during acute ischemia and premature ventricular contractions. *Circulation,* **46**:65–73.

15. Julian, D. G.; Valentine, P. A.; and Miller, G. C. (1964). Disturbances of rate, rhythm and conduction in acute myocardial infarction: a prospective study of 100 consecutive unselected patients with the aid of electrocardiographic monitoring. *Am. J. Med.*, **37**:915–27.

16. Pantridge, J. F., and Geddes, J. S. (1974). Primary ventricular fibrillation. *Eur. J. Cardiol.*, **1**:335–37.

17. Pantridge, J. F.; Webb, S. W.; Adgey, A. A. J.; et al. (1973). The first hour after the onset of acute myocardial infarction. *Prog. Cardiol.*, **3**:173–88.

CHAPTER 12

Applying a Clinical Trial

In Chapter 11 we discussed reading a clinical trial report in the literature. This chapter discusses using the trial as a guide to treating your patient.

You are a physician with a patient for whose disease the standard therapy is only sometimes successful. You have found a report of a recent clinical trial of a new treatment, and the trial showed it to be better than the standard. The methods were sound, and the result was statistically significant. What considerations should you employ in deciding whether to use the new therapy for your patient?

Let us sweep away some brush that would clutter and confuse the discussion. First, as your patient's physician, you would already have considered medical contraindications associated with the treatments studied in the trial. Strong contraindications will set certain treatments aside.

Second, some patients may have done poorly on the new therapy even though, on the average, it was better. How can the results be applied to your individual patient when you cannot be sure that he or she will not do poorly also? It is futile to demand certainty before new therapies are considered. Even among a group of apparently similar patients, some will do well and others poorly, unpredictably. Instead, attention should focus on the probability of success. The therapy with higher mean or median survival time or other relevant measure of effectiveness should, in general, be offered to the individual patient.

270

Third, we recognize that the expectations of success or life expectancy occasionally may not settle the decision problem because the choice can involve the patient's social situation and attitudes toward disability, complete recovery, life shortening, and risk. In this chapter we usually do not bring these considerations into the discussion formally, but you will, of course, have them in mind.

Fourth, the particular characteristics of your patient need to be considered to decide whether the results observed in the study patients apply to him or her. Your patient may be different in some important way. He or she may not be one who would have been eligible under the selection criteria used in the investigation. Or your patient may fit the eligibility requirements for admission and still have characteristics that influence the choice of therapy, such as unusual age, obesity, or symptoms.

Generalizing from study subjects to individual patients is a major focus of this chapter. Unfortunately, this problem has not received much formal attention and some of our discussion may seem oversimplified or incomplete.

Finally, when these hurdles are passed, we come to the question of prescribing the better treatment. You might ask how to tailor the treatment to the individual. Although much has been written about individualizing treatment, except for two or three patient categories, often based on staging of disease, we rarely have the luxury of much information from clinical trials about how treatment efficacy depends on patient characteristics. Costs alone usually prevent such extensive explorations. By and large, the trial reports the overall performance of patients assigned to each treatment and recommends the better therapy. Only occasionally does a trial recommend one treatment for one kind of patient and the other treatment for others. The point is that clinical trials ordinarily produce a modest number of recommendations, yet physicians frequently individualize therapy. Moreover, clinical trials may employ therapies that are difficult to mimic in practice. As a result, physicians may offer treatment that differs from that studied in a clinical trial. This is another focus of this chapter—generalizing from the therapy used in the study to the therapy you will give your patient.

12-1.   How Should You Treat Your Hypertensive Patient?

Clinical Problem 12-1.   *Is the HDFP Treatment Protocol Appropriate for Your Patient?*

You are reviewing the outpatient record of Ms. J.S. She is a 35-year-old black woman with diastolic blood pressures of 94, 98, and 96 mm Hg on three separate occasions. How should the results of the Hypertension Detection and Follow-up Program (HDFP) be applied to her? Should her mild hypertension be treated with drugs? What drugs should be used and how should treatment be given?

Background: The HDFP

We start to answer these questions by reviewing the summary of the HDFP* (Hypertension Detection and Follow-up Program Cooperative Group, 1979a).

> The Hypertension Detection and Follow-up Program (HDFP), in a community-based, randomized controlled trial involving 10,940 persons with high blood pressure (BP), compared the effects on five-year mortality of a systematic antihypertensive treatment program [Stepped Care (SC)] and referral to community medical therapy [Referred Care (RC)]. Participants, recruited by population-based screening of 158,906 people aged 30 to 69 years in 14 communities throughout the United States, were randomly assigned to SC or RC groups within each center and by entry diastolic blood pressure (DBP) (90 to 104, 105 to 114, and 115 + mm Hg). Over the five years of the study, more than two thirds of the SC participants continued to receive medication, and more than 50% achieved BP levels within the normotensive range, at or below the HDFP goal for DBP. Control of BP was consistently better for the SC than for the RC group. Five-year mortality from all causes was 17% lower for the SC group compared to the RC group (6.4 vs 7.7 per 100, $P < .01$) and 20% lower for the SC subgroup with entry DBP of 90 to 104 mm Hg compared to the corresponding RC subgroups (5.9 vs 7.4 per 100, $P < .01$). These findings of the HDFP indicate that the systematic effective management of hypertension has a great potential for reducing mortality for the large numbers of people with high BP in the population, including those with "mild" hypertension.

Using the approach outlined in Chapter 11, we observe that the opportunity for bias was minimal. Subjects were randomly assigned to treatment; the clear-cut endpoint, death, was specified in advance in this nonblinded trial; and 99.5% of the subjects were followed for the duration of the study. Chance is also an unlikely explanation, as the P value was low, $P < .01$. We conclude that SC therapy worked better than RC for the patients in the HDFP.

What are the implications of the HDFP for Ms. J.S.? Is she similar enough to the HDFP study subjects to make the results applicable?

12-2. Comparing Your Patient to the Study Subjects

Consider first the general problem: How can you evaluate the similarity of your patient to those studied?

1. *Admission characteristics.* Check whether your patient would have been admitted to the study had it been held in your institution. If not, a good deal of judgment may be required. Your patient may have a different type of illness or therapy may be contraindicated.

---

* All extracts in this chapter from Hypertension Detection and Follow-up Program Cooperative Group (1979a) are reprinted by permission from Hypertension Detection and Follow-up Program Cooperative Group, *JAMA*, **242**:2562–71, 1979. Copyright 1979, American Medical Association.

2. *Other important variables.* The study may report that certain variables have important effects on outcome, for example, stage of disease, physical status of patient, age, or sex. Ordinarily, the report discusses these issues one variable at a time, if at all, rather than simultaneously. If the report is carried out in this way, pay special attention to the variable that seems to make the most difference in performance.

Raise Specific Medical Doubts

Since the new treatment is effective, you would use it for your patient barring specific reasons to do otherwise. Being systematic about this issue may help resolve it.

> Write down the specific features of your patient that cause concern, and why they cause it (age? vital capacity? fasting glucose?). Do the doubts concern efficacy, side effects, or are they hard to state beyond a reluctance to overgeneralize?

We emphasize here and below that making doubts specific so that they can be affirmed or denied one by one is itself an important step in the process. Once you establish concrete concerns, rereading the report of the trial may harden or dissolve these doubts and settle the question. For example, the authors may say that they excluded patients above a certain age or those with high blood pressure because previous studies have shown that the new treatment or the standard was unsatisfactory for such patients. In addition, the report may cite useful literature that will resolve some issues.

Compare the Choices

After reviewing the paper and other relevant literature, some specific doubts may still remain. Now the situation is uncertain and resembles the problem of early use of a new treatment. It can be handled as a decision problem involving not only the likelihood of the success of the treatments, but also the losses and benefits that may occur. Although we cannot hope to know the best action for sure, it is worth giving effort to comparing the choices, so that we can choose the one we think is probably best.

Implications for Clinical Problem 12-1

Let us apply these ideas to our patient, Ms. J.S. Is she similar to the patients who were studied by the HDFP? We need to review the admission criteria of the HDFP.

[T]he . . . centers identified target populations on the basis of residential areas (census tracts, probability samples of larger areas, or entire housing projects); one center used employment rolls of entire divisions of large industries. With the exception of the industry-based program, an interviewer contacted individual households and constructed a census roster of all residents. As soon as possible thereafter, three consecutive BP readings (right arm, sitting) using a standard mercury manometer were taken on all persons aged 30 to 69 years. (Only bedfast and institutionalized persons were excluded.) If the mean of the second and third DBP (fifth phase) was 95 mm Hg or greater, the individual was referred to the HDFP clinical center regardless of current antihypertensive treatment status. At the center, if the mean of the second and fourth DBPs—taken with a special mercury manometer, the Hawksley random-zero device—was 90 mm Hg or greater, the person was considered a participant and randomized. . . .

The HDFP incorporated many unusual and important design features into its investigation. One of these was an attempt to include all people in the study provided that they had sufficient hypertension. Their method of subject selection resulted from a conscious attempt to make the study results generalizable to patients encountered in practice. Contrast this with methods of subject selection which are designed to select "good study subjects," namely reliable, cooperative, compliant individuals who frequent the institutions where the study is conducted. (See, for example, the Veterans Administration Cooperative Study Group, 1967, 1970, 1972.)

Although the HDFP recruited all patients with significant hypertension, the investigators did not define hypertension in a manner that can be easily reproduced in practice. Blood pressure was initially measured in the patient's home! [This is only alluded to in the admission criteria above. Another publication from the HDFP (Hypertension Detection and Follow-up Program Cooperative Group, 1977) explicitly states that the first blood pressure screen was conducted in the patient's home.] If home blood pressure is substantially lower than pressure in the doctor's office, as some have suggested, it is not possible to be sure that a patient seen only in the office fulfills the admission criteria.

Note also that pressure was measured with a special "random zero" manometer. This instrument conceals the exact blood pressure from the measurer until after the pressure is recorded. It seems unlikely, though, that this special manometer would yield pressures substantially different from a conventional sphygmomanometer.

The criteria in the screening and diagnosis of hypertension were based on the average of readings obtained at a single visit. Rarely is an averaging procedure meticulously followed in practice.

Finally, note that Ms. J.S. would not have been included in the study if her first blood pressure were an accurate reflection of her home pressure. Her first pressure was 94 mm Hg, below the HDFP cutoff.

The point is that clinical studies often employ elaborate diagnostic procedures that are difficult to reproduce in practice. This is true of the HDFP, a study designed to guide office practice. In other instances it may be extremely difficult to be sure that the patient in the office has the same extent or severity of disease as the study subjects.

What, then, are our concerns in generalizing from the HDFP to our patient Ms. J.S.? As suggested, we will write them down.

1. All HDFP patients had initial pressures measured at home that averaged 95 mm Hg or greater.
2. A special sphygmomanometer was used to confirm high blood pressure.
3. Formal averages of multiple measurements were the basis for diagnosis.

These concerns raise no specific reason to feel that SC therapy will not work for Ms. J.S., even though her hypertension may not be as sustained at home as that of the HDFP subjects. SC treatment was quite effective in reducing mortality in the study subjects. We recommend SC therapy be given to Ms. J.S.

So far, we have concentrated on the results observed for all HDFP patients. We now discuss how to review the results in a subgroup of study patients who are most like your patient.

12-3. Looking at Similar Patients

In general, you may be able to examine the performance of a subset of the trial's patients who have the key quality your patient possesses. If efficacy in the aged is of concern, you may be able to compare the two treatments for older patients in the trial. If a recent report does not discuss the matter, the authors may still have the data available in a form that could readily answer your query, and they may be happy to respond. This step addresses the specific question, because a favorable result for the new therapy increases confidence in its use even if subgroup results are not statistically significant.

> You can anticipate that subgroup analysis, because of small sample size, is unlikely to establish the efficacy of the new treatment using conventional tests of statistical significance.

For Ms. J.S., we are concerned that she is black and she is female. Some authors have suggested that the risks of hypertension depend on a patient's race and sex. Table 12-1 is taken from the report of the HDFP. It shows the mortality of the SC- and RC-treated patients in different race–sex and age groups.

Black women assigned to SC therapy had substantially lower mortality than those assigned to RC care. There is no reason to suspect that the benefits of SC therapy are any less for black females than for the entire HDFP population. If we want information about black women with mild hypertension (diastolic pressure between 90 and 104 mm Hg), we should consult the authors of the report for further information.

Table 12-1. Mortality from all causes for stepped care (SC) and referred care (RC) participants during five-year follow-up by race–sex or age at entry

| Race–sex or age (yr) | Sample size | | Deaths | | Life table death rates per 100 persons | | Reduction in mortality for |
	SC	RC	SC	RC	SC	RC	SC group
Black men	1064	1084	112	140	10.6	13.0	2.4
Black women	1344	1354	70	98	5.2	7.2	2.0
White men	1892	1861	109	126	5.8	6.8	1.0
White women	1185	1156	58	55	4.9	4.8	−.1
30–49	2429	2374	81	82	3.3	3.5	0.2
50–59	1852	1909	115	159	6.2	8.3	2.1
60–69	1203	1172	153	178	12.7	15.2	2.5

Source: Reprinted by permission from Hypertension Detection and Follow-up Program Cooperative Group, *JAMA,* **242:**2572–77, 1979. Copyright 1979, American Medical Association.

12-4.   Selection Biases

Part of the reason for writing out the specific concerns is to avoid the needle-in-the-haystack approach. In this method, one casts about for every possible way the patient may differ from those in the study whether or not the variables are believed to be medically relevant. Even if each of several variables had only 2 levels, together they soon produce many possible subgroups or strata; 2 for 1 variable, 4 for 2 variables, 1024 for 10 variables, or in general 2^k, where k is the number of variables. With several variables, the number of patients from the study falling into your patient's stratum will be vanishingly small. Also, in each stratum, we need patients from both treatments to compare results, and a small subsample size diminishes the reliability of the substudy.

For example, you might note that Ms. J.S. is a thin nonsmoker with a low serum cholesterol and no family history of cardiovascular disease. There may be very few black females among the 10,000 patients recruited for the HDFP who have an average diastolic pressure of 96 mm Hg, are thin nonsmokers with low cholesterol, and have no family history of heart disease. A query to the authors may produce no helpful information about such a narrowly defined subgroup.

Another hunt-and-peck approach is also unwise. This involves hunting out variables forming a stratum where the treatment tends to be less favorable for your patient. Instead, you should have specific concerns in advance about your patient. By merely hunting up the worst situation you can find by proliferating variables, you can produce a substantial bias.

You might have noticed in Table 12-1 that patients age 30 to 49 assigned to SC suffered about the same mortality as those assigned to RC care. SC treatment may be more costly and time consuming than RC. Is the apparently small benefit in this age group worth the price? Before answering this question, you should realize that you may have perused Table 12-1 and singled out this relevant subgroup because it has the worst results for SC therapy. Your perusal could bias your assessment. We illustrate by describing the consequences when we study an innovation in treatment and select the stratum with the least encouraging results.

12-5. Selection Bias When Choosing the Least Favorable Stratum

Treatment Effect May Be Reversed by Chance in the Least Favorable Stratum

To discuss this problem concretely, we need to choose a specific situation. We consider a two-treatment clinical trial comparing an innovative treatment (called simply treatment) to standard therapy (called control). Although the problem could be formulated in many ways, we specify the **true** gain (as opposed to the observed gain) from the innovative therapy and investigate the possibility that the direction of the treatment effect is estimated incorrectly in the stratum with results least favorable to the innovation.

Computing the Probability of a Chance Reversal

In terms of the HDFP, one might assume that the overall benefit of SC therapy is a reduction in 5-year mortality of 1.4 lives per 100 patients treated. (Note that 1.4 was the **observed** reduction in 5-year mortality computed from the mortality rates in the abstract.) The question is: What is the chance of observing that SC mortality was higher than RC mortality in one subgroup because of sampling variability?

To address the general problem, suppose that the true gain from treatment does not depend on the patient characteristics used to define the strata. If the overall true difference is μ_d, then the expected difference in mean outcome is μ_d in every stratum.

To simplify calculations, we assume that the n patients in **each** treatment group can be equally divided into k strata. Although this will never be exactly true, the results will be approximately correct for most situations. Suppose finally that the standard deviation of the response to treatment is σ, so that the estimated difference between the treatment groups, $\overline{d}$, has standard deviation equal to

$$\sigma_{\overline{d}} = \sigma\sqrt{\frac{1}{n} + \frac{1}{n}}$$

If we break the samples up into k strata of equal size, the standard deviation for difference in means for any subset is

$$\sigma\sqrt{\frac{k}{n} + \frac{k}{n}}$$

because of a reduction in sample size from n to n/k.

If we have 2 strata ($k = 2$), then on the average the values of the larger and smaller difference are, from special tables for ordered values drawn from the normal distribution,

$$\mu_d + .56\sigma\sqrt{\frac{4}{n}} \quad \text{(most favorable)}$$

$$\mu_d - .56\sigma\sqrt{\frac{4}{n}} \quad \text{(least favorable)}$$

If we were allowed to choose the stratum with the results appearing as least favorable to innovation, the difference in that stratum would, on the average, be more than half a standard deviation times $\sqrt{4/n}$ smaller than the true difference, μ_d. If there were 4 strata, the .56 would be replaced by 1.03, and for 8 strata, by 1.42. As we increased the number of strata, the standard deviation of the difference would grow, because of reduced sample size, to $\sigma\sqrt{2k/n}$.

As the number of strata increases, both the standard deviation of the estimated difference and the multiplier grow larger, increasing the chance for a reversed outcome for treatment in the worst stratum.

> Selection effects can lead to an estimated loss from innovation in the least favorable subgroup even when the true effect is a substantial gain.

EXAMPLE.   Suppose that the innovation under investigation has a true gain that equals 3 standard deviations of the estimated mean difference, so

$$\mu_d = 3\sigma\sqrt{\frac{2}{n}}$$

If we have k strata, then the bias from choosing the least favorable stratum is

$$-B_k\sigma\sqrt{\frac{2k}{n}}$$

where B_k is a tabled value, .56 for two strata ($k = 2$). Other values of B_k for k between 1 and 10 are given in Table 12-2. Suppose that $k = 8$. Then $B_k = 1.42$ and we get bias

$$-1.42\sigma\sqrt{\frac{2 \times 8}{n}} = -4.01\sigma\sqrt{\frac{2}{n}}$$

The average observed outcome in the worst of 8 strata is μ_d + bias or, in this example,

$$3\sigma\sqrt{\frac{2}{n}} - 4.01\sigma\sqrt{\frac{2}{n}}$$

Thus, when we select the worst of 8 differences, the expected difference is negative, even though only sampling error is at work. To summarize: a

Table 12-2.
Values of
B_k for
$k = 1$ to 10

k	B_k
1	.00
2	.56
3	.85
4	1.03
5	1.16
6	1.27
7	1.35
8	1.42
9	1.49
10	1.54

favorable effect of 3 standard deviations is likely to be overwhelmed by the selection of the least favorable difference from analysis of 8 strata.

We can also compute the probability of a negative difference. We do not discuss the details of this computation here, but it depends not only on the average value of the worst stratum but also on its variability. Table 12-3 tells, for various sizes of effect, μ_d compared to σ, the approximate probability that the favorable effect is reversed in the least favorable stratum. The table shows:

1. As we expect, the larger the original effect, μ_d, compared to σ, the smaller is the probability that the effect is reversed in the selected stratum.
2. The more strata we select from, the larger the probability of a reversal.

The table also shows that when the effects are modest in standard deviation units, say $\mu_d = 2\sigma$, even 2 or 4 strata produce high chances of reversal.

Just as choosing the least favorable stratum produces a bias downward, choosing the most favorable stratum will produce a bias upward. With the

Table 12-3.   If the least favorable of k strata is chosen when the overall difference μ_d is favorable to a new treatment at the 2, 3, or 4σ level, the entry shows the approximate probability that the effect is reversed in that stratum

k	μ_d/σ 2	3	4
2	.15	.03	.005
4	.50	.24	.09
8	.89	.71	.48

same assumptions, the bias will be of the same size. Thus there is a substantial danger that fishing expeditions can destroy the basic information gained from the whole sample.

Applying the Method in Clinical Problem 12-1

Returning to the HDFP, we note that one of the 4 race–sex subgroups shown in Table 12-1 had an outcome reversed from the overall result of SC benefit. White women given SC did slightly worse than white women assigned to RC. Although the 4 age–sex subgroups were not of equal size, we will assume that they were and try to use Table 12-2 to estimate whether a reversal by chance is a likely event.

If we assume that the true difference between SC and RC treatment is in fact the observed difference, we can estimate μ_d/σ from the P value reported for the results. When the P value is .05, we know the value of the ratio of the observed difference (here assumed to be equal to the true difference) to the standard deviation of the observed difference is approximately 2. The P value of the HDFP results is about .01; this implies that the ratio μ_d/σ is about 2.4. If the P value were .001, the ratio would be 3 (see Chapter 7).

Table 11-2 suggests that if the ratio μ_d/σ is 2 and $k = 4$, there is a 50% chance of a reversal in the least favorable subgroup. If the ratio is 3, there is a 24% chance of reversal. The HDFP results have a ratio of about 2.4 and we estimate that a reversal would occur by chance in 1 of 4 subgroups about 33% of the time. The unfavorable outcome observed in white women given SC therapy is not inconsistent with a chance effect.

12-6.   Implications for Evaluating Study Results by Stratum

We have recommended that you write down in advance your concerns about a patient. How will this help? Your written concerns flow from some medical question about the patient; they are not designed to choose the worst (or best) item from a cafeteria.

If you choose to examine a particular subgroup based on preexisting medical concerns, you will not be tempted to fit your patient into a subgroup with particularly poor results. You may even find an unexpectedly favorable outcome. Ms. J.S. has mild hypertension. There was some doubt prior to the HDFP whether patients with mild hypertension would benefit from therapy. The HDFP results show that this subgroup had striking benefits. Thus it is possible to address your concerns about an individual patient and still not indulge in a fishing expedition.

An exception arises when the only reason for your concern comes from the trial itself. Then you are stuck with the result unless you can find some literature to help further. If you have picked the result from a large number of possible results, you were fishing. Unless the result is extremely significant, you can discount it considerably. Looking everywhere for the worst can mislead, just as looking for the best misleads in the opposite direction.

Correcting the Bias in the Least Favorable Subgroup

Suppose that you do look at the worst of k strata, in spite of the discussion above. How bad does the worst stratum have to be before you decide that the treatment effect in that stratum differs from the effect overall? We give a rough-and-ready answer, again treating the groups as of equal size, but this time investigating the difference between the result in the least favorable group and $\bar{d}$, the **observed** overall result.

Let $\bar{d}_k$ be the observed difference in the least favorable stratum; $\bar{d}$, the overall difference; and n, the number of patients in **each** treatment group. As before, B_k is the bias factor for the expected outcome in the least favorable stratum. Its values are shown in Table 12-2. S_k is a new factor that adjusts the standard deviation to allow for the selection effect. Its values are given in Table 12-4.

The measure of departure in the least favorable stratum, the kth, is

$$z_k = \frac{(\bar{d} - \bar{d}_k) - B_k \sigma \sqrt{2k/n}}{S_k \sigma \sqrt{2k/n}}$$

Because the kth stratum has the least favorable results for innovation, $\bar{d}_k$ will be smaller than $\bar{d}$. If $\bar{d}_k$ is much smaller than $\bar{d}$ after correcting for selection bias, z_k will be substantially positive. The hypothesis of uniform treatment efficacy can be tested by referring z_k to a standard normal table. A similar approach can be used to evaluate results in the most favorable stratum.

EXAMPLE. The least favorable outcome among the 4 race–sex groups receiving SC therapy in the HDFP was an excess mortality of .1% among white women. What is the P value for this finding compared to the overall improvement in mortality on SC therapy of 1.3%? We will compute z_k by assuming that the 4 race–sex groups were of equal size. Then $\bar{d} = .013$, $\bar{d}_k = -.001, n = 10,940/2 = 5470$, and $k = 4$. We can estimate the standard deviation of the outcome (lived or died) using the expression for the standard deviation of a binomial response with $n = 1$ (Chapter 6). The average mortality rate in the two treatment groups was $\bar{p} = .07$, so the estimated stan-

Table 12-4. Values of S_k for $k = 2$ to 10

k	S_k
2	.43
3	.48
4	.49
5	.50
6	.50
7	.50
8	.50
9	.50
10	.49

dard deviation is

$$\sqrt{\bar{p}\bar{q}} = \sqrt{(.07)(.93)} = .255$$

We obtain B_k from Table 11-2 and S_k from Table 11-4 and

$$z_4 = \frac{.013 + .001 - 1.03(.255)\sqrt{(2 \times 4)/5470}}{.49(.255)\sqrt{(2 \times 4)/5470}}$$

$$= .83$$

Referring to Table III-1, we find that .83 has a one-sided P value of about .20. Thus the poor results observed among white women may well reflect a chance observation in a subgroup with a true gain of .013.

Comment: μ_d and $\bar{d}$

In our discussion of the probability of reversal in the least favorable stratum, we focused on the **true** difference, μ_d, rather than $\bar{d}$, the observed difference. We answered the question: What is the probability of reversal given a true positive difference? When reading a paper, we encounter only observed differences, and a more appropriate question might be: What is the probability of reversal given an **observed** positive difference? To answer this question, we need to focus on $\bar{d}$, but we found the calculations based on $\bar{d}$ too complex to include in this book. In our approach to correcting bias in the least favorable stratum, however, we were able to base our calculations on $\bar{d}$, the observed overall difference.

12-7.   Linear Relationships Between Patient Characteristics and Treatment Efficacy

We sometimes have reason to believe that response to treatment depends on one or more patient characteristics. We can describe some aspects of this situation by discussing the special case when that dependence is linear.

A Single Characteristic

Suppose that the difference between the innovative and standard therapies varies linearly with some patient characteristic, denoted x. Then we can estimate the gain from the innovative therapy, $\hat{d}_x$, by

$$\hat{d}_x = \bar{d} + b(x - \bar{x})$$

where $\bar{d}$ is the estimated overall gain, $\bar{x}$ the average value of x in the study population, and b the sample regression coefficient of gain on x. We use the symbol $\hat{d}_x$ to emphasize that the estimated gain for an individual patient depends on the patient's value for the characteristic x.

We consider the situation when $\bar{d}$ is positive, indicating overall superiority of the innovation. Then innovation is preferred for all values of x that satisfy

$$\bar{d} + b(x - \bar{x}) > 0$$

We can also assume that b is positive, changing the sign of x if necessary. This implies that the benefit of treatment is greatest for patients with large values of x, and least for patients with small values of x. Remembering that $\bar{d}$ and b are positive, $\hat{d}_x$ is positive unless x is so much smaller than $\bar{x}$ that

$$b(x - \bar{x}) < \bar{d}$$

When x is small enough so this inequality holds, then $\hat{d}_x$, the estimated gain, is negative, and the innovative treatment probably should not be given. From the inequality, we can draw three conclusions:

1. If $\bar{d}$ is large, representing a substantial average treatment effect, the estimated gain of treatment will not be negative except for patients whose x value is very far below $\bar{x}$.
2. If b is nearly equal to 0, the effect of x on treatment efficacy is small, and this variable can be ignored in choosing treatment.
3. If x is near $\bar{x}$, say within 1 standard deviation, the direction of treatment superiority will ordinarily not be reversed.

Several Characteristics

If more than one variable is involved in your concern, multiple regression might be used. The same considerations that encouraged writing down specific concerns apply here as well. You can fish in regression just as with subgroups. We encourage you to stick with variables that gave you specific concerns and not to hunt up every variable in the study.

We again turn to the HDFP and examine the relation between SC benefit and patient age to illustrate how regression may be used to apply results to the individual patient. Table 12-1 gives the difference in 5-year mortality between SC and RC treatment in three age groups. From the table we can construct three pairs of values for age and mortality-reduction pairs. Patients with mean age 40 (40 is assumed to be the mean age of the patients aged 30 to 49) had a reduction of .2, those age 55 had a reduction of 2.1, and those age 65 had 2.5. This is shown graphically in Figure 12-1. A line has been fitted by eye to the points.

Using the line rather than the points, we see that patients aged 50.8, the average age of HDFP subjects, have an estimated reduction of 1.3, the average mortality reduction. SC therapy provided more benefit for older people. However, the graph also suggests that the benefits might be reversed in younger subjects. Other studies of the effects of antihypertensive therapy have suggested greater benefits with older ages, so we feel fairly confident that benefits increase with age. We are less confident about the implications of this analysis for younger patients, since the relationship between age and response may not be linear, especially when extrapolated to age groups not included in the study.

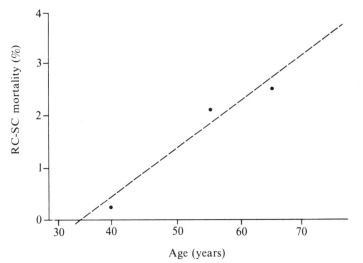

Figure 12-1.   Difference in 5-year mortality between RC and SC therapy by age.

12-8.   Is Your Patient from a Different Population?

A difficulty may arise if we cannot regard patients in the trial as comparable to our patient. For example, in the trial, patients suffering from an additional disease were excluded. Your patient has an additional disease. Now what?

   You have to decide whether the additional disease is related to the treatments, and how. The difficulty is that you probably cannot investigate this from the study results. Unless other literature can help, you are thrown back on your own theory of the effect of the additional disease.

> In applying a clinical trial to the individual patient, we have encountered two types of unresolved doubts: theoretically grounded and unexplained doubts. For the first, a cautious attitude toward the results of the trial seems appropriate because of justified doubts. For the second, excess caution makes unsupported fears override firm, favorable evidence about the new treatment.

12-9.   Favorable Nonsignificant Effects

What if the result were not statistically significant, but the new treatment did perform better than the standard? Doesn't this still suggest that the new treatment is slightly better? Yes. Why, then, do we tend to dismiss such a treatment?

   As a matter of medical policy, it may be undesirable to introduce slightly good looking treatments, since many will actually fail and the costs of introduction and change may be huge. But these concerns may not apply to your patient. If the treatment has no contraindications, is safe, and performed a little better than the standard, you may decide to try it, particularly if standard treatment has failed.

12-10. Choosing a Policy for Accepting Innovations

To illustrate the policy question, we imagine a stream of innovations, each being compared to its own standard. To be concrete about the distribution of gains, we assume that the average true gain from innovations is 0, but that the gains of individual innovations have a normal distribution centered at 0 with a standard deviation of 10. To simplify the problem further, we assume that the *estimated* gain in each study also has a standard deviation of 10. [This model might fit some surgery situations (Gilbert *et al.*, 1977).]

To make sure this model is clear, let us review. When an innovation occurs, its true improvement over the standard therapy is a random number drawn from a normal distribution with mean 0 and standard deviation 10. But we do not observe these; we observe the number (say -8) plus an error drawn from the same distribution, say an error of 12. Our observation is $-8 + 12 = 4$, giving in this instance a misleading favorable impression of the innovation. The probability that a treatment with a true gain of zero will produce an observed result larger than 20 is .025, and the probability that it will produce a positive value is .5.

We would like to winnow out the true zero and true negative values, and we can get help doing this by setting some limits. If we accept all positive observations, then $1/2$ of the trials are positive and $1/4$ of these come from the negative side (by a calculation not given here).

Retaining the same model, if we kept only treatments that gained 20 units on the average, we would keep, in all, .0785 (nearly 8%) of the innovations and .0754 would come from treatments with a true positive treatment effect. (The calculations are not shown.) Thus 96% of those accepted would have true positive effects.

The purpose of this discussion is to clarify the difficulty of accepting every positive appearance as favorable. We are not trying to settle the complicated problem of choosing cutoff levels or encouraging the use of significance levels for decisions. The discussion does illustrate that in some realistic circumstances one might accept 25% of bad therapies in order to accept every therapy producing a positive observation. Higher cutoffs will sacrifice some favorable therapies and can almost eliminate unfavorable ones.

12-11. Delivering the Therapy

Trial Results Are Obtained with a Specific Treatment Protocol

Let us now return to our patient, Ms. J.S. We decide that she should be treated even though she is considerably younger than the mean age of patients in the HDFP and her high blood pressure may not be sustained at home. How should she be treated? What drugs should be prescribed? How should she be followed?

The HDFP investigated only one treatment strategy, a complex stepped approach to drug therapy administered in special clinics to promote compliance. To be confident that your patient will receive the benefits of the SC therapy, you should follow the SC treatment regimen and follow-up strategy as precisely as possible.

We quote the entire section of the HDFP report (1979a) describing the SC therapy.

SC Treatment Regimen

Stepped Care participants were offered free a standardized program of antihypertensive therapy in HDFP centers. These centers differ from most traditional ambulatory care facilities in a number of ways. The participants were actively and intensively recruited. Uninterrupted antihypertensive drug therapy was attempted as far as possible by means of techniques presently believed to enhance adherence. Emphasis was placed on clinic attendance and adherence to medication schedules. Pill counts were used to help monitor drug adherence. Economic barriers to adherence were removed as much as possible; drugs, visits at the centers, laboratory tests, and, if necessary, transportation were provided at no cost to the participant. Waiting times were minimized by efficient operation and by use of various types of allied health personnel. Appointments were made at convenient hours, and a Program physician was on call at all times to deal with problems related to hypertension.

Medication was increased stepwise to bring patients to or below their goal DBP, defined as 90 mm Hg for those entering with DBP of 100 mm Hg or greater or already receiving the antihypertensive therapy, and a 10 mm decrease for those entering with DBP of 90 to 99 mm Hg. Antihypertensive drugs were prescribed in a broadly standardized stepwise sequence. Within each step, dosages were generally increased stepwise if DBP had not reached goal within a period specified by the HDFP *Manual of Operations*. The steps were as follows: *Step 1*—prescription of the diuretic, chlorthalidone (25 to 100 mg/day); triamterene (50 to 300 mg/day) or spironolactone (25 to 100 mg/day) could be prescribed as supplementary or alternative medication if indicated, *Step 2*—**addition** of an antiadrenergic drug, preferably reserpine (.1 to .25 mg/day); but alternatively methyldopa (500 to 2,000 mg/day), *Step 3*—**addition** of a vasodilator, hydralazine (30 to 200 mg/day), *Step 4*—**addition** of an antiadrenergic drug, guanethidine sulfate (10 to 200 mg/day), with or without discontinuation of medication added at Step 2 or 3, and *Step 5*—**addition** or **substitution** of other drugs. . . .

At all steps, the only drugs prescribed were those approved by the Food and Drug Administration for treatment of hypertension. Other antihypertensive drugs newly approved by the FDA during the course of the trial were then made available for Step 5 use.

As specified by the protocol, drug dosage and regimen could be changed as required to minimize side effects and maintain BP control.

Participants were generally informed of the desirability of avoiding high salt intake. The protocol also provided that for SC participants who were markedly overweight (40% or more above desirable weight), frankly hypercholesterolemic (250 mg/dL or greater), or heavy smokers (ten or more cigarettes per day), counseling was to be offered with regard to control of these risk factors. In all counseling, however, the primary emphasis was on drug therapy to achieve BP control.

Thus the HDFP treatment program consisted of:

1. Clinics designed to be pleasant, convenient, cheap, and to enhance compliance.
2. A complex regimen of antihypertensive drug therapy with most patients prescribed chlorthalidone, a thiazidelike diuretic. Reserpine and hydralazine were the next drugs prescribed, as needed, to reach each patient's goal blood pressure.
3. Supplemental advice concerning other cardiovascular risk factors.

What should you prescribe for Ms. J.S.? Does it matter that you usually prescribe the thiazide, hydrochlorothiazide, and not chlorthalidone? We have no specific reason to suspect the hydrochlorothiazide will be any less effective than chlorthalidone, even though it has a considerably shorter duration of action; but all other things being equal, why not prescribe the drug actually used in the trial? (The cost of drugs may be one factor in your decision.)

Note that propranolol was not extensively used in the HDFP, yet propranolol has become a popular antihypertensive drug. (The HDFP was planned in the early 1970s before propranolol was approved for hypertension.) Should propranolol be offered as a second agent instead of reserpine? The results of the HDFP tell us what we can expect from a strategy where reserpine is the step 2 agent. We can only speculate that results would not be substantially different with propranolol.

Can you provide the special, convenient, low-cost clinics and the attention to the other cardiovascular risk factors that were part of the HDFP strategy? Some physicians have remarked that the extra "tender loving care" provided by the HDFP is as likely to explain the SC results as is the antihypertensive therapy. In any case, the drug therapy was offered in this special setting and other approaches to delivering the same pharmacotherapy may result in diminished compliance, less satisfaction with the treatment, and loss of any benefit that may derive from substantial care. A trial tests therapy only in a specific setting. It is often difficult to know how important the setting is to the results.

Some trials are not really designed to guide practice, but to investigate whether a new therapy can benefit patients under ideal circumstances. The original Veterans Administration cooperative trial in hypertension can be considered such as investigation. The treatment strategy of this kind of trial may be impossible to mimic in practice.

Sometimes it is difficult to understand why a trial employed a therapy that cannot be reproduced in practice. For example, the Peterson ulcer trial discussed in Chapter 11 administered a specially formulated antacid which was not on the general market. From the point of view of a physician who wants to prescribe an antacid therapy that works, it is frustrating that the original trial to demonstrate the effectiveness of antacids in healing ulcers employed an antacid that cannot be prescribed.

Changing a Treatment Protocol

If you, as a physician, modify a clinical trial protocol in applying the results to your patient, you assume that you have captured, in your modified therapy, that aspect of the protocol which caused the beneficial results. Although you and many other physicians may think the modifications reasonable, your modified therapy has not been tested and this adds uncertainty to the expected outcome.

We see this problem as far reaching. Clinical trials test only one or two therapies at most; physicians prescribe a huge variety of different regimens. Here are some thoughts on a partial solution.

1. Pay close attention to the therapy actually prescribed in the clinical trial and copy it as closely as possible.
2. Realize that any modification of the therapy makes it an untested one.
3. Clinical investigators who intend to conduct trials to guide the therapy of individual patients should try to employ therapies that can be easily mimicked in practice.

12-12. Summary

The following is a checklist for applying the results of a clinical trial to your patient.

1. Review the trial to be confident that the therapy worked for the study patients (see Chapter 11).
2. Review the admission criteria to see if your patient is similar to the study patients.
3. Specify in advance if there is a subgroup of patients that is particularly appropriate to your patient and review the results of study patients in that subgroup to see if anything unusual occurred.
4. Do not search the study results for relevant subgroups with particularly favorable or unfavorable results. It will lead to a biased view of the effects of therapy in your patient.
5. Regression analysis can help determine whether the benefits of therapy might not be present in a patient who has some characteristic that is far from the average of the study subjects on a clinically important variable.
6. Mimic the treatment strategy employed in the trial as closely as possible since any modification could alter the expected results.

Problems

12-1. What is the problem of generalizing study results to your patient?

12-2. How can you tell whether your subject is similar to the study subjects? How do you handle possible differences?

12-3. How might selection bias enter the interpretation of a study? Is there a good procedure for identifying a subgroup of the study population that is particularly applicable to your patient?

12-4. How often will a trial with a P value of .05 favoring the new treatment have a reversed effect in one of 4 subgroups?

12-5. A trial shows that patients given a new therapy did slightly better than those given the standard, $P = .2$. When would you prescribe the new therapy for your patient? Would you change your hospital policy to provide the new therapy routinely?

12-6. Can you expect a clinical trial to assist in tailoring therapy to your patients' special characteristics? Why?

12-7. We believe that articles reporting clinical trials which bear titles such as "Antacid therapy for stress ulcers," "Antibiotic therapy for pancreatitis," or "Antihypertensive therapy for hypertension in pregnancy" are misleading. They suggest that a class of drugs have been evaluated, but usually only a particular drug has been used. Do you agree? Why?

12-8. Mr. F.L. is 65 years old. He has just been admitted to the intensive care unit with 2 hours of substernal chest pain and nonspecific T-wave changes on his electrocardiogram. He has no signs of heart failure and about 10 ventricular premature contractions (VPCs) a minute. Would you apply the results of the article on lidocaine prophylaxis (Lie *et al.*, 1974 reproduced in the Appendix to Chapter 10) to the patient? Why? How exactly would you treat F.L.? Would you alter the therapy if the VPCs did not decrease? Would you treat him if he were 75 years old, or if he had congestive heart failure?

References

Gilbert, J. P.; McPeek, B.; and Mosteller, F. (1977). Progress in surgery and anesthesia; benefits and risks of innovative therapy, in Bunker, J. P.; Barnes, B. A.; and Mosteller, F., eds., *Costs, Risks, and Benefits of Surgery*. New York: Oxford University Press.

Hypertension Detection and Follow-up Program Cooperative Group (1977). The hypertension detection and follow-up program: a progress report. *Circ. Res.*, **40**(Suppl. 1):106–9.

Hypertension Detection and Follow-up Program Cooperative Group (1979a). Five-year findings of the Hypertension Detection and Follow-up Program: I. Reduction in mortality of persons with high blood pressure, including mild hypertension. *JAMA*, **242**:2562–71.

Hypertension Detection and Follow-up Program Cooperative Group (1979b). Five-year findings of the Hypertension Detection and Follow-up Program: II. Mortality by race, sex and age. *JAMA*, **242**:2572–77.

Lie, K. I.; Wellens, H. J.; van Capell, F. J.; and Durrer, D. (1974). Lidocaine in the prevention of primary ventricular fibrillation. *N. Engl. J. Med.*, **291**:1324–26.

Veterans Administration Cooperative Study Group on Antihypertensive Agents (1967). Effects of treatment on morbidity in hypertension: results in patients with diastolic blood pressure averaging 115 through 129 mm Hg. *JAMA*, **202**:1028–34.

Veterans Administration Cooperative Study Group on Antihypertensive Agents (1970). Effects of treatment on morbidity in hypertension: II. Results in patients with diastolic blood pressure averaging 90 through 114 mm Hg. *JAMA*, **213**:1143–52.

Veterans Administration Cooperative Study Group on Antihypertensive Agents (1972). Effects of treatment on morbidity in hypertension: III. Influence of age, diastolic pressure, and prior cardiovascular disease; further analysis of side effects. *Circulation*, **45**:991–1004.

Additional Reading

See References at the end of Chapter 11.

APPENDIX I

Additional Problems

I-1. As stated in Chapter 2, 15% of the cases of hemophilia are from a new mutation. Ms. A. has 6 brothers, 1 of whom is hemophilic. What are the chances that Ms. A.'s hemophilic brother represents a "new" mutation?

I-2. A 56-year-old man has his serum cholesterol measured on 7 successive weekly visits. The results are:

260, 210, 240, 230, 260, 150, 230

   a. Find the mean $\bar{x}$, sample variance s^2, and standard deviation s of these values.
   b. Estimate the standard error of $\bar{x}$.

I-3. Blood pressure in 5 normal subjects was recorded as follows (pulse pressure = systolic pressure − diastolic pressure):

Subject	Systolic pressure	Diastolic pressure	Pulse pressure
1	145	90	55
2	120	70	50
3	110	65	45
4	120	55	65
5	105	70	35

a. Find the mean and variance of the systolic, diastolic, and pulse pressures.

b. Is the variance of the pulse pressure (the difference between the systolic and diastolic pressures) equal to the sum of the variance for the systolic and diastolic pressures? Why or why not?

I-4. Suppose that we have collected data on 96 families in which there are 3 male offspring with a normal father and a mother who is thought to be a carrier for hemophilia.

a. What is the probability that a given family has no affected male offspring? How many such families should we expect to see, on the average, in our sample of 96?

b. What is the standard deviation of the number of families with 0 affected male offspring in samples of 96 families with 3 male offspring?

c. Now suppose that the sample in fact includes 56 families in which all 3 boys are free of hemophilia. Do we have any reason to doubt whether the mothers are, in fact, carriers? Would your answer change if we had 76 families?

I-5. The following questions concern the diagnosis and treatment of sore throat due to group A streptococci (also called streptococcal or strep pharyngitis).

Group A streptococcus can be cultured from the throats of patients with sore throats due to streptococcal pharyngitis; however, about 10% of these patients will have negative cultures. A person can also be a "carrier of group A streptococci." Such a person will have a positive culture but no evidence of infection as measured by antibodies in the patient's serum. Some investigators have found that 20% of patients with sore throat not due to group A streptococci will have positive cultures with no antibodies in the serum. Usually, 20% of patients with sore throats have streptococcal pharyngitis.

a. Construct the 2×2 table of probabilities for culture results, positive or negative, by disease, streptococcal pharyngitis or not, for patients with sore thoat.

b. Calculate the sensitivity and specificity of throat culture for diagnosing streptococcal pharyngitis in patients with sore throat.

Streptococcal pharyngitis causes rheumatic fever, a severe disease which requires hospitalization, in 1% of patients who are not treated with penicillin and in virtually no patients who are treated with penicillin. Serious allergic reactions requiring hospitalization occur in .1% of patients given penicillin whether they have the disease or not.

c. Construct a decision tree comparing the following therapeutic strategies for the prevention of rheumatic fever. Strategy 1—treat all patients with penicillin who have a sore throat. Strategy 2—culture all patients with sore throat and treat patients with positive cultures with penicillin. Note that you will need the results of part a.

d. Is the choice of "best strategy" sensitive to the weighting of the severity of rheumatic fever compared to serious allergic reaction?

I-6. The following is part of an abstract of a paper in the *New England Journal of Medicine* (Vol. 298, No. 14, pp. 763–67, 1978) on the follow-up of diethylstilbestrol (DES)-treated mothers by Bibbo *et al.*

To assess the long term effects of diethylstilbestrol (DES) we conducted a health survey among 693 mothers who had taken the drug during pregnancy and a comparable group of 668 who had not. . . . There were 32 (4.6 percent) breast cancers among the 693 exposed and 21 (3.1 percent) among the 668 unexposed, but the difference was not statistically significant ($P = .16$).

What does $P = .16$ mean?

I-7. Calculate the mean, variance, and standard deviation of the following distribution.

x	0	1	2	3
$P(x)$	.3	.5	.1	.1

I-8. Calculate the sample mean, variance, and standard deviation of the 6 repeated plasma digoxin concentrations for patient 5 in Table I-1.

Table I-1. Digoxin concentration on 6 consecutive days in 10 patients on a continuous daily dose of .25 mg

Patient	Plasma concentration of digoxin (ng /ml)					
1	1.65	1.95	2.10	1.55	1.80	1.75
2	1.50	1.25	1.50	1.30	2.15	2.20
3	.80	.95	.95	.80	1.45	1.15
4	.70	.65	.80	1.10	1.05	1.00
5	.75	.75	1.00	.90	.80	.80
6	.55	.80	.80	.95	.80	.85
7	.70	.80	.80	.70	.80	.95
8	.75	.65	.75	.65	.50	.70
9	.55	.60	.75	.80	.55	.65
10	.60	.60	.75	.60	.60	.55

I-9. Table I-1 gives 6 repeated plasma digoxin concentrations for each of 10 patients. What are some potential sources of variability within a patient? Between patients?

I-10. Using the data in Table I-1, plot the interquartile range versus the median for patients 1 through 6. What interquartile range would you estimate for a patient with a median plasma digoxin concentration of 82 ng/ml? Are there any outlying observations in the data from patient 1?

I-11. Each of 3 tests for a certain disease has a sensitivity of .8 and a specificity of .9. Assume that the tests are statistically independent. Let x denote the number of positive tests of the 3 when all 3 are run on a patient. Given that the patient on which the tests are run has the disease, what is the distribution of x?

I-12. In the population tested in Problem I-11, 10% of the patients have the disease. If 10,000 patients are tested, write out the expected counts for the 4 × 2 table whose columns are disease status (D+, D−) and whose rows are outcomes of tests (number positive: 0, 1, 2, 3).

I-13. Use the data in the 4 × 2 table obtained in Problem I-12 to compute the odds favoring disease for each of the 4 outcomes. What is the probability of disease when one of three tests is positive?

I-14. A patient for whom our prior probability of disease was $^2/_3$ was given the three tests described in Problem I-11 above. We observed $x = 2$ (2 positive, 1 negative). We then calculated the odds of disease as

$$\frac{2}{1} \times \frac{3(.8)(.8)(.2)}{3(.1)(.1)(.9)} = 28.44 = \text{final odds}$$

a. Describe in detail the components of the final odds calculation. (Where did the numbers come from?)

b. What is the probability of disease corresponding to these odds?

I-15. Mr. W.R. has combined hyperlipidemia with a serum cholesterol of 320, 340, 380, 360, and 350 mg per 100 ml on 5 different occassions. As recommended, you start the patient on a low-cholesterol diet. Subsequent cholesterols in Mr. W.R. over the next few weeks are 300, 280, 320, 340, and 310 mg per 100.

a. Test whether cholesterol has decreased using the data from the two samples.

b. Set 90% confidence limits on the difference in the means.

c. How compatible is the second set of measurements with a cholesterol of 300 mg per 100 ml which is the goal of therapy?

d. Discuss the pros and cons of obtaining several pretreatment and posttreatment measurements as opposed to basing therapy on an initial observation.

I-16. It is currently believed that over half, say 60%, of males with symptoms of urethral discharge have nonspecific urethritis (NSU). The rest have gonorrhea (GC). A colleague has said that he has seen a lot of GC lately. That evening you see 10 male patients with urethral discharge. Of these, 9 have GC, the other NSU.

a. What is the probability of seeing at least 9 GC patients out of 10 patients with urethral discharge under the assumption of a 60% prevalence of NSU? What is the P value for these results assuming a 60% prevalence? (State whether you are performing a one-sided or two-sided test.)

b. What do you gain by this analysis beyond your clinical observation that you, too, saw a lot of GC?

I-17. A few percent of patients exposed to tuberculosis (TB) develop active TB. One year's treatment with prophylactic isoniazid (INH) has been recommended because patients treated with INH develop only $\frac{1}{3}$ as much active TB as did non-treated subjects. However, 1 to 2% of adults treated with isoniazid for a year develop hepatitis, which is fatal in 5% of cases; that is, the probability of death given hepatitis is equal to .05.

a. Construct a simple decision tree to compare isoniazid prophylaxis versus no prophylaxis in exposed patients (assume that patients do not get both TB and hepatitis).

b. There are certain "facts" which you have not been told: for instance, P(active TB|exposure) and P(severe morbidity|active TB). Is the decision sensitive to these or other unknowns? Would you treat yourself prophylactically if exposed?

c. What advantage does formulating the decision tree have over merely following the recommendations that all exposed patients (household contacts, recent PPD converters, and those with a positive PPD and abnormal chest x-ray) should receive a year's worth of isoniazid therapy?

Short Solutions to Odd-Numbered Problems

Chapter 1

1-1. a. $P(B|A) = \dfrac{.06}{.06 + .42}$

　　　　　　$= .13$

　　b.　　　　　A　not A

　　　　　B　| .06 | .24 |　.3

　　　not B　| .42 | .28 |　.7

1-3. The sensitivity is

$$P(T_1+ \text{ and } T_2+|D+) = (.90)(.80)$$
$$= .72$$

The specificity is

$$P(T_1- \text{ or } T_2-|D-) = 1 - P(T_1+ \text{ and } T_2+|D-)$$
$$= 1 - (.20)(.10)$$
$$= .98$$

1-5. T+ = two or more positive tests (ERP, US, PFT, ANG)
　　T− = more than two negative tests

a.

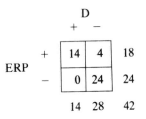

b. Sensitivity $P(T+|D+)$ = 1.00
   Specificity $P(T-|D-)$ = .75

   PV positive $P(D+|T+)$ = .67
   PV negative $P(D-|T-)$ = 1.00

   For ERP,

```
                   D
               +    −
        +  ┌────┬────┐
           │ 14 │  4 │  18
   ERP     ├────┼────┤
        −  │  0 │ 24 │  24
           └────┴────┘
             14   28    42
```

   Sensitivity $P(ERP+|D+)$ = 1.00
   Specificity $P(ERP-|D-)$ = .86

   PV positive $P(D+|ERP+)$ = .78
   PV negative $P(D-|ERP-)$ = 1.00

c. There is hardly any difference between the efficiency of these schemes, but the single-test ERP outperforms multiple tests in this sample.
d. The positive predictive value is slightly out of line.

Chapter 2

2-1. The family can have 0 to 5 hemophilic sons with probabilities

Number	0	1	2	3	4	5
Probability	1/32	5/32	10/32	10/32	5/32	1/32

2-3. If $T+$ and $T-$ are the possible test results, and $D+$ and $D-$ denote disease status, we define

Sensitivity: $P(T+|D+)$
Specificity: $P(T-|D-)$

The likelihood ratio for a negative test is

$$LR- = \frac{P(T-|D+)}{P(T-|D-)}$$

$$= \frac{1 - \text{sensitivity}}{\text{specificity}}$$

2-5. a.

	Fever ≥ 38.3°C	Cervical adenopathy	Pharyngeal exudate	History of exposure	Positive Gram stain
LR+	3.00	1.62	2.05	2.25	18.25
LR−	.75	.49	.71	.89	.28

b. Assuming a prior probability of .10, the posterior probability is .25.

c. Let LR_1+ be the likelihood ratio for presence of cervical adenopathy and LR_2+ be the likelihood ratio for history of exposure. Then, assuming independence, we have

$$\text{Posterior odds} = \text{prior odds} \times (LR_1+) \times (LR_2+)$$

$$= \frac{1}{9} \times 3.0 \times 2.25$$

$$= .75$$

The posterior probability is

$$\frac{.73}{1 + .73} = .42$$

d. Symptoms, history, and diagnostic tests may not be independent. In general, we have little information about independence and we have to reason from our clinical knowledge. In this case we might conclude that the signs of fever, exudate, and adenopathy are not independent of each other but are independent of the history of exposure and of the Gram stain.

e. Positive Gram stain is as sensitive as any test, more sensitive than most, and the most specific test.

2-7.

Number toxic	0	1	2	3	4
Probability	.32	.42	.21	.05	.00

Chapter 3

3-1.

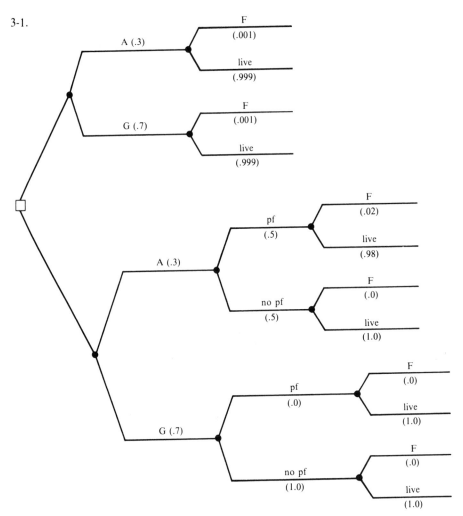

Decision tree for Problem 3-1. A = appendicitis, G = gastroenteritis, pf = perforation, and F = a fatal outcome.

3-3. When P(appendicitis) > .1, as in this case, immediate operation produces the lowest mortality; delay is preferred when P(appendicitis) < .01.

3-5. Considering death alone

$$P(\text{death}|\text{treatment}) = .0076$$
$$P(\text{death}|\text{no treatment}) = .025$$

However,

$$P(\text{morbidity}|\text{treatment}) = .057$$
$$P(\text{morbidity}|\text{no treatment}) = .0$$

Let us assign life a score of 0 and death a score of 1 (badness scoring) and morbidity a score of M. Then no treatment scores .025 and treatment scores .0076 +

.057M. The two results are equal when $M = .3$. As long as the value of M stays below 30% of that of death, treatment is preferred.

3-7. Another approach is to see at what probability of embolism the strategies produce equivalent death rates; at any $P(E)$ greater than this, treatment will produce fewer deaths.

$$P(F|\text{treatment}) = P(E)(.15)(.5) + P(E)(.15)(.5)(.0001)$$
$$+ P(E)(.85)(.0001) + [1 - P(E)](.0001)$$
$$= .0001 + .07499P(E)$$
$$P(F|\text{no treatment}) = P(E)(.5)(.5) = .25P(E)$$

These probabilities are equal when

$$.175P(E) = .0001$$

or

$$P(E) = .0006$$

We see that the estimate of the probability of embolism does not have to be precise. Therefore, if $P(E) \geq .0006$, treatment will minimize the probability of death.

Chapter 4

4-3. The patient ranges from 4 to 8, a range of 4 compared to the range of 7 in Problem 4-2.

4-5. a. The ordered differences are

$$-1, -1, 1, 2, 5, 8, 9, 10, 12, 14, 15, 21, 23, 26, 40, 112$$

b. The number of patients consuming less TNG while receiving propranolol is 14. The expected number is 8, because each treatment is equally likely to be more successful.

Chapter 5

5-1. a. The estimated mean is 86, the 80% confidence interval is (82.8, 89.2), and the 95% confidence interval is (81.1, 90.9). The confidence that his mean blood pressure is below 90 mm Hg is 94.5%.
b. The estimated standard deviation is 4.3.
c. The 80% critical value is 1.64 and the 80% confidence interval is (82.5, 89.5). The 95% confidence interval is (79.2, 92.8).

5-3. a. Solving for $n \geq [(1.28 \times 6)/5]^2$, we find that $n \geq 2.35$ or $n = 3$.
b. By the same method, $n = 59$—very large for a clinical situation.

5-7. A plot of the median against the range for these 10 patients gives evidence consistent with the hypothesis of constant variance.

5-9. The difference, $\bar{x}_2 - \bar{x}_1$, has variance 25. Thus we require a reduction of 8.3 mm Hg to ensure a 5% misclassification rate for persons not responding to treatment.

Chapter 5A

5A-1. Cum.

1	0	8
4	1	099
9	2	23679
(4)	3	0235
8	4	233
5	5	2
4	6	25
2	7	
2	8	58

$21 = 9 + 4 + 8$

5A-3. The range is sensitive to outliers. An outlying value will inflate the range and we will probably not detect the outlier.

5A-5. Although 2 of the 4 outliers occur in SL placebo, it would be difficult to construe this as a pattern.

Chapter 6

6-1. a. The one-sided P value is .194; the two-sided P value is .388.
b. The probability of 4 or more successes when $\pi = .10$ is .026.

6-3. The relapse rates are .00 for lithium and .55 for placebo. The pooled rate is .27. Thus the z score is 4.1, corresponding to a P value less than .001. We reject the hypothesis of equal rates.

6-5. Using the normal approximation, we obtain $z = 2.25$ for testing the equality of the two proportions. So the two-sided P value is .024. We conclude that the rate has changed.

6-7. The 80% confidence limits are .55 and .92.

6-9. Using the normal approximation, the 90% confidence interval is .52 to .68.

6-11. The observed proportion is .38 with $n = 100$. From Chart III-2 the 80% confidence interval is .31 to .44.

6-13. Since $m = np = 2$, the probability of 0 events is .135, from Table III-5.

6-15. In this problem, $m = 60 \times 6 = 360$, so the normal approximation gives a critical ratio of -1.58 for a count of 330 and a normal tail probability of .06.

Chapter 7

7-1. A calculation computes the probability that under given assumptions, the data would depart from the results expected as much as or more than those observed in the study. That calculated probability is a P value. Different assumptions would generate different P values. When the P value is small, it casts doubt on the assumptions because other possibilities are likely to be true.

7-3. The probability that the null hypothesis, such as that $\pi = .50000$ . . . , is exactly true is zero. Thus the probability that the results happened by chance would ordinarily be zero. It is the probability **given** the null hypothesis that we compute.

7-5. The physician must take many other matters into account, such as ability to deliver the therapy, danger to the patient, and patient's preferences.

7-7. When we have a very special issue which we can make quantitative and a statistical test appropriate for it, the P value helps us reach **conclusions** about the situation, even though they may not settle treatment policy.

Chapter 8

8-1. The chi-square statistic measures the departure between counts and their expectations by summing

$$\frac{(O - E)^2}{E}$$

where O is the observed count and E the expected. Large discrepancies between O and E lead to large numerators, although they can be discounted if E is large.

8-3. The idea of a degree of freedom generally is that it is a free dimension. Thus the degrees of freedom in a contingency table tell how many free dimensions are available. The dimensions are related to the cells, but because of the constraints in totals, not every cell is free.

8-5. 20%.

8-7. a.

Observed counts

	KEPT APPOINTMENT	MISSED APPOINTMENT	*Total*
Reminded	20	5	25
Not reminded	8	12	20
Total	28	17	45

Expected counts

	KEPT APPOINTMENT	MISSED APPOINTMENT	*Total*
Reminded	15.6	9.4	25
Not reminded	12.4	7.6	20
Total	28.0	17.0	45

b. $\chi^2 = \left[(20 - 15.6) - \frac{1}{2} \right]^2 \left(\frac{1}{15.6} + \frac{1}{12.4} + \frac{1}{9.4} + \frac{1}{7.6} \right)$

$= 5.8$

The P value is between .01 and .05.

Chapter 9

9-1. The plot shows no trend unless we set aside period 7, the placebo-treated response. Then the plot, especially through periods of 1 to 6 of increasing dose, shows a declining rate.

9-3. The plot of migraine rate against log dose suggests an approximately linear relation.

9-5. Among the doses tested, the highest, 160 mg, was most effective. Higher doses may be more effective, but this requires extrapolation of our results beyond the range studied. The study provides no basis for that extrapolation. The migraine rate appears to be substantially reduced by 40 mg three times a day.

9-7. The observation of Paradise *et al.* that an undocumented history of frequent sore throat is a poor predictor of future sore throats may reflect regression effects rather than the unreliability of an "undocumented history." Regression effect predicts that patients who have an unusually high number of colds and sore throats one year will have fewer the next.

Chapter 9A

9A-1. Sometimes transforming the x values or y values by logarithms, square roots, or reciprocals, for example, may straighten them.

9A-3. The F test tests the significance of the line in reducing the sum of squares of deviations. Is the line a better fit than just the grand mean of the y's?

9A-5. $r^2 = 100/400 = .25$.

Chapter 9B

9B-1. R correlates y with $\hat{y}$, r correlates y with x. Except possibly for sign, r also is the correlation of y with $\hat{y}$.

9B-3. If we can choose b_0 to be zero as well, then $2^{k+1} = 2^{12} = 4096$.

9B-5. Data often become sparse when many models must be separately fitted.

9B-7. The adjusted R^2 is 18.4%, substantially smaller than the R^2 of 76.7% achieved using $1/Cr$.

9B-9. The predicted values for the first two men are 116.4 and 190.5, respectively. Using the second equation, the predicted values are 125.8 and 198.0. The two sets of predicted values agree reasonably well, as would be expected if height and weight are highly correlated in the 180 men.

Chapter 10

10-1. 0.85.

10-3. Twelve deaths in the surgical group and 33 deaths in the medical group. The resulting 2×2 table is:

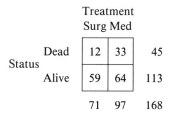

Treatment
Surg Med

		Surg	Med	
Status	Dead	12	33	45
	Alive	59	64	113
		71	97	168

The chi-square statistic for this table (See Chapter 8) is 6.67, which implies $P < .01$. Thus, the chi-square test for association gives a smaller P-value than the life table analysis. The life table analysis differs from the chi-square test for association in that it considers the order in which the deaths occur. Here, we infer that the deaths of the surgically treated patients occurred somewhat earlier in the 5-year interval than the deaths in the medically treated patients.

Chapter 11

11-1. In parallel studies, independent samples receive each treatment, and the patients are ordinarily treated concurrently. The assignment of the treatment is what is randomized. For example, with three treatments, each individual ordinarily has a one-third chance of receiving each treatment. In a crossover design, the patient undergoes two or more treatments usually at different times. What is randomized is the order in which the treatment is given.

11-3. In randomized, unblinded trials, bias can arise from different ascertainment of endpoints and from different interpretation of clinical events. Ascertainment bias can be minimized with a uniform follow-up program. Unambiguous endpoints such as death, hospitalization, or operation are probably not subject to bias. Sometimes, ancillary therapy differs between groups in an unblinded trial. Such treatment differences can be regarded as part of the different treatment strategies, rather than bias.

11-5. When endpoints are not specified in advance, many endpoints can be considered in exploratory analysis, increasing the probability that significant relationships will be detected even when the null hypothesis is correct.

11-7. Treatment was randomized and blinded. Complete follow-up was reported. A P value was reported, but power was not discussed. Primary ventricular fibrillation was the main endpoint, and the effect in the entire study population was emphasized. This is a "good" study, but some design features were incompletely described. The slightly smaller number of deaths in the treatment group is consistent with our view that lidocaine reduces death.

Chapter 12

12-1. My patient may not have been eligible for the study, and also may be rather different from those in the study even if he or she would have passed the admission requirements. Also, I may not be able to provide the exact therapy used in the study.

12-3. We recommend writing down in advance the variables that concern you, explaining to yourself why they are definite concerns, and then seeing if something is to be learned from the study about this subgroup of patients, or from other literature.

12-5. If there is no contraindication, there is no reason for not trying it if it can be provided as described in the study. At the same time, it may not be wise to change hospital policy on the basis of such slender evidence, because it leads to the introduction of too many poor therapies, with all the attendant costs and mistakes that accompany hospital changes.

12-7. We believe that the title should reflect the exact treatments used so that successful therapy can be mimicked in practice.

Statistical Tables and Charts

This section provides a set of statistical tables and charts. We have extended and supplemented the tables required in the book so that these tables will be sufficient for most commonly encountered problems in clinical research.

At the top of each page, we list all the tables in order with the table on that page in boldface type. This makes it easy to find an individual table once the book is open to this section because it tells you whether to leaf forward, backward, or stay.

Table III-1.   Probabilities, $A(z)$, for the Standard Gaussian or Normal Distribution

Area under the standard normal curve from 0 to z, shown shaded, is $A(z)$.

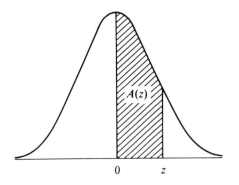

EXAMPLE.   If Z is the standard normal random variable and $z = 1.54$, then

$$A(z) = P(0 < Z < z) = .4382,$$
$$P(Z > z) = .0618$$
$$P(Z < z) = .9382,$$
$$P(|Z| < z) = .8764$$

Table III-1. Probabilities, $A(z)$, for the standard gaussian or normal distribution

z	.00	.01	.02	.03	.04	.05	.06	.07	.08	.09
0.0	.0000	.0040	.0080	.0120	.0160	.0199	.0239	.0279	.0319	.0359
0.1	.0398	.0438	.0478	.0517	.0557	.0596	.0636	.0675	.0714	.0753
0.2	.0793	.0832	.0871	.0910	.0948	.0987	.1026	.1064	.1103	.1141
0.3	.1179	.1217	.1255	.1293	.1331	.1368	.1406	.1443	.1480	.1517
0.4	.1554	.1591	.1628	.1664	.1700	.1736	.1772	.1808	.1844	.1879
0.5	.1915	.1950	.1985	.2019	.2054	.2088	.2123	.2157	.2190	.2224
0.6	.2257	.2291	.2324	.2357	.2389	.2422	.2454	.2486	.2517	.2549
0.7	.2580	.2611	.2642	.2673	.2704	.2734	.2764	.2794	.2823	.2852
0.8	.2881	.2910	.2939	.2967	.2995	.3023	.3051	.3078	.3106	.3133
0.9	.3159	.3186	.3212	.3238	.3264	.3289	.3315	.3340	.3365	.3389
1.0	.3413	.3438	.3461	.3485	.3508	.3531	.3554	.3577	.3599	.3621
1.1	.3643	.3665	.3686	.3708	.3729	.3749	.3770	.3790	.3810	.3830
1.2	.3849	.3869	.3888	.3907	.3925	.3944	.3962	.3980	.3997	.4015
1.3	.4032	.4049	.4066	.4082	.4099	.4115	.4131	.4147	.4162	.4177
1.4	.4192	.4207	.4222	.4236	.4251	.4265	.4279	.4292	.4306	.4319
1.5	.4332	.4345	.4357	.4370	.4382	.4394	.4406	.4418	.4429	.4441
1.6	.4452	.4463	.4474	.4484	.4495	.4505	.4515	.4525	.4535	.4545
1.7	.4554	.4564	.4573	.4582	.4591	.4599	.4608	.4616	.4625	.4633
1.8	.4641	.4649	.4656	.4664	.4671	.4678	.4686	.4693	.4699	.4706
1.9	.4713	.4719	.4726	.4732	.4738	.4744	.4750	.4756	.4761	.4767
2.0	.4772	.4778	.4783	.4788	.4793	.4798	.4803	.4808	.4812	.4817
2.1	.4821	.4826	.4830	.4834	.4838	.4842	.4846	.4850	.4854	.4857
2.2	.4861	.4864	.4868	.4871	.4875	.4878	.4881	.4884	.4887	.4890
2.3	.4893	.4896	.4898	.4901	.4904	.4906	.4909	.4911	.4913	.4916
2.4	.4918	.4920	.4922	.4925	.4927	.4929	.4931	.4932	.4934	.4936
2.5	.4938	.4940	.4941	.4943	.4945	.4946	.4948	.4949	.4951	.4952
2.6	.4953	.4955	.4956	.4957	.4959	.4960	.4961	.4962	.4963	.4964
2.7	.4965	.4966	.4967	.4968	.4969	.4970	.4971	.4972	.4973	.4974
2.8	.4974	.4975	.4976	.4977	.4977	.4978	.4979	.4979	.4980	.4981
2.9	.4981	.4982	.4982	.4983	.4984	.4984	.4985	.4985	.4986	.4986
3.0	.4987	.4987	.4987	.4988	.4988	.4989	.4989	.4989	.4990	.4990

Source: F. Mosteller, R. E. K. Rourke, and G. B. Thomas, Jr., *Probability with Statistical Applications,* 2nd ed., © 1970, Addison-Wesley, Reading, Mass. (Table III); reprinted with permission.

Table III-2. Values of $t_{df,1-\alpha}$ for Student *t* Distributions

EXAMPLE. For a *t* distribution with 9 degrees of freedom, the critical value corresponding to a two-sided confidence level of $1 - \alpha$ = .95 or a two-sided significance level of α = .05 is 2.26.

Table III-2. Values of $t_{df,1-\alpha}$ for Student *t* distributions

Degrees of freedom	(Two-sided) probability level $(1 - \alpha)$					
	.50	.80	.90	.95	.98	.99
1	1.00	3.08	6.31	12.71	31.82	63.66
2	.82	1.89	2.92	4.30	6.96	9.93
3	.76	1.64	2.35	3.18	4.54	5.84
4	.74	1.53	2.13	2.78	3.75	4.60
5	.73	1.48	2.02	2.57	3.36	4.03
6	.72	1.44	1.94	2.45	3.14	3.71
7	.71	1.41	1.89	2.36	3.00	3.50
8	.71	1.40	1.86	2.31	2.90	3.36
9	.70	1.38	1.83	2.26	2.82	3.25
10	.70	1.37	1.81	2.23	2.76	3.17
15	.69	1.34	1.75	2.13	2.60	2.95
30	.68	1.31	1.70	2.04	2.46	2.75
50	.68	1.30	1.68	2.01	2.40	2.68
100	.68	1.29	1.66	1.98	2.36	2.63
∞[a]	.67	1.28	1.64	1.96	2.33	2.58

[a] For infinite degrees of freedom, the *t* distribution is the normal distribution.

Table III-3. Three-Place Tables of the Binomial Distribution

This table gives the values of the function

$$b(d; n, \pi) = \binom{n}{d} \pi^d(1 - \pi)^{n-d}$$

$$= \frac{n!}{d!(n - d)!} \pi^d(1 - \pi)^{n-d}$$

This is the probability of exactly d successes in n independent binomial trials with probability of success on a single trial equal to π.

Values of the functions are given for $d = 0, 1, \ldots, n; n = 2, 3, \ldots, 17$; and $\pi = .01, .05, .10, .20, .30, .40, .50, .60, .70, .80, .90, .95$, and $.99$.

In this table, each three-digit entry should be read with a decimal preceding it. For entries $0+$, the probability is less than $.0005$ but greater than 0.

EXAMPLE. If $n = 6$ and $\pi = .3$, the probability of $d = 3$ successes is $.185$. The probability of 3 or fewer successes ($d \leq 3$) is $.118 + .303 + .324 + .185 = .930$.

Table III-3. Three-place tables of the binomial distribution

n	d	.01	.05	.10	.20	.30	.40	.50	.60	.70	.80	.90	.95	.99	d
2	0	980	902	810	640	490	360	250	160	090	040	010	002	0+	0
	1	020	095	180	320	420	480	500	480	420	320	180	095	020	1
	2	0+	002	010	040	090	160	250	360	490	640	810	902	980	2
3	0	970	857	729	512	343	216	125	064	027	008	001	0+	0+	0
	1	029	135	243	384	441	432	375	288	189	096	027	007	0+	1
	2	0+	007	027	096	189	288	375	432	441	384	243	135	029	2
	3	0+	0+	001	008	027	064	125	216	343	512	729	857	970	3
4	0	961	815	656	410	240	130	062	026	008	002	0+	0+	0+	0
	1	039	171	292	410	412	346	250	154	076	026	004	0+	0+	1
	2	001	014	049	154	265	346	375	346	265	154	049	014	001	2
	3	0+	0+	004	026	076	154	250	346	412	410	292	171	039	3
	4	0+	0+	0+	002	008	026	062	130	240	410	656	815	961	4
5	0	951	774	590	328	168	078	031	010	002	0+	0+	0+	0+	0
	1	048	204	328	410	360	259	156	077	028	006	0+	0+	0+	1
	2	001	021	073	205	309	346	312	230	132	051	008	001	0+	2
	3	0+	001	008	051	132	230	312	346	309	205	073	021	001	3
	4	0+	0+	0+	006	028	077	156	259	360	410	328	204	048	4
	5	0+	0+	0+	0+	002	010	031	078	168	328	590	774	951	5
6	0	941	735	531	262	118	047	016	004	001	0+	0+	0+	0+	0
	1	057	232	354	393	303	187	094	037	010	002	0+	0+	0+	1
	2	001	031	098	246	324	311	234	138	060	015	001	0+	0+	2
	3	0+	002	015	082	185	276	312	276	185	082	015	002	0+	3
	4	0+	0+	001	015	060	138	234	311	324	246	098	031	001	4
	5	0+	0+	0+	002	010	037	094	187	303	393	354	232	057	5
	6	0+	0+	0+	0+	001	004	016	047	118	262	531	735	941	6

Table III-3. *Continued*

n	d	.01	.05	.10	.20	.30	.40	.50	.60	.70	.80	.90	.95	.99	d
7	0	932	698	478	210	082	028	008	002	0+	0+	0+	0+	0+	0
	1	066	257	372	367	247	131	055	017	004	0+	0+	0+	0+	1
	2	002	041	124	275	318	261	164	077	025	004	0+	0+	0+	2
	3	0+	004	023	115	227	290	273	194	097	029	003	0+	0+	3
	4	0+	0+	003	029	097	194	273	290	227	115	023	004	0+	4
	5	0+	0+	0+	004	025	077	164	261	318	275	124	041	002	5
	6	0+	0+	0+	0+	004	017	055	131	247	367	372	257	066	6
	7	0+	0+	0+	0+	0+	002	008	028	082	210	478	698	932	7
8	0	923	663	430	168	058	017	004	001	0+	0+	0+	0+	0+	0
	1	075	279	383	336	198	090	031	008	001	0+	0+	0+	0+	1
	2	003	051	149	294	296	209	109	041	010	001	0+	0+	0+	2
	3	0+	005	033	147	254	279	219	124	047	009	0+	0+	0+	3
	4	0+	0+	005	046	136	232	273	232	136	046	005	0+	0+	4
	5	0+	0+	0+	009	047	124	219	279	254	147	033	005	0+	5
	6	0+	0+	0+	001	010	041	109	209	296	294	149	051	003	6
	7	0+	0+	0+	0+	001	008	031	090	198	336	383	279	075	7
	8	0+	0+	0+	0+	0+	001	004	017	058	168	430	663	923	8
9	0	914	630	387	134	040	010	002	0+	0+	0+	0+	0+	0+	0
	1	083	299	387	302	156	060	018	004	0+	0+	0+	0+	0+	1
	2	003	063	172	302	267	161	070	021	004	0+	0+	0+	0+	2
	3	0+	008	045	176	267	251	164	074	021	003	0+	0+	0+	3
	4	0+	001	007	066	172	251	246	167	074	017	001	0+	0+	4
	5	0+	0+	001	017	074	167	246	251	172	066	007	001	0+	5
	6	0+	0+	0+	003	021	074	164	251	267	176	045	008	0+	6
	7	0+	0+	0+	0+	004	021	070	161	267	302	172	063	003	7
	8	0+	0+	0+	0+	0+	004	018	060	156	302	387	299	083	8
	9	0+	0+	0+	0+	0+	0+	002	010	040	134	387	630	914	9
10	0	904	599	349	107	028	006	001	0+	0+	0+	0+	0+	0+	0
	1	091	315	387	268	121	040	010	002	0+	0+	0+	0+	0+	1
	2	004	075	194	302	233	121	044	011	001	0+	0+	0+	0+	2
	3	0+	010	057	201	267	215	117	042	009	001	0+	0+	0+	3
	4	0+	001	011	088	200	251	205	111	037	006	0+	0+	0+	4
	5	0+	0+	001	026	103	201	246	201	103	026	001	0+	0+	5
	6	0+	0+	0+	006	037	111	205	251	200	088	011	001	0+	6
	7	0+	0+	0+	001	009	042	117	215	267	201	057	010	0+	7
	8	0+	0+	0+	0+	001	011	044	121	233	302	194	075	004	8
	9	0+	0+	0+	0+	0+	002	010	040	121	268	387	315	091	9
	10	0+	0+	0+	0+	0+	0+	001	006	028	107	349	599	904	10
11	0	895	569	314	086	020	004	0+	0+	0+	0+	0+	0+	0+	0
	1	099	329	384	236	093	027	005	001	0+	0+	0+	0+	0+	1
	2	005	087	213	295	200	089	027	005	001	0+	0+	0+	0+	2
	3	0+	014	071	221	257	177	081	023	004	0+	0+	0+	0+	3
	4	0+	001	016	111	220	236	161	070	017	002	0+	0+	0+	4
	5	0+	0+	002	039	132	221	226	147	057	010	0+	0+	0+	5
	6	0+	0+	0+	010	057	147	226	221	132	039	002	0+	0+	6
	7	0+	0+	0+	002	017	070	161	236	220	111	016	001	0+	7
	8	0+	0+	0+	0+	004	023	081	177	257	221	071	014	0+	8
	9	0+	0+	0+	0+	001	005	027	089	200	295	213	087	005	9

Table III-3. *Continued*

n	d	.01	.05	.10	.20	.30	.40	.50	.60	.70	.80	.90	.95	.99	d
11	10	0+	0+	0+	0+	0+	001	005	027	093	236	384	329	099	10
	11	0+	0+	0+	0+	0+	0+	0+	004	020	086	314	569	895	11
12	0	886	540	282	069	014	002	0+	0+	0+	0+	0+	0+	0+	0
	1	107	341	377	206	071	017	003	0+	0+	0+	0+	0+	0+	1
	2	006	099	230	283	168	064	016	002	0+	0+	0+	0+	0+	2
	3	0+	017	085	236	240	142	054	012	001	0+	0+	0+	0+	3
	4	0+	002	021	133	231	213	121	042	008	001	0+	0+	0+	4
	5	0+	0+	004	053	158	227	193	101	029	003	0+	0+	0+	5
	6	0+	0+	0+	016	079	177	226	177	079	016	0+	0+	0+	6
	7	0+	0+	0+	003	029	101	193	227	158	053	004	0+	0+	7
	8	0+	0+	0+	001	008	042	121	213	231	133	021	002	0+	8
	9	0+	0+	0+	0+	001	012	054	142	240	236	085	017	0+	9
	10	0+	0+	0+	0+	0+	002	016	064	168	283	230	099	006	10
	11	0+	0+	0+	0+	0+	0+	003	017	071	206	377	341	107	11
	12	0+	0+	0+	0+	0+	0+	0+	002	014	069	282	540	886	12
13	0	878	513	254	055	010	001	0+	0+	0+	0+	0+	0+	0+	0
	1	115	351	367	179	054	011	002	0+	0+	0+	0+	0+	0+	1
	2	007	111	245	268	139	045	010	001	0+	0+	0+	0+	0+	2
	3	0+	021	100	246	218	111	035	006	001	0+	0+	0+	0+	3
	4	0+	003	028	154	234	184	087	024	003	0+	0+	0+	0+	4
	5	0+	0+	006	069	180	221	157	066	014	001	0+	0+	0+	5
	6	0+	0+	001	023	103	197	209	131	044	006	0+	0+	0+	6
	7	0+	0+	0+	006	044	131	209	197	103	023	001	0+	0+	7
	8	0+	0+	0+	001	014	066	157	221	180	069	006	0+	0+	8
	9	0+	0+	0+	0+	003	024	087	184	234	154	028	003	0+	9
	10	0+	0+	0+	0+	001	006	035	111	218	246	100	021	0+	10
	11	0+	0+	0+	0+	0+	001	010	045	139	268	245	111	007	11
	12	0+	0+	0+	0+	0+	0+	002	011	054	179	367	351	155	12
	13	0+	0+	0+	0+	0+	0+	0+	001	010	055	254	513	878	13
14	0	869	488	229	044	007	001	0+	0+	0+	0+	0+	0+	0+	0
	1	123	359	356	154	041	007	001	0+	0+	0+	0+	0+	0+	1
	2	008	123	257	250	113	032	006	001	0+	0+	0+	0+	0+	2
	3	0+	026	114	250	194	085	022	003	0+	0+	0+	0+	0+	3
	4	0+	004	035	172	229	155	061	014	001	0+	0+	0+	0+	4
	5	0+	0+	008	086	196	207	122	041	007	0+	0+	0+	0+	5
	6	0+	0+	001	032	126	207	183	092	023	002	0+	0+	0+	6
	7	0+	0+	0+	009	062	157	209	157	062	009	0+	0+	0+	7
	8	0+	0+	0+	002	023	092	183	207	126	032	001	0+	0+	8
	9	0+	0+	0+	0+	007	041	122	207	196	086	008	0+	0+	9
	10	0+	0+	0+	0+	001	014	061	155	229	172	035	004	0+	10
	11	0+	0+	0+	0+	0+	003	022	085	194	250	114	026	0+	11
	12	0+	0+	0+	0+	0+	001	006	032	113	250	257	123	008	12
	13	0+	0+	0+	0+	0+	0+	001	007	041	154	356	359	123	13
	14	0+	0+	0+	0+	0+	0+	0+	001	007	044	229	488	869	14
15	0	860	463	206	035	005	0+	0+	0+	0+	0+	0+	0+	0+	0
	1	130	366	343	132	031	005	0+	0+	0+	0+	0+	0+	0+	1
	2	009	135	267	231	092	022	003	0+	0+	0+	0+	0+	0+	2
	3	0+	031	129	250	170	063	014	002	0+	0+	0+	0+	0+	3

Normal / *t* / **Binomial** / Binomial confidence limits / Poisson /
F-5% / *F*-1% / Poisson confidence limits / χ^2 / Binomial, π = ½
/ Random digits

Table III-3. *Continued*

								π								
n	*d*	.01	.05	.10	.20	.30	.40	.50	.60	.70	.80	.90	.95	.99	*d*
15	4	0+	005	043	188	219	127	042	007	001	0+	0+	0+	0+	4
	5	0+	001	010	103	206	186	092	024	003	0+	0+	0+	0+	5
	6	0+	0+	002	043	147	207	153	061	012	001	0+	0+	0+	6
	7	0+	0+	0+	014	081	177	196	118	035	003	0+	0+	0+	7
	8	0+	0+	0+	003	035	118	196	177	081	014	0+	0+	0+	8
	9	0+	0+	0+	001	012	061	153	207	147	043	002	0+	0+	9
	10	0+	0+	0+	0+	003	024	092	186	206	103	010	001	0+	10
	11	0+	0+	0+	0+	001	007	042	127	219	188	043	005	0+	11
	12	0+	0+	0+	0+	0+	002	014	063	170	250	129	031	0+	12
	13	0+	0+	0+	0+	0+	0+	003	022	092	231	267	135	009	13
	14	0+	0+	0+	0+	0+	0+	0+	005	031	132	343	366	130	14
	15	0+	0+	0+	0+	0+	0+	0+	0+	005	035	206	463	860	15
16	0	851	440	185	028	003	0+	0+	0+	0+	0+	0+	0+	0+	0
	1	138	371	329	113	023	003	0+	0+	0+	0+	0+	0+	0+	1
	2	010	146	275	211	073	015	002	0+	0+	0+	0+	0+	0+	2
	3	0+	036	142	246	146	047	009	001	0+	0+	0+	0+	0+	3
	4	0+	006	051	200	204	101	028	004	0+	0+	0+	0+	0+	4
	5	0+	001	014	120	210	162	067	014	001	0+	0+	0+	0+	5
	6	0+	0+	003	055	165	198	122	039	006	0+	0+	0+	0+	6
	7	0+	0+	0+	020	101	189	175	084	019	001	0+	0+	0+	7
	8	0+	0+	0+	006	049	142	196	142	049	006	0+	0+	0+	8
	9	0+	0+	0+	001	019	084	175	189	101	020	0+	0+	0+	9
	10	0+	0+	0+	0+	006	039	122	198	165	055	003	0+	0+	10
	11	0+	0+	0+	0+	001	014	067	162	210	120	014	001	0+	11
	12	0+	0+	0+	0+	0+	004	028	101	204	200	051	006	0+	12
	13	0+	0+	0+	0+	0+	001	009	047	146	246	142	036	0+	13
	14	0+	0+	0+	0+	0+	0+	002	015	073	211	275	146	010	14
	15	0+	0+	0+	0+	0+	0+	0+	003	023	113	329	371	138	15
	16	0+	0+	0+	0+	0+	0+	0+	0+	003	028	185	440	851	16
17	0	843	418	167	023	002	0+	0+	0+	0+	0+	0+	0+	0+	0
	1	145	374	315	096	017	002	0+	0+	0+	0+	0+	0+	0+	1
	2	012	158	280	191	058	010	001	0+	0+	0+	0+	0+	0+	2
	3	001	041	156	239	125	034	005	0+	0+	0+	0+	0+	0+	3
	4	0+	008	060	209	187	080	018	002	0+	0+	0+	0+	0+	4
	5	0+	001	017	136	208	138	047	008	001	0+	0+	0+	0+	5
	6	0+	0+	004	068	178	184	094	024	003	0+	0+	0+	0+	6
	7	0+	0+	001	027	120	193	148	057	009	0+	0+	0+	0+	7
	8	0+	0+	0+	008	064	161	185	107	028	002	0+	0+	0+	8
	9	0+	0+	0+	002	028	107	185	161	064	008	0+	0+	0+	9
	10	0+	0+	0+	0+	009	057	148	193	120	027	001	0+	0+	10
	11	0+	0+	0+	0+	003	024	094	184	178	068	004	0+	0+	11
	12	0+	0+	0+	0+	001	008	047	138	208	136	017	001	0+	12
	13	0+	0+	0+	0+	0+	002	018	080	187	209	060	008	0+	13
	14	0+	0+	0+	0+	0+	0+	005	034	125	239	156	041	001	14
	15	0+	0+	0+	0+	0+	0+	001	010	058	191	280	158	012	15
	16	0+	0+	0+	0+	0+	0+	0+	002	017	096	315	374	145	16
	17	0+	0+	0+	0+	0+	0+	0+	0+	002	023	167	418	843	17

Source: F. Mosteller, R. E. K. Rourke, and G. B. Thomas, Jr., *Probability with Statistical Applications*, 2nd ed., © 1970, Addison-Wesley, Reading, Mass. (Table IV); reprinted by permission.

Table III-4. Table of 80% and 95% Confidence Limits for a Binomial Proportion

This table gives the lower limits, π_L, and the upper limits, π_U, of the 80% and 95% confidence intervals for π, the probability of success on a single trial. Values are given for n, the number of trials, from 1 to 20, and all values of d, the number of successes.

EXAMPLE. For $n = 9$ and $d = 3$, the 80% confidence interval is (.17, .56) and the 95% confidence interval is (.10, .71).

Table III-4. Table of 80% and 95% confidence limits for a binomial proportion

n	d	80% confidence limits π_L	π_U	95% confidence limits π_L	π_U	n	d	80% confidence limits π_L	π_U	95% confidence limits π_L	π_U
1	0	.00	.80	.00	.95	7	3	.22	.70	.13	.77
	1	.20	1.00	.05	1.00		4	.30	.78	.23	.87
2	0	.00	.55	.00	.78		5	.42	.88	.34	.95
	1	.11	.89	,03	.97		6	.58	.97	.45	.99
	2	.45	1.00	.22	1.00		7	.78	1.00	.62	1.00
3	0	.00	.50	.00	.63	8	0	.00	.25	.00	.31
	1	.07	.72	.02	.86		1	.02	.38	.01	.50
	2	.28	.93	.14	.98		2	.10	.50	.05	.68
	3	.50	1.00	.37	1.00		3	.19	.62	.11	.71
4	0	.00	.41	.00	.53		4	.25	.75	.19	.81
	1	.05	.59	.01	.75		5	.38	.81	.29	.89
	2	.21	.79	.10	.90		6	.50	.90	.32	.95
	3	.41	.95	.25	.99		7	.62	.98	.50	.99
	4	.59	1.00	.47	1.00		8	.75	1.00	.69	1.00
5	0	.00	.32	.00	.50	9	0	.00	.22	.00	.29
	1	.04	.53	.01	.66		1	.02	.34	.01	.44
	2	.16	.68	.08	.81		2	.09	.44	.04	.56
	3	.32	.84	.19	.92		3	.17	.56	.10	.71
	4	.47	.96	.34	.99		4	.22	.66	.17	.75
	5	.68	1.00	.50	1.00		5	.34	.78	.25	.83
6	0	.00	.28	.00	.40		6	.44	.83	.29	.90
	1	.03	.50	.01	.60		7	.56	.91	.44	.96
	2	.13	.64	.06	.73		8	.66	.98	.56	.99
	3	.26	.74	.15	.85		9	.78	1.00	.71	1.00
	4	.36	.87	.27	.94	10	0	.00	.20	.00	.27
	5	.50	.97	.40	.99		1	.02	.31	.01	.40
	6	.72	1.00	.60	1.00		2	.08	.45	.04	.60
7	0	.00	.22	.00	.38		3	.15	.55	.09	.62
	1	.03	.42	.01	.55		4	.20	.62	.15	.73
	2	.12	.58	.05	.66		5	.31	.69	.22	.78
							6	.38	.80	.27	.85

313

Table III-4. *Continued*

n	d	80% confidence limits π_L	π_U	95% confidence limits π_L	π_U	n	d	80% confidence limits π_L	π_U	95% confidence limits π_L	π_U
11	7	.45	.85	.38	.91	14	2	.06	.32	.03	.39
	8	.55	.92	.40	.96		3	.12	.39	.06	.50
	9	.69	.98	.60	.99		4	.16	.46	.10	.61
	10	.80	1.00	.73	1.00		5	.22	.53	.15	.63
							6	.26	.60	.21	.69
11	0	.00	.19	.00	.25		7	.33	.67	.21	.79
	1	.03	.28	.01	.37		8	.40	.74	.31	.79
	2	.08	.41	.03	.50		9	.47	.78	.37	.85
	3	.15	.50	.08	.63		10	.54	.84	.39	.90
	4	.20	.58	.14	.67		11	.61	.88	.50	.94
	5	.29	.66	.20	.75		12	.68	.94	.61	.97
	6	.34	.71	.25	.80		13	.75	.95	.69	1.00
	7	.42	.80	.33	.86		14	.85	1.00	.79	1.00
	8	.50	.85	.37	.92	15	0	.00	.14	.00	.19
	9	.59	.92	.50	.97		1	.02	.23	.00	.30
	10	.72	.97	.63	.99		2	.06	.30	.02	.37
	11	.81	1.00	.75	1.00		3	.11	.37	.06	.45
12	0	.00	.17	.00	.24		4	.15	.43	.10	.55
	1	.01	.27	.00	.35		5	.21	.50	.14	.63
	2	.06	.37	.03	.45		6	.24	.57	.19	.67
	3	.13	.45	.07	.55		7	.31	.62	.19	.71
	4	.17	.54	.12	.65		8	.38	.69	.29	.81
	5	.26	.63	.18	.71		9	.43	.76	.33	.81
	6	.30	.70	.24	.76		10	.50	.79	.37	.86
	7	.37	.74	.29	.82		11	.57	.85	.45	.90
	8	.46	.83	.35	.88		12	.63	.88	.55	.94
	9	.55	.87	.45	.93		13	.70	.94	.63	.98
	10	.63	.94	.55	.97		14	.77	.98	.70	1.00
	11	.73	.99	.65	1.00		15	.86	1.00	.81	1.00
	12	.83	1.00	.76	1.00	16	0	.00	.13	.00	.18
13	0	.00	.15	.00	.22		1	.01	.22	.00	.27
	1	.02	.25	.00	.33		2	.06	.25	.02	.35
	2	.07	.37	.03	.43		3	.10	.35	.05	.43
	3	.13	.42	.07	.52		4	.14	.43	.09	.50
	4	.16	.50	.11	.59		5	.19	.50	.13	.57
	5	.24	.58	.17	.67		6	.23	.56	.18	.65
	6	.26	.62	.22	.74		7	.29	.62	.18	.73
	7	.38	.74	.26	.78		8	.35	.65	.27	.73
	8	.42	.76	.33	.83		9	.38	.71	.27	.82
	9	.50	.84	.41	.89		10	.44	.77	.35	.82
	10	.58	.87	.48	.93		11	.50	.81	.43	.87
	11	.63	.93	.57	.97		12	.57	.86	.50	.91
	12	.75	.98	.67	1.00		13	.65	.90	.57	.95
	13	.85	1.00	.78	1.00		14	.75	.94	.65	.98
14	0	.00	.15	.00	.21		15	.78	.95	.73	1.00
	1	.01	.25	.00	.31		16	.87	1.00	.82	1.00

Normal / *t* / Binomial / **Binomial confidence limits** / Poisson /
F-5% / *F*-1% / Poisson confidence limits / χ^2 / Binomial, $\pi = \frac{1}{2}$
/ Random digits

Table III-4. *Continued*

n	*d*	80% confidence limits		95% confidence limits		*n*	*d*	80% confidence limits		95% confidence limits	
		π_L	π_U	π_L	π_U			π_L	π_U	π_L	π_U
17	0	.00	.10	.00	.17	19	2	.05	.24	.02	.32
	1	.02	.20	.00	.25		3	.09	.31	.04	.36
	2	.06	.25	.02	.34		4	.12	.37	.08	.43
	3	.10	.35	.05	.42		5	.16	.42	.11	.50
	4	.11	.41	.09	.49		6	.19	.47	.15	.57
	5	.18	.47	.12	.54		7	.25	.52	.15	.63
	6	.21	.52	.17	.59		8	.28	.57	.22	.65
	7	.26	.58	.17	.66		9	.32	.62	.23	.69
	8	.33	.64	.25	.75		10	.38	.68	.31	.77
	9	.36	.67	.25	.75		11	.43	.72	.35	.78
	10	.42	.74	.34	.83		12	.48	.75	.37	.85
	11	.48	.79	.41	.83		13	.53	.81	.43	.85
	12	.53	.82	.46	.88		14	.58	.84	.50	.89
	13	.59	.89	.51	.91		15	.63	.88	.57	.92
	14	.65	.90	.58	.95		16	.69	.91	.64	.96
	15	.75	.94	.66	.98		17	.76	.95	.68	.98
	16	.80	.98	.75	1.00		18	.82	.98	.77	1.00
	17	.90	1.00	.83	1.00		19	.89	1.00	.85	1.00
18	0	.00	.11	.00	.16	20	0	.00	.10	.00	.14
	1	.01	.19	.00	.24		1	.01	.18	.00	.22
	2	.05	.25	.02	.34		2	.05	.23	.02	.29
	3	.09	.33	.05	.38		3	.08	.30	.04	.35
	4	.12	.39	.08	.44		4	.11	.35	.07	.41
	5	.17	.44	.12	.56		5	.15	.40	.10	.47
	6	.20	.50	.16	.62		6	.19	.45	.14	.53
	7	.26	.55	.16	.62		7	.24	.50	.14	.59
	8	.30	.60	.24	.67		8	.27	.54	.21	.65
	9	.34	.66	.24	.76		9	.31	.59	.22	.71
	10	.40	.70	.33	.76		10	.36	.64	.29	.71
	11	.45	.74	.38	.84		11	.41	.69	.29	.78
	12	.50	.80	.38	.84		12	.46	.73	.35	.79
	13	.56	.83	.44	.88		13	.50	.76	.41	.86
	14	.61	.88	.56	.92		14	.55	.81	.47	.86
	15	.67	.91	.62	.95		15	.60	.85	.53	.90
	16	.75	.95	.68	.98		16	.65	.89	.59	.93
	17	.81	.99	.76	1.00		17	.70	.92	.65	.96
	18	.85	1.00	.84	1.00		18	.77	.95	.71	.98
19	0	.00	.11	.00	.15		19	.82	.99	.78	1.00
	1	.02	.18	.00	.23		20	.90	1.00	.86	1.00

Table III-5. Three-Place Tables for the Poisson Distribution

For each value of m, the first column gives values of the individual terms

$$P(X = x) = \frac{e^{-m}m^x}{x!}$$

and the second column gives the cumulative

$$P(X \geq x) = \sum_{r=x}^{x} \frac{e^{-m}m^r}{r!}$$

Values of probabilities are given for $x = 0, 1, 2, \ldots$, and for

$$m = .1(.1)3.5(.5)7(1)14 .$$

Each three-digit entry should be read with a decimal point preceding it. If no entry appears in a column for a particular value of x, then the value of the corresponding individual term or cumulative is less than .0005. For entries $1-$, the probability is larger than .9995 but less than 1.

EXAMPLES. For $m = 8$,

$P(X = 7)\ \ = .140$
$P(X = 19) = .000$, or less than .0005
$P(X \geq 0)\ \ = 1$ (exactly)
$P(X \geq 1)\ \ = 1-$, or more than .9995
$P(X \geq 12) = .112$

Table III-5. Three-place tables for the Poisson distribution

x	IND. m=0.1	CUM. m=0.1	IND. m=0.2	CUM. m=0.2	IND. m=0.3	CUM. m=0.3	IND. m=0.4	CUM. m=0.4	IND. m=0.5	CUM. m=0.5	IND. m=0.6	CUM. m=0.6	IND. m=0.7	CUM. m=0.7	x
0	905	1.	819	1.	741	1.	670	1.	607	1.	549	1.	497	1.	0
1	090	095	164	181	222	259	268	330	303	393	329	451	348	503	1
2	005	005	016	018	033	037	054	062	076	090	099	122	122	156	2
3			001	001	003	004	007	008	013	014	020	023	028	034	3
4							001	001	002	002	003	003	005	006	4
5													001	001	5

x	IND. m=0.8	CUM. m=0.8	IND. m=0.9	CUM. m=0.9	IND. m=1.0	CUM. m=1.0	IND. m=1.1	CUM. m=1.1	IND. m=1.2	CUM. m=1.2	IND. m=1.3	CUM. m=1.3	IND. m=1.4	CUM. m=1.4	x
0	449	1.	407	1.	368	1.	333	1.	301	1.	273	1.	247	1.	0
1	359	551	366	593	368	632	366	667	361	699	354	727	345	753	1
2	144	191	165	228	184	264	201	301	217	337	230	373	242	408	2
3	038	047	049	063	061	080	074	100	087	121	100	143	113	167	3
4	008	009	011	013	015	019	020	026	026	034	032	043	039	054	4
5	001	001	002	002	003	004	004	005	006	008	008	011	011	014	5
6					001	001	001	001	001	002	002	002	003	003	6
7													001	001	7

x	IND. m=1.5	CUM. m=1.5	IND. m=1.6	CUM. m=1.6	IND. m=1.7	CUM. m=1.7	IND. m=1.8	CUM. m=1.8	IND. m=1.9	CUM. m=1.9	IND. m=2.0	CUM. m=2.0	IND. m=2.1	CUM. m=2.1	x
0	223	1.	202	1.	183	1.	165	1.	150	1.	135	1.	122	1.	0
1	335	777	323	798	311	817	298	835	284	850	271	865	257	878	1
2	251	442	258	475	264	507	268	537	270	566	271	594	270	620	2
3	126	191	138	217	150	243	161	269	171	296	180	323	189	350	3
4	047	066	055	079	064	093	072	109	081	125	090	143	099	161	4
5	014	019	018	024	022	030	026	036	031	044	036	053	042	062	5
6	004	004	005	006	006	008	008	010	010	013	012	017	015	020	6
7	001	001	001	001	001	002	002	003	003	003	003	005	004	006	7
8								001	001	001	001	001	001	001	8

Normal / *t* / Binomial / Binomial confidence limits / **Poisson** /
F-5% / *F*-1% / Poisson confidence limits / χ^2 / Binomial, $\pi = \frac{1}{2}$
/ Random digits

Table III-5. *Continued*

x	IND. m=2.2	CUM. m=2.2	IND. m=2.3	CUM. m=2.3	IND. m=2.4	CUM. m=2.4	IND. m=2.5	CUM. m=2.5	IND. m=2.6	CUM. m=2.6	IND. m=2.7	CUM. m=2.7	IND. m=2.8	CUM. m=2.8
0	111	1.	100	1.	091	1.	082	1.	074	1.	067	1.	061	1.
1	244	889	231	900	218	909	205	918	193	926	181	933	170	939
2	268	645	265	669	261	692	257	713	251	733	245	751	238	769
3	197	377	203	404	209	430	214	456	218	482	220	506	222	531
4	108	181	117	201	125	221	134	242	141	264	149	286	156	308
5	048	072	054	084	060	096	067	109	074	123	080	137	087	152
6	017	025	021	030	024	036	028	042	032	049	036	057	041	065
7	005	007	007	009	008	012	010	014	012	017	014	021	016	024
8	002	002	002	003	002	003	003	004	004	005	005	007	006	008
9				001	001	001	001	001	001	001	001	002	002	002
10												001		001

x	IND. m=2.9	CUM. m=2.9	IND. m=3.0	CUM. m=3.0	IND. m=3.1	CUM. m=3.1	IND. m=3.2	CUM. m=3.2	IND. m=3.3	CUM. m=3.3	IND. m=3.4	CUM. m=3.4	IND. m=3.5	CUM. m=3.5
0	055	1.	050	1.	045	1.	041	1.	037	1.	033	1.	030	1.
1	160	945	149	950	140	955	130	959	122	963	113	967	106	970
2	231	785	224	801	216	815	209	829	201	841	193	853	185	864
3	224	554	224	577	224	599	223	620	221	641	219	660	216	679
4	162	330	168	353	173	375	178	397	182	420	186	442	189	463
5	094	168	101	185	107	202	114	219	120	237	126	256	132	275
6	045	074	050	084	056	094	061	105	066	117	072	129	077	142
7	019	029	022	034	025	039	028	045	031	051	035	058	039	065
8	007	010	008	012	010	014	011	017	013	020	015	023	017	027
9	002	003	003	004	003	005	004	006	005	007	006	008	007	010
10	001	001	001	001	001	001	001	002	002	002	002	003	002	003
11										001	001	001	001	001

318

Normal / *t* / Binomial / Binomial confidence limits / **Poisson** /
F-5% / F-1% / Poisson confidence limits / χ^2 / Binomial, $\pi = \tfrac{1}{2}$
/ Random digits

Each value of m has two columns: the first is the individual term $P(X=x)$, the second (headed "1.") is the cumulative $P(X \ge x)$. (Values are given in units of 0.001.)

x	m=4.0 P	≥	m=4.5 P	≥	m=5.0 P	≥	m=5.5 P	≥	m=6.0 P	≥	m=6.5 P	≥	m=7.0 P	≥
0	018	1.	011	1.	007	1.	004	1.	002	1.	002	1.	001	1.
1	073	982	050	989	034	993	022	996	015	998	010	998	006	999
2	147	908	112	939	084	960	062	973	045	983	032	989	022	993
3	195	762	169	826	140	875	113	912	089	938	069	957	052	970
4	195	567	190	658	175	735	156	798	134	849	112	888	091	918
5	156	371	171	468	175	560	171	642	161	715	145	776	128	827
6	104	215	128	297	146	384	157	471	161	554	157	631	149	699
7	060	111	082	169	104	238	123	314	138	394	146	473	149	550
8	030	051	046	087	065	133	085	191	103	256	119	327	130	401
9	013	021	023	040	036	068	052	106	069	153	086	208	101	271
10	005	008	010	017	018	032	029	054	041	084	056	123	071	170
11	002	003	004	007	008	014	014	025	023	043	033	067	045	099
12	001	001	002	002	003	005	007	011	011	020	018	034	026	053
13			001	001	001	002	003	004	005	009	009	016	014	027
14						001	001	002	002	004	004	007	007	013
15								001	001	001	002	003	003	006
16										001	001	001	001	002
17													001	001

x	m=8.0 P	≥	m=9.0 P	≥	m=10.0 P	≥	m=11.0 P	≥	m=12.0 P	≥	m=13.0 P	≥	m=14.0 P	≥
0		1.		1.		1.		1.		1.		1.		1.
1	003	1.–	001	1.–		1.–		1.–		1.–		1.–		1.–
2	011	997	005	999	002	1.–	001	1.–		1.–		1.–		1.–
3	029	986	015	994	008	997	004	999	002	999	001	1.–		1.–
4	057	958	034	979	019	990	010	995	005	998	003	999	001	1.–
5	092	900	061	945	038	971	022	985	013	992	007	996	004	998
6	122	809	091	884	063	933	041	962	025	980	015	989	009	994
7	140	687	117	793	090	870	065	921	044	954	028	974	017	986
8	140	547	132	676	113	780	089	857	066	910	046	946	030	968
9	124	407	132	544	125	667	109	768	087	845	066	900	047	938

Table III-5. *Continued*

x	IND. m = 8.0	CUM. m = 8.0	IND. m = 9.0	CUM. m = 9.0	IND. m = 10.0	CUM. m = 10.0	IND. m = 11.0	CUM. m = 11.0	IND. m = 12.0	CUM. m = 12.0	IND. m = 13.0	CUM. m = 13.0	IND. m = 14.0	CUM. m = 14.0	x
10	099	283	119	413	125	542	119	659	105	758	086	834	066	891	10
11	072	184	097	294	114	417	119	540	114	653	101	748	084	824	11
12	048	112	073	197	095	303	109	421	114	538	110	647	098	740	12
13	030	064	050	124	073	208	093	311	106	424	110	537	106	642	13
14	017	034	032	074	052	136	073	219	090	318	102	427	106	536	14
15	009	017	019	041	035	083	053	146	072	228	088	325	099	430	15
16	005	008	011	022	022	049	037	093	054	156	072	236	087	331	16
17	002	004	006	011	013	027	024	056	038	101	055	165	071	244	17
18	001	002	003	005	007	014	015	032	026	063	040	110	055	173	18
19		001	001	002	004	007	008	018	016	037	027	070	041	117	19
20			001	001	002	003	005	009	010	021	018	043	029	077	20
21					001	002	002	005	006	012	011	025	019	048	21
22						001	001	002	003	006	006	014	012	029	22
23							001	001	002	003	004	008	007	017	23
24									001	001	002	004	004	009	24
25										001	001	002	002	005	25
26											001	001	001	003	26
27													001	001	27
28														001	28

Source: F. Mosteller and R. E. K. Rourke, *Sturdy Statistics.* © 1973, Addison-Wesley, Reading, Mass. (Table A-8); reprinted by permission.

Normal / *t* / Binomial / Binomial confidence limits / Poisson /
F-5% / *F*-1% / Poisson confidence limits / χ^2 / Binomial, $\pi = \frac{1}{2}$
/ Random digits

Table III-6. Table of Critical Values of the *F* Distribution for Significance Levels of .05 and .01

Table III-6*A* gives the critical values for a significance level of .05, and Table III-6*B* gives the critical value for .01. Choose the column corresponding to the numerator degrees of freedom and the row corresponding to the denominator degrees of freedom.

EXAMPLE. If an *F* variable with 1 and 17 degrees of freedom equals 4.80, we find that the critical values for significance levels of .05 and .01 are 4.45 and 8.40, respectively, so $.01 < P < .05$.

Table III-6*A*. Critical values of the *F* distribution for a significance level of .05

ν_2 \ ν_1	1	2	3	4	5	6	7	8	9
1	161.45	199.50	215.71	224.58	230.16	233.99	236.77	238.88	240.54
2	18.51	19.00	19.17	19.25	19.30	19.33	19.35	19.37	19.39
3	10.13	9.55	9.28	9.12	9.01	8,94	8.89	8.85	8.81
4	7.71	6.94	6.59	6.39	6.26	6.16	6.09	6.04	6.00
5	6.61	5.79	5.41	5.19	5.05	4.95	4.88	4.82	4.77
6	5.99	5.14	4.76	4.53	4.39	4.28	4.21	4.15	4.10
7	5.59	4.74	4.35	4.12	3.97	3.87	3.79	3.73	3.68
8	5.32	4.46	4.07	3.84	3.69	3.58	3.50	3.44	3.39
9	5.12	4.26	3.86	3.63	3.48	3.37	3.29	3.23	3.18
10	4.96	4.10	3.71	3.48	3.33	3.22	3.14	3.07	3.02
11	4.84	3.98	3.59	3.36	3.20	3.09	3.01	2.95	2.90
12	4.75	3.89	3.49	3.26	3.11	3.00	2.91	2.85	2.80
13	4.67	3.81	3.41	3.18	3.03	2.92	2.83	2.77	2.71
14	4.60	3.74	3.34	3.11	2.96	2.85	2.76	2.70	2.65
15	4.54	3.68	3.29	3.06	2.90	2.79	2.71	2.64	2.59
16	4.49	3.63	3.24	3.01	2.85	2.74	2.66	2.59	2.54
17	4.45	3.59	3.20	2.96	2.81	2.70	2.61	2.55	2.49
18	4.41	3.55	3.16	2.93	2.77	2.66	2.58	2.51	2.46
19	4.38	3.52	3.13	2.90	2.74	2.63	2.54	2.48	2.42
20	4.35	3.49	3.10	2.87	2.71	2.60	2.51	2.45	2.39
21	4.32	3.47	3.07	2.84	2.68	2.57	2.49	2.42	2.37
22	4.30	3.44	3.05	2.82	2.66	2.55	2.46	2.40	2.34
23	4.28	3.42	3.03	2.80	2.64	2.53	2.44	2.37	2.32
24	4.26	3.40	3.01	2.78	2.62	2.51	2.42	2.36	2.30
25	4.24	3.39	2.99	2.76	2.60	2.49	2.40	2.34	2.28
26	4.23	3.37	2.98	2.74	2.59	2.47	2.39	2.32	2.27
27	4.21	3.35	2.96	2.73	2.57	2.46	2.37	2.31	2.25
28	4.20	3.34	2.95	2.71	2.56	2.45	2.36	2.29	2.24
29	4.18	3.33	2.93	2.70	2.55	2.43	2.35	2.28	2.22
30	4.17	3.32	2.92	2.69	2.53	2.42	2.33	2.27	2.21
40	4.08	3.23	2.84	2.61	2.45	2.34	2.25	2.18	2.12
60	4.00	3.15	2.76	2.53	2.37	2.25	2.17	2.10	2.04
120	3.92	3.07	2.68	2.45	2.29	2.18	2.09	2.02	1.96
∞	3.84	3.00	2.60	2.37	2.21	2.10	2.01	1.94	1.88

Normal / *t* / Binomial / Binomial confidence limits / Poisson /
F-5% / **F-1%** / Poisson confidence limits / χ^2 / Binomial, $\pi = \frac{1}{2}$
/ Random digits

Table III-6B. Critical values of the *F* distribution for a significance level of .01

ν_2 \ ν_1	1	2	3	4	5	6	7	8	9
1	4052.20	4999.50	5403.30	5624.60	5763.70	5859.00	5928.30	5981.60	6022.50
2	98.50	99.00	99.17	99.25	99.30	99.33	99.36	99.37	99.39
3	34.12	30.82	29.46	28.71	28.24	27.91	27.67	27.49	27.35
4	21.20	18.00	16.70	15.98	15.52	15.21	14.98	14.80	14.66
5	16.26	13.27	12.06	11.39	10.97	10.67	10.46	10.29	10.16
6	13.75	10.93	9.78	9.15	8.75	8.47	8.26	8.10	7.98
7	12.25	9.55	8.45	7.85	7.46	7.19	6.99	6.84	6.72
8	11.26	8.65	7.59	7.01	6.63	6.37	6.18	6.03	5.91
9	10.56	8.02	6.99	6.42	6.06	5.80	5.61	5.47	5.35
10	10.04	7.56	6.55	5.99	5.64	5.39	5.20	5.06	4.94
11	9.65	7.21	6.22	5.67	5.32	5.07	4.89	4.74	4.63
12	9.33	6.93	5.95	5.41	5.06	4.82	4.64	4.50	4.39
13	9.07	6.70	5.74	5.21	4.86	4.62	4.44	4.30	4.19
14	8.86	6.51	5.56	5.04	4.70	4.46	4.28	4.14	4.03
15	8.68	6.36	5.42	4.89	4.56	4.32	4.14	4.00	3.89
16	8.53	6.23	5.29	4.77	4.44	4.20	4.03	3.89	3.78
17	8.40	6.11	5.19	4.67	4.34	4.10	3.93	3.79	3.68
18	8.29	6.01	5.09	4.58	4.25	4.01	3.84	3.71	3.60
19	8.19	5.93	5.01	4.50	4.17	3.94	3.77	3.63	3.52
20	8.10	5.85	4.94	4.43	4.10	3.87	3.70	3.56	3.46
21	8.02	5.78	4.87	4.37	4.04	3.81	3.64	3.51	3.40
22	7.95	5.72	4.82	4.31	3.99	3.76	3.59	3.45	3.35
23	7.88	5.66	4.76	4.26	3.94	3.71	3.54	3.41	3.30
24	7.82	5.61	4.72	4.22	3.90	3.67	3.50	3.36	3.26
25	7.77	5.57	4.68	4.18	3.86	3.63	3.46	3.32	3.22
26	7.72	5.53	4.64	4.14	3.82	3.59	3.42	3.29	3.18
27	7.68	5.49	4.60	4.11	3.78	3.56	3.39	3.26	3.15
28	7.64	5.45	4.57	4.07	3.75	3.53	3.36	3.23	3.12
29	7.60	5.42	4.54	4.04	3.73	3.50	3.33	3.20	3.09
30	7.56	5.39	4.51	4.02	3.70	3.47	3.30	3.17	3.07
40	7.31	5.18	4.31	3.83	3.51	3.29	3.12	2.99	2.89
60	7.08	4.98	4.13	3.65	3.34	3.12	2.95	2.82	2.72
120	6.85	4.79	3.95	3.48	3.17	2.96	2.79	2.66	2.56
∞	6.63	4.61	3.78	3.32	3.02	2.80	2.64	2.51	2.41

Table III-7. Table of 80% and 95% Confidence Limits for the Expectation of a Poisson Variable

EXAMPLES. If $x = 7$, the 80% confidence interval for m, the expectation, is

(3.91, 10.97)

and the 95% confidence interval is

(2.81, 14.42)

Table III-7. Table of 80% and 95% confidence limits for the expectation of a Poisson variable

	80%		95%			80%		95%	
x	LOWER	UPPER	LOWER	UPPER	*x*	LOWER	UPPER	LOWER	UPPER
0	.00	1.82	.00	3.69	26	18.94	33.03	16.77	37.67
1	.22	3.55	.03	5.57	27	20.41	34.42	17.63	38.16
2	.82	4.56	.24	7.22	28	22.04	34.58	19.05	39.76
3	1.54	5.88	.62	8.77	29	22.04	35.92	19.05	40.94
4	1.82	7.56	1.09	10.24	30	22.33	37.39	20.33	41.75
5	2.65	8.53	1.62	11.67					
					31	23.74	39.07	21.36	43.45
6	3.55	9.92	2.20	13.06	32	25.71	39.07	21.36	44.26
7	3.91	10.97	2.81	14.42	33	25.71	40.23	22.94	45.28
8	4.76	12.48	3.45	15.76	34	25.71	41.62	23.76	47.02
9	5.70	13.24	4.12	17.08	35	26.97	43.25	23.76	47.69
10	5.88	15.21	4.80	18.39					
					36	28.47	44.20	25.40	48.74
11	7.56	15.44	5.49	19.68	37	29.98	44.48	26.31	50.42
12	7.56	16.91	6.20	20.96	38	29.98	45.79	26.31	51.29
13	8.53	18.54	6.92	22.23	39	30.15	47.20	27.73	52.15
14	9.92	18.94	7.65	23.49	40	31.51	48.99	28.97	53.72
15	9.92	20.41	8.40	24.74					
					41	33.03	49.46	28.97	54.99
16	10.97	22.04	9.15	25.98	42	34.42	49.94	30.02	55.51
17	12.48	22.33	9.90	27.22	43	34.42	51.25	31.67	56.99
18	12.48	23.74	10.67	28.45	44	34.58	52.64	31.67	58.72
19	13.24	25.71	11.44	29.67	45	35.92	54.29	32.28	58.84
20	15.21	25.71	12.22	30.89	46	37.39	55.16	34.05	60.24
21	15.21	26.97	12.82	31.68	47	39.07	55.33	34.66	61.90
22	15.44	28.47	13.77	32.28	48	39.07	56.61	34.66	62.81
23	16.91	29.98	14.92	34.05	49	39.07	57.95	36.03	63.49
24	18.54	30.15	14.92	34.67	50	40.23	59.44	37.67	64.95
25	18.54	31.51	16.77	36.03					

Source: Reprinted by permission from E. L. Crow and R. S. Gardner, *Biometrika,* **46**:441–53, 1959.

Normal / t / Binomial / Binomial confidence limits / Poisson / F-5% / F-1% / Poisson confidence limits / χ^2 / Binomial, $\pi = \frac{1}{2}$ / Random digits

Table III-8.   Critical Values of χ^2 for Various Degrees of Freedom

EXAMPLES.   (a) The probability that a χ^2 random variable with 1 degree of freedom exceeds 3.84 is $P(\chi^2_1 \geq 3.84) = .05$. (b) The table shows that $P(\chi^2_{10} \geq 15)$ lies between .10 and .20. Interpolation gives $P(\chi^2_{10} \geq 15) = .14$, approximately.

Normal / *t* / Binomial / Binomial confidence limits / Poisson /
F-5% / *F*-1% / Poisson confidence limits / χ^2 / Binomial, $\pi = \frac{1}{2}$
/ Random digits

Table III-8. Critical values of χ^2 for various degrees of freedom

Degrees of freedom	Probability levels						
	.99	.95	.50	.20	.10	.05	.01
1	.00	.00	.45	1.64	2.71	3.84	6.63
2	.02	.10	1.39	3.22	4.61	5.99	9.21
3	.11	.35	2.37	4.64	6.25	7.81	11.34
4	.30	.71	3.36	5.99	7.78	9.49	13.28
5	.55	1.15	4.35	7.29	9.24	11.07	15.09
6	.87	1.64	5.35	8.56	10.64	12.59	16.81
7	1.24	2.17	6.35	9.80	12.02	14.07	18.48
8	1.65	2.73	7.34	11.03	13.36	15.51	20.09
9	2.09	3.33	8.34	12.24	14.68	16.92	21.67
10	2.56	3.94	9.34	13.44	15.99	18.31	23.21
11	3.05	4.57	10.34	14.63	17.28	19.68	24.72
12	3.57	5.23	11.34	15.81	18.55	21.03	26.22
13	4.11	5.89	12.34	16.98	19.81	22.36	27.69
14	4.66	6.57	13.34	18.15	21.06	23.68	29.14
15	5.23	7.26	14.34	19.31	22.31	25.00	30.58
16	5.81	7.96	15.34	20.47	23.54	26.30	32.00
17	6.41	8.67	16.34	21.61	24.77	27.59	33.41
18	7.01	9.39	17.34	22.76	25.99	28.87	34.81
19	7.63	10.12	18.34	23.90	27.20	30.14	36.19
20	8.26	10.85	19.34	25.04	28.41	31.41	37.57
21	8.90	11.59	20.34	26.17	29.62	32.67	38.93
22	9.54	12.34	21.34	27.30	30.81	33.92	40.29
23	10.20	13.09	22.34	28.43	32.01	35.17	41.64
24	10.86	13.85	23.34	29.55	33.20	36.42	42.98
25	11.52	14.61	24.34	30.68	34.38	37.65	44.31
26	12.20	15.38	25.34	31.79	35.56	38.89	45.64
27	12.88	16.15	26.34	32.91	36.74	40.11	46.96
28	13.56	16.93	27.34	34.03	37.92	41.34	48.28
29	14.26	17.71	28.34	35.14	39.09	42.56	49.59
30	14.95	18.49	29.34	36.25	40.26	43.77	50.89
50	29.71	34.76	49.33	58.16	63.17	67.50	76.15

Table III-9. Three-Place Tables of the Cumulative Binomial for $\pi = \frac{1}{2}$

The sample size is n, and the number in the less frequent class is r. The probability table is

$$P(X \le r) = \sum_{x=0}^{r} \binom{n}{x}\left(\frac{1}{2}\right)^n$$

Each three-digit entry should be read with a decimal point preceding it. The one-tailed P value for a sign test is the cumulative probability corresponding to the number in the less frequent class. Double the table entry to get the two-tailed P-value.

EXAMPLE. A sign test with sample size 15 gave 11 positive observations and 4 negative observations. The number in the less frequent class is 4. Thus $P = .059$.

Table III-9. Three-place tables of the cumulative binomial for $\pi = \frac{1}{2}$

n	0	1	2	3	4								
1	500												
2	250												
3	125	500											
4	062	312											
5	031	188	500										
6	016	109	344										
7	008	062	227	500									
8	004	035	145	363									
9	002	020	090	254	500								
10	001	011	055	172	377								

	1	2	3	4	5	6	7	8	9				
11	006	033	113	274	500								
12	003	019	073	194	387								
13	002	011	046	133	291	500							
14	001	006	029	090	212	395							
15		004	018	059	151	304	500						
16		002	011	038	105	227	402						
17		001	006	025	072	166	315	500					
18		001	004	015	048	119	240	407					
19			002	010	032	084	180	324	500				
20			001	006	021	058	132	252	412				

	3	4	5	6	7	8	9	10	11	12	13	14	
21	001	004	013	039	095	192	332	500					
22		002	008	026	067	143	262	416					
23		001	005	017	047	105	202	339	500				
24		001	003	011	032	076	154	271	419				
25			002	007	022	054	115	212	345	500			

Table III-9. *Continued*

n	3	4	5	6	7	8	9	10	11	12	13	14	
26			001	005	014	038	084	163	279	423			
27			001	003	010	026	061	124	221	351	500		
28				002	006	018	044	092	172	286	425		
29				001	004	012	031	068	132	229	356	500	
30				001	003	008	021	049	100	181	292	428	

n	7	8	9	10	11	12	13	14	15	16	17		
31	002	005	015	035	075	141	237	360	500				
32	001	004	010	025	055	108	189	298	430				
33	001	002	007	018	040	081	148	243	364	500			
34		001	005	012	029	061	115	196	304	432			
35		001	003	008	020	045	088	155	250	368	500		

n	8	9	10	11	12	13	14	15	16	17	18	19	
36	001	002	006	014	033	066	121	203	309	434			
37		001	004	010	024	049	094	162	256	371	500		
38		001	003	007	017	036	072	128	209	314	436		
39		001	002	005	012	027	054	100	168	261	375	500	
40			001	003	008	019	040	077	134	215	318	437	

n	10	11	12	13	14	15	16	17	18	19	20	21	22
41	001	002	006	014	030	059	106	174	266	378	500		
42		001	004	010	022	044	082	140	220	322	439		
43		001	003	007	016	033	063	111	180	271	380	500	
44		001	002	005	011	024	048	087	146	226	326	440	
45			001	003	008	018	036	068	116	186	276	383	500

n	12	13	14	15	16	17	18	19	20	21	22	23	24
46	001	002	006	013	027	052	092	151	231	329	441		
47	001	002	004	009	020	039	072	121	191	280	385	500	
48		001	003	007	015	030	056	097	156	235	333	443	
49		001	002	005	011	022	043	076	126	196	284	388	500
50			001	003	008	016	032	059	101	161	240	336	444

n	14	15	16	17	18	19	20	21	22	23	24	25	26	27
51	001	002	005	012	024	046	080	131	201	288	390	500		
52	001	002	004	009	018	035	063	106	166	244	339	445		
53		001	003	006	014	027	049	084	136	205	292	392	500	
54		001	002	005	010	020	038	067	110	170	248	342	446	
55		001	001	003	007	015	029	052	089	140	209	295	394	500

Source: F. Mosteller and R. E. K. Rourke, *Sturdy Statistics,* © 1973, Addison-Wesley, Reading, Mass. (Table A-6); reprinted by permission.

Normal / *t* / Binomial / Binomial confidence limits / Poisson /
F-5% / *F*-1% / Poisson confidence limits / χ^2 / Binomial, $\pi = \frac{1}{2}$
/ **Random digits**

Table III-10.   1000 random digits[a]

Line number	Column number									
	00–09		10–19		20–29		30–39		40–49	
00	15544	80712	97742	21500	97081	42451	50623	56071	28882	28739
01	01011	21285	04729	39986	73150	31548	30168	76189	56996	19210
02	47435	53308	40718	29050	74858	64517	93573	51058	68501	42723
03	91312	75137	86274	59834	69844	19853	06917	17413	44474	86530
04	12775	08768	80791	16298	22934	09630	98862	39746	64623	32768
05	31466	43761	94872	92230	52367	13205	38634	55882	77518	36252
06	09300	43847	40881	51243	97810	18903	53914	31688	06220	40422
07	73582	13810	57784	72454	68997	72229	30340	08844	53924	89630
08	11092	81392	58189	22697	41063	09451	09789	00637	06450	85990
09	93322	98567	00116	35605	66790	52965	62877	21740	56476	49296
10	80134	12484	67089	08674	70753	90959	45842	59844	45214	36505
11	97888	31797	95037	84400	76041	96668	75920	68482	56855	97417
12	92612	27082	59459	69380	98654	20407	88151	56263	27126	63797
13	72744	45586	43279	44218	83638	05422	00995	70217	78925	39097
14	96256	70653	45285	26293	78305	80252	03625	40159	68760	84716
15	07851	47452	66742	83331	54701	06573	98169	37499	67756	68301
16	25594	41552	96475	56151	02089	33748	65239	89956	89559	33687
17	65358	15155	59374	80940	03411	94656	69440	47156	77115	99463
18	09402	31008	53424	21928	02198	61201	02457	87214	59750	51330
19	97424	90765	01634	37328	41243	33564	17884	94747	93650	77668

[a] A good source of random digits and random normal deviates is The Rand Corporation, *A Million Random Digits with 100,000 Normal Deviates*. Glencoe, Ill.: Free Press of Glencoe, 1955.

Source: F. Mosteller, and R. E. K. Rourke, *Sturdy Statistics,* © 1973, Addison-Wesley, Reading, Mass. (Table A-22); reprinted by permission.

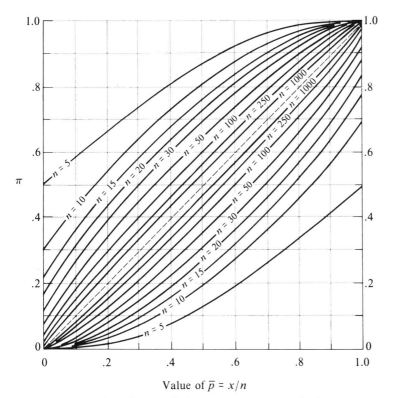

Chart III-1. Chart for obtaining 95% confidence limits on π, the probability of success on a single binomial trial when a sample of n trials gives the proportion $\bar{p}$ of successes. To obtain confidence limits for π, enter the horizontal axis at the observed value of $\bar{p}$. Read the vertical axis at the two points where the two curves for n cross the vertical line erected from $\bar{p}$, for example, $\bar{p} = .4$, $n = 30$, lower confidence limit = .22, upper confidence limit = .60. [Reprinted by permission from Clopper, C. J., and Pearson, E. S. (1934). *Biometrika,* **26**:410.]

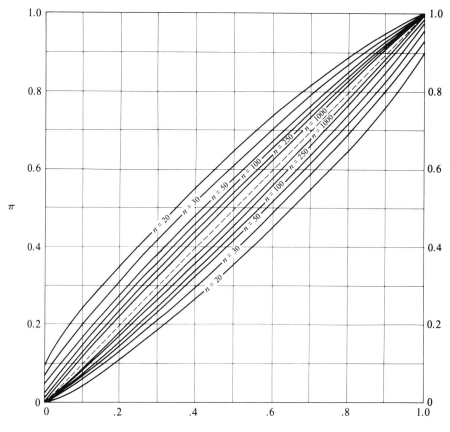

Value of $\bar{p} = x/n$

Chart III-2. Chart for obtaining 80% confidence limits on π, the probability of success on a single binomial trial when a sample of n trials gives the proportion $\bar{p}$ of successes. To obtain confidence limits for π, enter the horizontal axis at the observed value of $\bar{p}$. Read the vertical axis at the two points where the two curves for n cross the vertical line erected from $\bar{p}$, for example, $\bar{p} = .4$, $n = 30$, lower confidence limit = .30, upper confidence limit = .51. To obtain 80% confidence limits when the sample size is 20 or less, refer to Table III-4.

330

Index